Industrial Drying of Foods

Industrial Drying of Foods

Edited by

CHRISTOPHER G.J. BAKER

ArvaTec (UK) Ltd, Wantage, UK
and
Chemical Engineering Department,
Kuwait University, Kuwait

BLACKIE ACADEMIC & PROFESSIONAL
An Imprint of Chapman & Hall
London · Weinheim · New York · Tokyo · Melbourne · Madras

**Published by Blackie Academic and Professional, an imprint of
Chapman & Hall, 2–6 Boundary Row, London SE1 8HN, UK**

Chapman & Hall, 2–6 Boundary Row, London SE1 8HN, UK

Chapman & Hall, GmbH, Pappelallee 3, 69469 Weinheim, Germany

Chapman & Hall USA, 115 Fifth Avenue, New York, NY 10003, USA

Chapman & Hall Japan, ITP-Japan, Kyowa Building, 3F, 2-2-1 Hirakawacho,
Chiyoda-ku, Tokyo 102, Japan

DA Book (Aust.) Pty Ltd, 648 Whitehorse Road, Mitcham 3132, Victoria, Australia

Chapman & Hall India, R. Seshadri, 32 Second Main Road, CIT East, Madras
600 035, India

First edition 1997

© 1997 Chapman & Hall

Typeset in 10/12 Times by Blackpool Typesetting Services Limited, UK
Printed in Great Britain by St Edmundsbury Press, Bury St Edmunds

ISBN   0 7514 0384 9

A catalogue record for this book is available from the British Library

Library of Congress Catalog Card Number: 97-72361

♾ Printed on acid-free text paper, manufactured in accordance with ANSI/NISO
Z39.48-1992 (Permanence of Paper)

Contents

Contributors   xi
Preface   xiii
List of symbols   xv

1 The industrial drying of foods: an overview   1
   R.B. KEEY

2 Drying fundamentals   7
   A.S. MUJUMDAR

   2.1 Introduction   7
   2.2 Basic dryer types   9
   2.3 Thermodynamic properties of air–water mixtures and moist solids   11
        2.3.1 Psychrometry   11
        2.3.2 Equilibrium moisture content   16
        2.3.3 Water activity   19
   2.4 Drying kinetics   21
   2.5 Food quality parameters   26
   2.6 Summary   29
   Acknowledgements   29
   References   29

3 Through-flow dryers for agricultural crops   31
   S. SOKHANSANJ

   3.1 Introduction   31
   3.2 On-farm dryers   32
        3.2.1 Heated-air dryers   32
        3.2.2 Unheated-air dryers   38
        3.2.3 Combination heated- and unheated-air dryers   42
   3.3 Commercial dryers   43
        3.3.1 Commercial grain dryers   43
        3.3.2 Malt dryers   44
   3.4 Dryer control   46
   3.5 Design of through-circulation dryers   47
        3.5.1 Design calculations   47
        3.5.2 Grain properties   53
        3.5.3 Verification of design calculation   54
   3.6 Crop quality and drying   56
        3.6.1 Quality models   57
        3.6.2 Stress cracks and broken kernels   59
   3.7 Crop drying costs   59
        3.7.1 Fixed and variable costs   59
        3.7.2 Risk costs   61
        3.7.3 Total costs   61
   3.8 Summary   62
   References   63

4   Fluidized bed dryers   **65**
     R.E. BAHU

     4.1   Introduction   65
     4.2   Basics of fluidization   65
     4.3   Features of construction   68
           4.3.1   Distributor   68
           4.3.2   Plenum chamber   70
           4.3.3   Freeboard region and gas cleaning systems   71
           4.3.4   Exhaust air recycle   71
     4.4   Types of fluidized bed dryer   72
           4.4.1   Introduction   72
           4.4.2   Batch fluidized bed dryers   72
           4.4.3   Well-mixed fluidized bed dryers   73
           4.4.4   Plug-flow fluidized bed dryers   74
           4.4.5   Multi-stage fluidized bed dryers   74
           4.4.6   Vibro-fluidized bed dryers   76
           4.4.7   Internally heated fluidized bed dryers   76
           4.4.8   Fluidized bed granulators/coaters   78
           4.4.9   Mechanically agitated fluidized bed dryers   78
           4.4.10 Centrifugal fluidized bed dryers   80
           4.4.11 The Jetzone dryer   80
     4.5   Operating considerations   80
           4.5.1   Control   81
           4.5.2   Fire and explosion hazards   81
           4.5.3   Energy conservation   81
     4.6   Applications   81
     4.7   Test procedures   82
     4.8   Design methods   84
           4.8.1   Scoping design   84
           4.8.2   Detailed design calculations   86
     Acknowledgements   88
     References   88

5   Spray dryers   **90**
     K. MASTERS

     5.1   Introduction   90
     5.2   Spray drying principles   92
           5.2.1   Atomization   92
           5.2.2   Spray-air contact and flow   94
           5.2.3   Evaporation, drying and particle formation   96
           5.2.4   Powder handling   99
     5.3   Spray dryer layouts   101
           5.3.1   Cocurrent drying chamber with rotary atomizer   101
           5.3.2   Cocurrent drying chamber with nozzle atomization   102
           5.3.3   Cocurrent drying chamber with integrated fluid bed 103
           5.3.4   Cocurrent drying chamber with integrated belt   104
           5.3.5   Mixed flow drying chamber with integrated fluid bed 105
     5.4   Meeting powder specifications   106
           5.4.1   Particulate structure and size   106
           5.4.2   Moisture content   107
           5.4.3   Bulk density   109
           5.4.4   Hygroscopicity   109
     5.5   Special design features for hygiene and safety   110
           5.5.1   Drying air filtration   110
           5.5.2   Insulation and cladding of drying chambers   110
           5.5.3   Prevention of fires and explosions   112

6 Contact dryers **115**

D. OAKLEY

6.1 Introduction 115
6.2 Industrial equipment and applications 116
   6.2.1 Introduction 116
   6.2.2 Contact dryer selection 116
   6.2.3 Vacuum tray dryers 118
   6.2.4 Vacuum band dryers 118
   6.2.5 Plate dryers 120
   6.2.6 Thin-film dryers 121
   6.2.7 Drum dryers 122
   6.2.8 Rotating batch vacuum dryers 124
   6.2.9 Horizontally agitated dryers 125
   6.2.10 Indirectly heated rotary dryers 127
   6.2.11 Vertically agitated dryers 128
6.3 Theoretical overview of contact drying 129
6.4 Design of contact dryers 132
Acknowledgements 133
References 133

7 Freeze dryers **134**

J.W. SNOWMAN

7.1 Introduction 134
7.2 Process overview 134
7.3 Description of the freeze drying process 138
   7.3.1 Pretreatment of food 138
   7.3.2 Freezing 139
   7.3.3 Sublimation (primary drying) 141
   7.3.4 Desorption (secondary drying) 144
   7.3.5 Storage after drying 145
   7.3.6 Rehydration and use 145
7.4 Equipment 146
7.5 The cost of freeze drying 149
7.6 Freeze drying procedures 150
   7.6.1 Coffee and tea 150
   7.6.2 Raw meat products 151
   7.6.3 Cooked meat products 152
   7.6.4 Beef 152
   7.6.5 Cooked pork and lamb 152
   7.6.6 Cooked chicken meat 152
   7.6.7 Offal 153
   7.6.8 Shrimp 153
   7.6.9 Egg 153
   7.6.10 Fish 153
   7.6.11 Mushrooms 153
   7.6.12 Strawberries 153
   7.6.13 Milled vegetables 154
   7.6.14 Difficult materials 154
7.7 The packaging of freeze dried food 154
7.8 Conclusion 155
References 155

8   Dielectric dryers   **156**
P.L. JONES and A.T. ROWLEY

8.1   Introduction   156
8.2   The fundamentals of dielectric heating   156
      8.2.1   Radio frequency dryers   157
      8.2.2   Microwave heating systems   160
      8.2.3   Safety of dielectric heating equipment   162
8.3   Dielectric drying   162
      8.3.1   The power dissipated with a dielectric   162
      8.3.2   The effect of a dielectric on an electric field   165
      8.3.3   Variation of dielectric properties with moisture content   166
      8.3.4   Moisture levelling   168
      8.3.5   Heat and mass transfer within drying solids   169
8.4   Applications of dielectric heat in drying   173
      8.4.1   Microwave or radio frequency drying?   173
      8.4.2   Microwave applications   174
      8.4.3   Radio frequency applications   175
      8.4.4   Application of combination dryers   176
8.5   Cost of RF and microwave dryers   177
8.6   Conclusions   177
References   177

9   Specialized drying systems   **179**
D.J. BARR and C.G.J. BAKER

9.1   Introduction   179
9.2   Pneumatic-conveying dryers   179
      9.2.1   Introduction   179
      9.2.2   Equipment   180
      9.2.3   Hygiene   185
      9.2.4   Applications   186
      9.2.5   Design of pneumatic-conveying dryers   187
      9.2.6   Comparison of different drying systems   191
9.3   Spin-flash dryer   192
      9.3.1   Introduction   192
      9.3.2   Equipment   192
      9.3.3   Applications   193
9.4   Rotary dryers   194
      9.4.1   Introduction   194
      9.4.2   Equipment   194
      9.4.3   Applications   198
      9.4.4   Design of rotary dryers   198
9.5   Tray and tunnel dryers   200
      9.5.1   Introduction   200
      9.5.2   Tray dryers   201
      9.5.3   Tunnel dryers   202
      9.5.4   Design of tunnel dryers   205
9.6   Band dryers   206
      9.6.1   Introduction   206
      9.6.2   Equipment   207
      9.6.3   Applications   208
      9.6.4   Design of band dryers   208
References   209

10 Solar dryers   **210**
L. IMRE

10.1   Introduction   210
10.2   Construction of solar dryers   211
       10.2.1   Functional parts of solar dryers   211
       10.2.2   Main types of solar dryers   211
10.3   Solar natural dryers   212
       10.3.1   Cabinet-type dryers   212
       10.3.2   Cabinet dryers fitted with a chimney   214
       10.3.3   Cabinet dryers fitted with a chimney and heat storage 215
10.4   Semi-artificial solar dryers   216
       10.4.1   Solar-tunnel dryers   216
       10.4.2   Greenhouse-type solar dryers   218
       10.4.3   Solar room dryers   218
10.5   Solar-assisted dryers   219
       10.5.1   Solar-assisted dryer for seeds and herbs   219
       10.5.2   Solar-assisted dryers with heat storage   221
       10.5.3   Solar-assisted dryers with rock-bed heat storage   225
       10.5.4   Solar-assisted dryer combined with adsorbent units   225
       10.5.5   Solar-assisted dryers combined with heat pumps   226
10.6   Economic evaluation of solar dryers   227
       10.6.1   Savings   227
       10.6.2   Investment costs   228
10.7   Design of solar dryers   228
       10.7.1   Definition of the drying process   229
       10.7.2   Selection of solar dryer type   230
       10.7.3   Structural design of solar dryers   231
       10.7.4   Modelling and simulation of solar dryers   233
10.8   Dryer operation and process control strategies   234
       10.8.1   Operational aspects   234
       10.8.2   Less sophisticated solar dryers   235
       10.8.3   Solar-assisted dryers   235
       10.8.4   Solar dryers with recirculation   237
       10.8.5   Solar dryers fitted with a water storage tank and auxiliary heater 237
References   238

11 Dryer selection   **242**
C.G.J. BAKER

11.1   Introduction   242
11.2   Specification of the drying process   244
       11.2.1   Form of feed   244
       11.2.2   Upstream and downstream processing operations   246
       11.2.3   Moisture content of feed and product   246
       11.2.4   Dryer throughput   248
       11.2.5   Physical properties of the feed   249
       11.2.6   Physical properties of the product   250
       11.2.7   Quality changes during drying   251
       11.2.8   Drying kinetics   253
11.3   Preliminary dryer selection   254
       11.3.1   Nature of the feed   254
       11.3.2   Nature of the product   255
       11.3.3   Throughput   256
       11.3.4   Mode of heating   256
11.4   Bench-scale tests   257
       11.4.1   Role of bench-scale tests   257
       11.4.2   Details of testing methods   261

	11.5	Economic comparison of alternatives	264
		11.5.1 Introduction	264
		11.5.2 Vendor-generated capital cost estimates	264
		11.5.3 In-house generated capital cost estimates	265
		11.5.4 Cost comparison	267
	11.6	Pilot-plant trials and final selection	267
	11.7	Typical examples of the use of the dryer selection algorithm	268
		11.7.1 Example 1	268
		11.7.2 Example 2	269
	11.8	Summary	270
	References		271

12 Dryer operation and control   **272**
S.P. GARDINER

	12.1	Introduction	272
	12.2	Dryer operation	272
		12.2.1 Introduction	272
		12.2.2 Operating conditions	272
		12.2.3 Energy savings	274
	12.3	Safety, health and the environment	276
		12.3.1 Introduction	276
		12.3.2 Safety	277
		12.3.3 Environmental issues	281
	12.4	Dryer control and instrumentation	282
		12.4.1 Operating and control strategy	282
		12.4.2 Alarms and interlocks	283
		12.4.3 Dryer control strategies	284
		12.4.4 Types of process control system	287
		12.4.5 Process measurements	291
	References		297

Index   **299**

Contributors

R.E. Bahu   Barr-Rosin Ltd, 48 Bell Street, Maidenhead, Berks
                     SL6 1BR, UK

C.G.J. Baker   *ArvaTec* (UK) Ltd, Long Acre, East Hanney, Wantage,
                     Oxfordshire OX12 0HP, UK

D.J. Barr   Barr-Rosin Ltd, 48 Bell Street, Maidenhead, Berks
                     SL6 1BR, UK

S.P. Gardiner   SPG Consultants, 5 Greenacres, Frodsham, Cheshire
                     WA6 6BU, UK

L. Imre   Faculty of Mechanical Engineering, Technical University
                     of Budapest, Muegyetem-Rkp 3, H-1521 Budapest,
                     Hungary

P.L. Jones   Peal Jay Consultants, 2 Newton Lane, Chester CH2 3RP,
                     UK

R.B. Keey   Department of Chemical and Process Engineering,
                     University of Canterbury, Private Bag 4800, Christchurch,
                     New Zealand

K. Masters   Niro A/S, Gadsaxevej 305, PO Box 45, DK-2860
                     Soeborg, Denmark

A.S. Mujumdar   Department of Chemical Engineering, McGill University,
                     3480 University Street, Montreal, Quebec, Canada
                     H3A 2A7

D. Oakley   Separation Processes Service, AEA Technology, 404
                     Harwell, Didcot, Oxfordshire OX11 0RA, UK

A.T. Rowley   EA Technology Ltd, Capenhurst, Chester CH1 6ES, UK

J.W. Snowman   Northcote, Stockcroft Road, Balcombe, W. Sussex
                     RH17 6LH, UK

S. Sokhansanj   Department of Agricultural and Bioresource Engineering,
                     University of Saskatchewan, 57 Campus Drive,
                     Saskatoon, Saskatchewan, Canada S7N 0W0

Preface

Drying is traditionally defined as that unit operation which converts a liquid, solid or semi-solid feed material into a solid product of significantly lower moisture content. In most, although not all, cases it involves the application of thermal energy, which causes water to evaporate into the vapour phase. In practice, this definition encompasses a number of technologies which differ markedly in, for example, the manner in which energy is supplied to the foodstuff and in which product is transported through the dryer. Depending on the dryer type, the residence time may vary from a few seconds to several hours. Dryers designed to handle liquid feedstocks are naturally quite different from those intended to process moist solids. Even within these two broad categories, however, many distinct varieties of dryer have evolved to meet specific processing needs.

The dryer is frequently the last processing stage in the manufacture of a dehydrated food product. As such, it may not only bring about the desired reduction in moisture content but may also have a significant effect on a number of other properties, such as flavour, colour, texture, viability, and nutrient retention, for example. These properties, which are generally considered to affect the perceived quality of the end product, are often influenced by the temperature–moisture content–time profiles experienced by the foodstuff as it moves through the dryer. The underlying chemistry and physics are highly complex and, broadly speaking, only poorly understood. Thus, in practice, a dryer is considerably more complex than a device that merely removes moisture.

This book has been written primarily to provide practical assistance on the drying of foodstuffs to process engineers and technologists working in the food-manufacturing industry. As such, considerable emphasis has been placed on providing relevant practical detail, backed up, where appropriate, by underlying theory. Its contents include descriptions of the fundamentals of drying, and of the large number of dryer types to be found in the food industry. Advice on dryer selection is also given and, finally, issues relating to the operation and control of dryers discussed. The considerable volume of work devoted to elucidating a better understanding of moisture transport in solids is not covered in depth. The reason for this is that, in practice, drying-rate curves have to be measured experimentally, rather than calculated from fundamentals. Theoretically based models that can be used with confidence to interpolate between data obtained under different drying conditions are used in practice and are described in appropriate chapters of the book.

I would like to express my appreciation to Blackie Academic & Professional for the opportunity to produce this book; to the contributors, all of whom are leading experts in their particular fields and are therefore very busy people, for the many hours they must have devoted to preparing their texts; and finally to my wife and family for their patience and support during the editing process. Several of the contributors are or have been associated with AEA Technology's Separation Processes Service in one capacity or another. This organization has made a notable contribution to drying research over many years, which has undoubtedly influenced the content of this book. This is gratefully acknowledged.

Drying is a technology that has provided me with many exciting scientific and technical challenges over the years. I hope that the readers of this book will not only benefit in an immediate sense from the accumulated wisdom of the contributors, but will be stimulated to advance our collective knowledge of drying in different areas of the food industry.

Christopher G.J. Baker
Wantage, February 1997

List of symbols

Symbol	Definition	Units
a	Coefficient	K^{-1}
a	Half-thickness of slab (Chapter 2)	m
a_w	Water activity	–
a_x	Specific surface area for heat transfer at x	m^{-1}
a'	Coefficient	K^{-1}
a_1	Coefficient	$kJ\ kg^{-1}$
a_2	Coefficient	K^{-1}
a_3	Exponent	–
a_4	Coefficient	K^{-1}
a_5	Coefficient	K
a_6	Exponent	–
A	Area	m^2
A_b	Area of bed	m^2
A_c	Area of solar collector	m^2
A_h	Heat exchange area	m^2
A_p	Plate area	m^2
A_x	Cross-sectional area	m^2
b	Coefficient	K^{-1}
b'	Coefficient	–
B	Axial dispersion number	–
c_a	Specific heat of dry air	$kJ\ kg^{-1}\ K^{-1}$
c_{base}	Capacity of base-case dryer	$kg\ s^{-1}$
c_d	Specific heat of dry solids	$kJ\ kg^{-1}\ K^{-1}$
c_1	Specific heat of water (liquid)	$kJ\ kg^{-1}\ K^{-1}$
c_{new}	Capacity of new dryer	$kg\ s^{-1}$
c_s	Humid heat	$kJ\ kg^{-1}\ K^{-1}$
c_v	Specific heat of water (vapour)	$kJ\ kg^{-1}\ K^{-1}$
C_{base}	Purchase cost of base-case dryer	$
C_c	Total initial cost	$
C_e	Unit price of electricity	$\ (kWh)^{-1}$
C_f	Unit price of fuel	$\ (kWh)^{-1}$
C_g	Grain cost	$\ t^{-1}$
C_{new}	Purchase cost of new dryer	$
C_s	Total annual savings	$\ y^{-1}$
$C_{s,k}$	k^{th} component of annual savings	$\ y^{-1}$
C_0	Initial cost	$
C_0	Capacitance of empty applicator (Chapter 8)	F
C_1	Annual fixed cost	$\ y^{-1}$
C_2	Inspection cost	$\ y^{-1}$

Symbol	Definition	Units
C_3	Energy cost	$\$\,y^{-1}$
C_4	Over-drying cost	$\$\,y^{-1}$
C_5	Inventory cost	$\$\,y^{-1}$
C_6	Loss-of-quality cost	$\$\,y^{-1}$
C_7	Late-harvest cost	$\$\,y^{-1}$
C'	Effective capacitance	F
d	Plate separation	m
D_L	Diffusion coefficient	$m^2\,s^{-1}$
D_{LO}	Pre-exponential factor	$m^2\,s^{-1}$
e_c	Collection efficiency	–
e_d	Drying efficiency	–
$e_{d.avr}$	Average drying effectiveness	–
e_t	Total energy effectiveness	–
$\mathbf{E}_{applied}$	Applied field	$V\,m^{-1}$
E_a	Activation energy for diffusion	$kJ\,kmol^{-1}$
E_e	Electrical energy consumption	$kW\,hy^{-1}$
E_f	Fuel energy consumption	$kW\,hy^{-1}$
$E(t)$	Residence time distribution function	–
$\mathbf{E}$	Electric field strength (filled applicator)	$V\,m^{-1}$
$\mathbf{E}_{rms}$	Root-mean-square electric field	$V\,m^{-1}$
$\mathbf{E}_0$	Electric field strength (empty applicator)	$V\,m^{-1}$
$\mathbf{E}'$	Electric field in air gaps surrounding dielectric	$V\,m^{-1}$
$\mathbf{E}^*$	Complex conjugate of E	$V\,m^{-1}$
f	Frequency	s^{-1}
f_p	Probability factor	–
G	Air mass flow rate (dry basis)	$kg\,s^{-1}$
G_v	Air volumetric flow rate	$m^3\,s^{-1}$
G'	Air mass flow rate (dry basis) per unit area	$kg\,m^{-2}\,s^{-1}$
h	Enthalpy of wet solids	$kJ\,kg^{-1}$
h	Convective heat transfer coefficient (Chapter 2)	$kW\,m^{-2}\,K^{-1}$
h_i	Enthalpy of inlet solids	$kJ\,kg^{-1}$
h_o	Enthalpy of outlet solids	$kJ\,kg^{-1}$
H	Enthalpy of air–water vapour mixture	$kJ\,kg^{-1}$
H_i	Enthalpy of inlet air	$kJ\,kg^{-1}$
H_o	Enthalpy of outlet air	$kJ\,kg^{-1}$
i	Interest rate	–
i_{en}	Energy price inflation rate	–
i_{eq}	Equipment price inflation rate	–
I	Solar energy flux	$W\,m^{-2}$
$I_{t.base}$	Value of cost index at purchase of base-case dryer	–
$I_{t.new}$	Current value of cost index	–

Symbol	Definition	Units
$\mathbf{j}$	Complex operator	–
j	Integer	–
$\mathbf{J}$	Current	A
$\mathbf{J_D}$	Displacement current	A
k_v	Mass transfer coefficient	$\text{kmol m}^{-2}\,\text{s}^{-1}$
k	Propagation constant (Chapter 8)	m^{-1}
k	Coefficient	s^{-1}
$K(t)$	Coefficient	–
k'	Coefficient	s^{-1}
k''	Rate constant	–
K	Parameter	–
K_o	Parameter	–
L	Bed length	m
m	Order of reaction	–
M_{air}	Molecular mass of air	–
M_B	Bed mass per unit area	kg m^{-2}
M_s	Mass of dry solids	kg
M_w	Mass of water	kg
n	Exponent	–
n_l	Economic life	y
n_o	Annual operating hours	h y^{-1}
n'	Payback time	y
N	Drying rate	$\text{kg m}^{-2}\,\text{s}^{-1}$
N_c	Drying rate in constant-rate period	$\text{kg m}^{-2}\,\text{s}^{-1}$
p	Partial pressure of water	Pa
p_a	Atmospheric pressure	Pa
p_i	Partial pressure of water in inlet air	Pa
p_w	Vapour pressure of pure water	Pa
P	Electric power	W
P_v	Power density	W m^{-3}
q	Quality parameter	–
Q	Heat flux	kW m^{-2}
Q_{bed}	Total heat absorbed by bed	kW
Q_{ev}	Latent heat component of Q_{bed}	kW
Q_l	Heat loss	kW m^{-2}
Q_{sen}	Sensible heat component of Q_{bed}	kW
r	Ratio of salvage value to initial cost	–
R	Electrical resistance	Ω
R	Dimensionless parameter (Chapter 3)	–
R_f	Ratio of annual capital cost to initial cost	–
R_g	Universal gas constant	$\text{kJ kmol}^{-1}\,\text{K}^{-1}$
R_m	Ratio of maintenance cost to initial cost	–

Symbol	Definition	Units
s	Grain loading	$\mathrm{kg\ m^{-2}}$
S	Solids mass flowrate (dry basis)	$\mathrm{kg\ s^{-1}}$
S_i	Storage index	–
S'	Solids mass flowrate (dry basis) per unit area	$\mathrm{kg\ m^{-2}\ s^{-1}}$
t	Time	s
t_c	Drying time in the constant-rate period	s
t_d	Drying time	s
t_f	Drying time in the falling-rate period	s
t_g	Germination-loss time	d
t_m	Mean residence time	s
t_u	Time that product remains unsold	h
t'	Time over which greenness drops to half its initial value	s
T	Temperature	K
T_a	Air temperature	K
T_{abs}	Temperature of absorber	K
T_{ai}	Inlet air temperature	K
T_{ao}	Outlet air temperature	K
T_{as}	Adiabatic saturation temperature	K
T_{bed}	Temperature of bed	K
T_{bp}	Boiling point temperature	K
T_c	Collapse temperature	K
T_D	Inlet air temperature required for drying	K
T_{db}	Dry-bulb temperature	K
T_E	Temperature of lowest eutectic	K
T_G	Glass-transition temperature	K
T_{hm}	Temperature of heating medium	K
T_L	Collector outlet temperature	K
T_r	Reference temperature	K
T_s	Solids temperature	K
T_{si}	Inlet solids temperature	K
T_{so}	Outlet solids temperature	K
T_T	Required water temperature in tank	K
T_{TS}	Set-point temperature	K
T_w	Temperature of wall	K
T_{wb}	Wet-bulb temperature	K
$T(t)$	Temperature function	K
u	Gas velocity	$\mathrm{m\ s^{-1}}$
u_{mf}	Minimum fluidizing velocity	$\mathrm{m\ s^{-1}}$
U	Overall heat transfer coefficient	$\mathrm{kW\ m^{-2}\ K^{-1}}$
U_m	Overall heat transfer coefficient (modified)	$\mathrm{kW\ m^{-2}\ K^{-1}}$
U_x	Overall heat transfer coefficient at x	$\mathrm{kW\ m^{-2}\ K^{-1}}$

Symbol	Definition	Units
v	Volume	m^3
v_H	Humid volume	$m^3\,kg^{-1}$
v_s	Volumetric flow per unit area	$m^3\,m^{-2}\,s^{-1}$
V	Voltage	V
w	Gain mass	t
W_{ev}	Evaporation rate	$kg\,s^{-1}$
x	Linear dimension	m
X	Moisture content (dry basis)	$kg\,kg^{-1}$
X_{av}	Average moisture content (dry basis)	$kg\,kg^{-1}$
X_c	Critical moisture content (dry basis)	$kg\,kg^{-1}$
X_f	Free moisture content	$kg\,kg^{-1}$
X_i	Inlet or initial moisture content	$kg\,kg^{-1}$
X_o	Outlet moisture content (dry basis)	$kg\,kg^{-1}$
$X_{o.avr}$	Average outlet moisture content (dry basis)	$kg\,kg^{-1}$
X_w	Moisture content (wet basis)	$kg\,kg^{-1}$
X_{wi}	Inlet moisture content (wet basis)	$kg\,kg^{-1}$
X_{wo}	Outlet moisture content (wet basis)	$kg\,kg^{-1}$
X_{ws}	Moisture content at which grain, if harvested, would not be exposed to bad weather	$kg\,kg^{-1}$
$X(t)$	Drying curve function	$kg\,kg^{-1}$
X_1	Moisture content (dry basis) at start of drying	$kg\,kg^{-1}$
X_2	Moisture content (dry basis) at end of drying	$kg\,kg^{-1}$
X^*	Equilibrium moisture content	$kg\,kg^{-1}$
Y	Absolute humidity	$kg\,kg^{-1}$
Y_{as}	Absolute humidity at adiabatic saturation temperature	$kg\,kg^{-1}$
Y_i	Absolute humidity of inlet air	$kg\,kg^{-1}$
Y_o	Absolute humidity of outlet air	$kg\,kg^{-1}$
Y_p	Percent humidity	−
Y_s	Absolute humidity at saturation	$kg\,kg^{-1}$
Y_{wb}	Absolute humidity at wet-bulb temperature	$kg\,kg^{-1}$
$\mathbf{Y}_c$	Complex electrical admittance (empty applicator)	Ω^{-1}
$\mathbf{Y}_c'$	Complex electrical admittance (filled applicator)	Ω^{-1}
z	Length	m
$\mathbf{Z}_c$	Complex electrical impedance (empty applicator)	Ω
$\mathbf{Z}_c'$	Complex electrical impedance (filled applicator)	Ω
α	Coefficient	K^{-1}
α_c	Absorbance of absorber coating	−
γ	Greenness	−
Γ_p	Particle diffusivity	$m2\,s^{-1}$
Δh	Difference between inlet and outlet solids enthalpies	$kJ\,kg^{-1}$

Symbol	Definition	Units
Δt	Time interval	s
Δt_b	Length of break period	s
Δt_d	Length of drying period	s
$\Delta t_{i.b}$	Length of jth break period	s
$\Delta t_{i.d}$	Length of jth drying period	s
Δt_m	Time interval	s
Δx	Layer thickness	m
Δy	Incremental length in direction of solids flow	m
Δt_1	Time interval on batch drying curve	s
Δt_2	Time interval on batch drying curve	s
ΔC	Price differential	$ t^{-1}$
ΔH_w	Heat of wetting	kJ kg^{-1}
ΔP_b	Pressure drop across bed	N m^{-2}
ΔP_{mf}	Pressure drop across bed at minimum fluidization conditions	N m^{-2}
ΔT_{lm}	Log mean temperature difference	K
ΔX	Difference between inlet and outlet moisture contents of solids	kg kg^{-1}
$\Delta Y(t)$	Increase in absolute humidity of air as a function of time	Kg kg^{-1}
ε_0	Permittivity of free space	F m^{-1}
ε_r	Relative permittivity	–
ε_r'	Dielectric constant	–
ε_r''	Dielectric loss factor	–
η	Normalized free moisture content	–
η'	Normalized free moisture content	–
ξ	Effectiveness of solar collector	–
ξ_{max}	Maximum effectiveness of solar collector	–
λ	Latent heat of vaporization of water	kJ kg^{-1}
λ_{as}	Latent heat of vaporization of water at T_{as}	kJ kg^{-1}
λ_r	Latent heat of vaporization of water at T_r	kJ kg^{-1}
λ_s	Latent heat of vaporization of water at T_s	kJ kg^{-1}
λ_{wb}	Latent heat of vaporization of water at T_{wb}	kJ kg^{-1}
υ	Reduced drying rate	–
ρ_a	Density of inlet air	kg m^{-3}
ρ_s	Density of solids	kg m^{-3}
ψ	Relative humidity	–
σ	Electrical conductivity	Ω^{-1} m^{-1}
σ_γ	Standard deviation of greenness	–
τ	Transparency	–
ϕ_a	Solar energy flux transferred to air	W
ϕ_e	Effective solar energy flux used for drying	W

Symbol	Definition	Units
ϕ_u	Solar energy flux required by collector	W
ϕ_O	Incident solar energy flux	W
ω	Angular frequency	rad s^{-1}

1 The Industrial Drying of Foods: An Overview

R.B. KEEY

In 1985, McKeon of the Irish Industrial Development Authority concluded that the use of dehydrated food was declining in Europe and North America, while increasing elsewhere. It is likely that this situation still pertains a decade later. McKeon noted that the increased level of freezer ownership has resulted in a static or dwindling demand for many dried foods in the markets of the developed world. Nevertheless, several product categories were still experiencing rapid growth at the time; amongst other items, these included meat substitutes, breakfast cereals and fruit juices.

Worldwide, however, food dehydration has an important place. There is a substantial world trade in grains and dairy powders which require large-scale drying operations. Currently (1997), for example, a single plant is being built in New Zealand to produce 20 tonnes h^{-1} of dried milk powder.

This book is concerned with the theory of food drying operations and the equipment that is available to undertake them. Inevitably, in such a wide-ranging work with a multiple authorship, there is a diversity of viewpoints and a generalization of comment. Further, as Mujumdar writes in Chapter 2, food materials themselves are diverse, and include liquid solutions and gels, capillary–porous rigid bodies and capillary–colloidal materials. The range of inherent biophysical and biochemical properties has generated a host of technical solutions to the drying of foodstuffs.

On the other hand, one must not overemphasize the significance of this diversity. Irrespective of the nature of the food material, basic and common psychrometric and thermodynamic relationships underlie any drying operation; there are universal principles concerning the mode of operating dryers, whether batch or continuously worked units with a progressive flow of moist material through the plant. The nature of the food material essentially sets boundaries for the operating conditions and the types of processing equipment that might be suitable.

Hallström and co-authors (1988) point out that food properties are of two kinds: there are engineering properties which influence the heat and mass transfer, and other properties which are important in assessing the safety and quality of the products. In general, engineering properties are known or may be estimated to sufficient accuracy to make a preliminary assessment of energy demands and equipment needs, even if the properties are based on those of similar materials to the ones of interest. However, safety and quality issues are more material-specific. The susceptibility to biodeterioration may be determined by the structure of the material being dried (Strumillo *et al.*, 1996), but generally it is influenced to a greater or lesser extent by the chosen process temperature. A useful list of kinetic data for chemical changes, vitamin destruction, enzyme inactivation and micro-organism lethality is given by Hallström *et al.* (1988) and provides some indicators of the these changes with certain materials as they are dried at particular temperatures. A thin-layer, cross-circulation rig for analysing the deterioration kinetics of biomaterials has been described recently by Huber and Menner (1996). Less elaborate tests, however, may provide guidelines for temperature limits.

Fundamentals of the drying process are reviewed by Mujumdar in Chapter 2. A number of empirical equations have been proposed to correlate moisture desorption, and care must be exercised in extrapolating any correlation beyond the tested range of the embedded data as the moisture–equilibrium relationship is highly non-linear. Some understanding of the nature of moisture–solid bonding is advantageous in choosing an appropriate expression, although a simple exponential fit may be suitable in the high relative humidity range. The traditional description of drying kinetics, a constant rate period followed by one or more falling rate periods, may not be followed, particularly if the material is heat sensitive and some thermal transitions take place. Frequently, a Fickian moisture diffusion process is assumed, although the actual moisture movement in capillary–cellular material which shrinks on drying can be complex and composed of several mechanisms. For aqueous sugar solutions and skim milk in water, for example, the apparent moisture diffusion coefficient is a marked exponential function of water content, while the temperature dependence may be described by means of an Arrhenius-like equation (van der Lijn, 1976). Short-cut methods of describing the drying kinetics (such as the use of the characteristic drying curve) have certain advantages, if used judiciously, particularly in assessing the drying behaviour on full-scale plant. In this regard, Kerkhof's (1994) comparative evaluation of lumped-parameter methods for describing the drying rate of bioproducts in a fluidized bed dryer is useful in demonstrating the scope and weaknesses of various short-cut approaches.

Chapter 2 concludes with a brief summary of quality related aspects of food drying. Such issues are of overriding importance in food manufacture. While the presence of water is a necessary precondition for the enzymic and microbiological spoilage of foods, and the removal of water minimizes enzyme activity and microbial growth, drying can also be an effective means of preserving micro-organisms in a viable state. To minimize organoleptic changes in foods on drying, times and temperatures are kept as low as possible. Thus microbes are not exposed to highly lethal conditions. Some foods (such as pasta) can be preheated before drying for security; in other cases, extreme care with hygiene has to be undertaken with the selection and pre-processing of ingredients to avoid the ingress of pathogens. A useful discussion of the microbial resistance to drying is given by Gibbs (1985).

A number of authors discuss the use of particular kinds of dryers. Sokhansanj, in Chapter 3, considers through-flow dryers for agricultural crops. He notes that specific energy demands for these dryers range between 3600 and 4900 kJ kg^{-1} water evaporated. This is not unexpected, as the ideal heat demand for even an adiabatic dryer without solids heating is likely to be considerably greater than the latent heat of vaporization at the operating temperatures of these dryers (Keey, 1992). Besides providing a process design basis, Sokhansanj describes the use of a numerical quality index, which for grains might be the germination rate, relative gluten content before and after drying or the baking characteristics of the resultant flour. He also includes a probate model to account for the possibility that the thermal degradation kinetics might not be entirely deterministic. Such an approach is also likely to be useful in following the thermal destruction of micro-organisms which inherently have a variable resistance to heat. In this chapter, there is also a section on crop drying costs, which presumably refer to Canadian practice, and should be applied with caution to gain an insight into costs in other environments.

Bahu (Chapter 4) considers fluidized bed dryers which are commonly employed to produce dry particulate material because of their design simplicity and the confidence which can be placed in their process design. Various configurations are in use, including multistaged units and a cooling stage for heat sensitive foods. Hybrid dryers would be employed, with the fluidized bed as the second stage, when the feed material is very wet and sticky. Fluidized beds have also been used to create certain puffed food products (Shilton and Naranjan, 1994).

Spray drying was first introduced as a means of preserving fresh foods in powdered form. Masters, in Chapter 5, notes that today spray drying is a mature technology, often combining atomization, fluidization and agglomeration in a single system to meet end-product specifications of quality within a safe, hygienic and environment friendly process. The hazards of spray dryers are well-known, particularly with the recycle of combustible particles to hot air zones, and modern units incorporate suitable explosion relief systems. Spray dryers can now be designed to handle high fat, and sugar-containing foods, with particle temperatures being maintained relatively low even with high inlet air temperatures. A cocurrently worked chamber with an integrated belt for product removal would be used with high fat and sugary products when extended dwell times are needed for crystallization and drying.

Contact drying, as noted by Oakley in Chapter 6, is also regarded as environment friendly, because contact dryers can be operated with very low gas flows leading to high thermal efficiency and ease of exhaust gas clean-up. Because of the generally high capital costs of contact dryers compared with a possible alternative convective unit, the use of contact dryers is confined to the vacuum drying of heat sensitive, high value materials. Various kinds of agitated and scraped film dryers have been designed to handle materials which are very sticky or are likely to cause baking on drying.

Freeze drying is surveyed by Snowman in Chapter 7. This method of drying minimizes the physical changes that normally accompany drying, with a good retention of aroma, flavour and nutrients. (Theoretical and practical aspects of aroma retention in freeze and spray drying are considered in a review by Coumans *et al.*, 1994.) The technique of freeze drying involves subliming moisture that has been prefrozen and leaves behind a porous structure which can be readily rehydrated in subsequent use. The process, as a method of preservation, has to compete with direct refrigeration, which is unsuitable for certain products, such as instant beverages. Some freeze dried products are employed for convenience as constituents in other dry formulations, such as instant soups, or in special circumstances where minimal product weight is important, as in mountaineering and military use. As fats cannot be freeze dried, any product with a high fat content is likely to go rancid if this approach to dehydration is attempted.

Jones and Rowley, in Chapter 8, consider the place of dielectric dryers. These dryers are still limited in application to special circumstances, primarily because of the capital and operating costs. One example cited is the postbaking of biscuits, when convective drying in the falling rate period is enhanced by dielectric heating. Particular care has to be exercised with these methods in the drying of potentially incendive materials, such as granular dairy products, to prevent dielectric breakdown and consequential ignition.

In Chapter 9, which deals with specialized drying systems, Barr and Baker describe various dryers not considered in the previous chapters. These include batch tray dryers, the traditional workhorse. Although very flexible in use, and ideal for the small volume production of a variety of goods, these dryers have intensive labour requirements, as pointed out by the authors. Batch drying may lead to large moisture content variations in the solids unless the material is dried to low moisture levels. Perforated band dryers also lead to similar variability in drying (Keey, 1992), unless the solids are spread very thinly over the band.

Imre notes in Chapter 10 that open air drying in the sun was probably the earliest drying process employed by mankind. Solar drying is seen as an appropriate technology alternative to open air drying in regions of high insolation. Other developments include the use of solar assisted dryers, in conjunction with heat pump systems, to provide energy efficient units (Alves-Filho and Strømmen, 1996).

Baker provides some guidelines for dryer selection in Chapter 11. Knowledge, experience and science all play important roles, the author notes. The throughput requirements and the feed material's characteristics may exclude certain options, but there is usually more than one solution that is both technically feasible and economically viable. A given plant manufacturer will normally provide one particular solution based on the vendor's own expertise and experience. Choice between competing bids will be determined by price and the purchaser's assessment of the technical skill of the vendor. Although the literature contains some selection flowcharts, these provide only very rough and generalized indicators. However, more sophisticated guides are being developed, based on expert system and ranking methods for example (Kemp and Bahu, 1994), which will ultimately provide better selection tools.

The operation and control of dryers are considered by Gardiner in Chapter 12. He rightly points out that dryers frequently have a service life of 20 years or more. During their lifetime, it is therefore not uncommon for them to be used to dry several different products for which they were probably not designed. Operating flexibility is therefore very important. The effects of the principal process variables on product quality and throughput, energy economy, environmental impact, and safety are addressed in this chapter.

Concurrent dryers are normally easier to operate than those in which there is a countercurrent movement of goods and drying gas. Concurrent operation is self-regulating: a small deviation in feed conditions is attenuated as the two streams pass through the dryer converging towards equilibrium. On the other hand, an upset in either of the inlet streams flowing in a countercurrent direction is immediately felt in the adjacent outlet stream. Difficulties with condensation can also be experienced when cold feed material comes into contact with cooled, humidified gas.

Process control difficulties in drying are of two extreme kinds. With pneumatic conveying and spray drying installations, gas and particle residence times are extremely short and are of similar order to the time constants of the process-control loops. In marked contrast, the solids residence time in many band and rotary cascading dryers is lengthy; dead times are also long. Furthermore, except under special circumstances, it is not normally practical to measure the product moisture content on-line. Frequently then, the control of a dryer has consisted of simply regulating process variables, such as inlet air temperature and feedrate, rather than attempting feedback control of product variables. As discussed by Gardiner, newer control methodologies, based on expert systems and neural networks, are being developed. In another development, Kaminski *et al.* (1996) show how neural networks may be used to gain an insight into process behaviour. The example they use relates to the thermal degradation of vitamin C during the drying of a biomaterial and does not require any prior knowledge of the relationship between process variables and a quality index.

A number of non-conventional techniques have certain attractions. Animal feed materials such as sliced spent sugar beet and chopped alfalfa have been dried in a fluidized bed of superheated steam, with the evaporated moisture being used elsewhere in the factory (Jensen, 1994). Osmotic dehydration is the process whereby the moist solid is soaked in a concentrated sugar solution. Significant water removal takes place as well as some impregnation of the material with solute. Currently the process is used for producing candied fruits, but other possibilities include vegetable dehydration (Raoult-Wack, 1994). Another immersion process is the use of hot oil which acts as the heating medium for drying. The process is akin to deep fat frying and is believed to be used in the small scale drying of edible coconut.

The commercial success of any new food drying installation depends upon the dryer meeting the specified performance in all respects: product quality, production rate, safe and sanitary operation. This book shows that the design of many dryers can have a sound scientific base to give confidence in the operation of these units. However, small changes in process conditions may have a significant bearing on the nature of the product. An example is spray drying, where operation above the solvent's boiling point induces skin formation and puffing of the resultant dried particle. Increasingly, we are able to explain such phenomena, and thus tell processors of the likely boundaries of operation that are required to achieve satisfactory products. This book is one bridge spanning the gap between product knowledge and process insight on the one hand and equipment capability on the other.

References

Alves-Filho, O. and Strømmen, I. (1996). The application of heat pump in drying of biomaterials. *Drying Technol.*, **14**(9), 2061–90.

Coumans, W.J., Kerkhof, P.J.A.M. and Bruin, S. (1994) Theoretical and practical aspects of aroma retention in spray drying and freeze drying. *Drying Technol.*, **12**(1&2), 99–150.

Gibbs, P.A. (1985) Microbiological quality of dried foods, in *Concentration and Drying of Foods*, (ed. D. MacCarthy), Elsevier, London & New York, pp. 89–111.

Hallström, B., Skjöldebrand, C. and Trägårdh, C. (1988) *Heat Transfer and Food Products*, Elsevier, London & New York.

Huber, S. and Menner, M. (1996) A new laboratory dryer for analyzing the deterioration kinetics of biomaterials. *Drying Technol.*, **14**(9), 1947–66.

Jensen, S.A. (1994) Industrial experience in superheated steam fluid bed drying under pressure of beet pulp, sewage sludge and wood chips, in *Drying '94* (eds V. Rudolph and R.B. Keey), University of Queensland, pp. 519–26.

Kaminski, W., Strumillo, P. and Tomczak, E. (1996) Genetic algorithms and artificial neural networks for description of thermal degradation kinetics, *Drying Technol.*, **14**(9), 2117–34.

Keey, R.B. (1992) *Drying of Loose and Particulate Materials*, Taylor & Francis, New York, Washington.

Kemp, I.C. and Bahu, R.G. (1994) A new algorithm for dryer selection, in *Drying '94* (eds V. Rudolph and R.B. Keey), University of Queensland, pp. 439–46.

Kerkhof, P.J.A.M. (1994) A test of lumped-parameter methods for the drying rate in fluidized bed dryers for bioproducts, in *Drying '94* (eds V. Rudolph and R.B. Keey), University of Queensland, pp. 131–40.

McKeon, J. (1985) Market trends in dehydrated foods, in *Concentration and Drying of Foods* (ed. D. MacCarthy), Elsevier, London and New York, pp. 1–10.

Raoult-Wack, A.L. (1994) Recent advances in the osmotic dehydration of foods, *Trends in Food Science and Technology*, **5**, 255–60.

Shilton, N.C. and Niranjan, K. (1994) Puff drying of foods in a fluidized bed, in *Developments in Food Engineering* (eds. T. Yano *et al.*), Chapman & Hall, London, pp. 337–9.

Strumillo, C., Zbicinski, I. and Liu, X.D. (1996) Effect of particle structure on quality retention of bioproducts during thermal drying. *Drying Technol.*, **14**(9), 1921–46.

Van der Lijn, J. (1976) Simulation of heat and mass transfer in drying. Agricultural University, Wageningen, PhD thesis.

2 Drying Fundamentals

A.S. MUJUMDAR

2.1 Introduction

Drying is traditionally defined as that unit operation which converts a liquid, solid or semi-solid feed material into a solid product of significantly lower moisture content. In most cases, drying involves the application of thermal energy, which causes water to evaporate into the vapour phase. Freeze drying provides an exception to this definition, since this process is carried out below the triple point, and water vapour is formed directly through the sublimation of ice (King, 1971). The requirements of thermal energy, phase change and a solid final product distinguish this operation from mechanical dewatering, evaporation, extractive distillation, adsorption and osmotic dewatering.

Drying is a complex process involving simultaneous coupled, transient heat, mass and momentum transport. These are often accompanied by chemical or biochemical reactions and phase transformations, such as glass transition and crystallization, along with the shrinkage.

Foods are dried commercially, starting either from their natural state (e.g. vegetables, fruits, milk, spices, grains) or after processing (e.g. instant coffee, whey, soup mixes, non-dairy creamers). The production of a processed food may sometimes involve drying at several stages in the operation. In some cases, pre-treatment of the food product may be necessary prior to drying.

In addition to preserving the product and extending its shelf life, drying may be carried out to accomplish one or more of the following additional objectives:

- obtain desired physical form (e.g. powder, flakes, granules);
- obtain desired colour, flavour or texture;
- reduce volume or weight for transportation;
- produce new products which would not otherwise be feasible.

Drying is important to the food industry as it consumes up to 10% of the total energy used in that sector. The selection of a dryer is, however, driven more by product quality considerations than by energy saving potential. Environmental impact and safety of operation are additional factors which influence the selection of a drying system.

While over 200 different types of dryer have found various applications in industry, only about 20 basic types and their variants are commonly used in practice. This wide range of dryers is due to the diverse physical forms of the products to be dried, the production rates desired, and the quality constraints on the dried product. The wet feedstock may be in the form of a liquid (slurry, suspension or solution), a solid (particulate, sheet-like, pelletized, extruded forms) or pasty. Further, one may distinguish between three main types of food material: liquid solutions and gels, capillary-porous rigid materials and capillary-porous colloidal materials.

Drying of food materials is further complicated by the fact that physical, chemical and biochemical transformations may occur during drying, some of which may be desirable and others undesirable. Physical changes such as glass transitions or crystallization during drying can result in changes in mechanisms of mass transfer and rates of heat transfer within the material, often in an unpredictable manner.

Figure 2.1 shows a schematic flow chart for the overall process design of dryers. It is important to note that both feed and product characteristics as well as equipment features are essential parts of the design process. Further, laboratory or pilot-scale tests are essential unless there is relevant prior experience with the product as well as the type of dryer selected. Dryer design calculations require thermodynamic information, e.g. solid–moisture equilibrium data and gas–vapour thermodynamic properties (for direct dryers), as well as kinetics information (e.g. drying rates as functions of relevant parameters). The final design choice is often a compromise between the requirements of thermodynamics (for the highest thermal and mass transfer efficiency) and kinetics

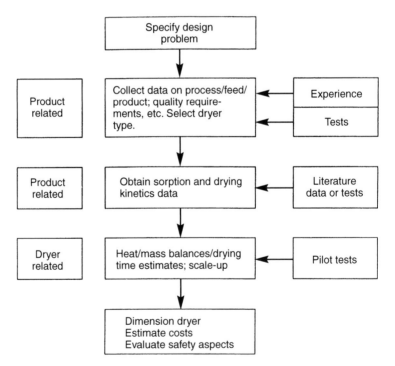

Figure 2.1 Flow chart for process design of dryer.

(for the smallest dryer dimensions). Appropriate scale-up criteria need to be developed for the dryer selected.

The object of this chapter is to provide a brief summary of the basic concepts of the drying of solids, with special reference to food materials. Topics covered include:

- a discussion of basic dryer types and selection criteria;
- thermodynamics of moist solids (e.g. sorption phenomena, equilibrium moisture content, water activity);
- psychrometric relations for the air–water system;
- kinetics of drying (e.g. drying rate curves, characteristic drying rate);
- hygrothermal and quality properties of foods.

2.2 Basic Dryer Types

Figure 2.2 indicates the basic dryer types based on the mode of heat input e.g. convection, conduction, radiation, dielectric heating or combinations of one or more of these modes. Convective drying accounts perhaps for over 90% of dehydrated food production, despite the fact that the other dryer types provide some important advantages in terms of energy efficiency, product quality, and environmental impact.

Dryers can be further classified in a number of ways, for example, on the basis of pressure (vacuum or near-atmospheric), temperature of the product

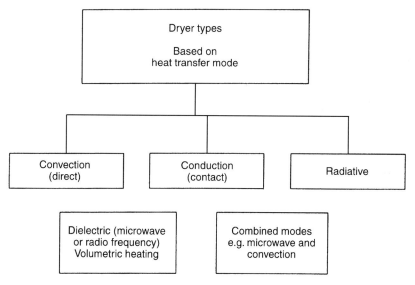

Convection, conduction and radiant heating supply heat at the
surface of the material and RF heating supplies heat volumetrically

Figure 2.2 Basic dryer types classified according to the mode of heat transfer.

during drying (low or high; below or above the boiling point of water), mode of operation (batch or continuous), and the method of material handling within the dryer (stationary, agitated, dispersed, fluidized, converged, falling under gravity). In each category, finer sub-classification is possible. For instance, at least 30 variants of the fluidized bed dryer can be identified for drying applications involving particles, slurries, planks, and continuous sheets.

Dryers can be classified broadly as batch or continuous types, depending on their mode of operation. In general, batch dryers are preferred for smaller scale operations (under 50 kg h^{-1}), long residence times (e.g. several hours), where batch equipment is located upstream and downstream, and also when quality integrity within a batch is essential (e.g. in pharmaceuticals). Not all dryers can be operated in a batch mode. For example, spray, rotary, flash, and drum dryers can only be operated in the continuous mode. Fluidized and through-circulation dryers, on the other hand, can operate as batch or continuous dryers.

The drying time required for a given product is also a key factor influencing the choice of dryer. Only flash and spray dryers have inherently short residence times (under 1 min), while only a batch tray dryer can economically provide residence times exceeding several hours. Most other continuous dryers have residence times ranging from several minutes to 2 h. When longer dwell times are required due to the special characteristics of the material (such as a wide particle size distribution), it may be necessary to separate and recycle the larger, wetter product back into the dryer in order to provide additional drying time. Table 2.1 lists the various criteria used to classify dryers. The most commonly used types are indicated with an asterisk.

Table 2.1 Dryer classification according to various criteria

Criterion	Types
Mode of operation	Batch, continuous*
Physical form of feed	Liquid, paste-like, powder, chips, continuous sheet, large planks
State of material in dryer	Stationary, moving, agitated, vibrated, converged, dispersed, fluidized
Heat input	Convection,* conduction, radiation, electromagnetic field, combination of various modes, adiabatic/non-adiabatic
Operating pressure	Vacuum; atmospheric;* high pressure
Drying medium	Hot air;* superheated steam; flue gases
Drying temperature	Below triple point, below boiling point,* above boiling point
Relative flow direction of material and gas in dryer	Cocurrent, countercurrent, mixed flow, cross-flow
Time-variation of heat input	Continuous,* intermittent
Number of stages	Single, multiple
Residence time	Short (<1 min), medium (1–30 min), long (>30 min)

*Most common types.

One of the most important factors to be considered in the design and operation of food dryers is the inherent heat sensitivity of foods. In process design calculations, it is essential to ensure that the material does not exceed its maximum permissible temperature. In some special cases, sterilization of the product may also be accomplished in the dryer. Dryer selection is discussed in depth in Chapter 11. The following list includes some of the key criteria that are particularly relevant for food materials and should be considered in this process:

- Ability to handle the material at the required rate and to deliver a product of specified quality;
- Ease of hygienic operation without contamination;
- Turndown ratio;
- Flexibility (to process other similar products);
- Safe operation (no fire/explosion risk, toxicologically safe);
- Ease of control;
- Energy consumption;
- Low cost, small space requirements;
- Multi-processing capability (e.g. cooling and granulation/instantizing, in the same unit).

A special comment concerning toxicological safety in the drying of foods is essential. Micro-organisms can form exotoxins during the early stages of drying, due to high moisture levels, and may convert nitrates in foods (e.g. meat, spinach, cheese) into toxic nitrite. Highly toxic oxycholesterol products may be formed even under mild conditions when drying cholesterol-containing foods such as milk, meat and eggs.

When foods are dried in direct contact with products of combustion containing carcinogenic polycyclic aromatic hydrocarbons (PAHs), these may be adsorbed on to the foods. Further, the NO_x compounds in combustion gases used in direct-fired dryers, as well as the nitrites present in some foods as additives (e.g. cheeses, cured meats), can react with secondary amines in the food to produce powerful carcinogens such as nitrosamines. For the aforementioned reasons, in the absence of definitive, extensive scientific studies, the use of direct-fired dryers is not recommended for the convective drying of food materials. The air used for such drying operations must be heated indirectly by using suitable heat exchangers.

2.3 Thermodynamic properties of air–water mixtures and moist solids

2.3.1 Psychrometry

As noted earlier, a majority of dryers in the food industry are of the direct (or convective) type. In other words, hot air is used both to supply the heat for evaporation and to carry away the evaporated moisture from the product.

Notable exceptions are freeze and vacuum dryers, which are used almost exclusively for drying highly heat-sensitive products because they tend to be significantly more expensive than dryers that operate near to atmospheric pressure. Another exception is the emerging technology of superheated steam drying (Mujumdar, 1995). In certain cases, such as the drum drying of pasty foods, some or all of the heat is supplied indirectly by conduction.

Drying with heated air implies humidification and cooling of the air in a well-insulated (adiabatic) dryer. Thus hygrothermal properties of humid air are required for the design calculations of such dryers. Pakowski *et al.* (1991) have presented a comprehensive summary of the engineering properties of humid air. Both thermodynamic (e.g. adiabatic saturation temperature, humid heat, humid enthalpy) as well as transport properties (e.g. thermal conductivity, moisture diffusivity, permeability, inter-phase heat/mass transfer coefficients) are essential in dryer calculations. Table 2.2 provides a listing of brief definitions of various terms encountered in drying and psychrometry. It includes several terms not explicitly discussed in the text.

Figure 2.3 is a psychrometric chart for the air–water system, plotted using the dryPAK software developed by Pakowski *et al.* (1996). This figure shows the relationship between the temperature (abscissa) and absolute humidity (ordinate, in g water per kg dry air) of humid air at 1 atmosphere total pressure over the range 0° to 180°C. Lines representing percent humidity and adiabatic saturation are drawn according to the thermodynamic definitions of these terms. Most handbooks of engineering provide more detailed psychrometric charts including additional information and extended temperature ranges. Mujumdar (1995) includes psychrometric charts for several gas–organic vapour systems as well.

Numerous empirical correlations are available in the literature which could be used to compute the psychrometric properties of air and steam. Schmidt (1969) provides detailed thermodynamic and transport property data for air, water and steam. Pakowski *et al.* (1991) have evaluated the various correlations with regard to their applicability in dryer calculations. They caution strongly against the extrapolation of empirical correlations beyond the ranges of parameters for which they were derived.

Table 2.3 summarizes the essential thermodynamic relationships for humid air. Equations for the adiabatic saturation- and wet-bulb temperature lines on the psychrometric chart are as follows (Geankoplis, 1983):

$$\frac{Y - Y_{as}}{T - T_{as}} = -\frac{c_s}{\lambda_{as}} = -\frac{1.005 + 1.88Y}{\lambda_{as}} \qquad (2.1)$$

and

$$\frac{Y - Y_{wb}}{T - T_{wb}} = -\frac{h/M_{air} k_y}{\lambda_{wb}} \qquad (2.2)$$

The ratio $(h/M_{air} k_y)$, termed the psychrometric ratio, lies between 0.96 and 1.005 for air–water vapour mixtures; thus it is nearly equal to the value of humid heat

c_s. If the effect of humidity is neglected, the adiabatic saturation- and wet-bulb temperatures (T_{as} and T_{wb}, respectively) are almost equal for the air–water system. Note, however, that T_{as} and T_{wb} are conceptually quite different. The adiabatic saturation temperature is a gas temperature and a thermodynamic entity. Plots of Y_{as} versus T_{as} on a psychrometric chart are straight lines and represent the path followed by the air in an adiabatic dryer. In contrast, the wet-bulb temperature is a heat and mass transfer rate-based parameter and refers to the temperature of the liquid phase. Under constant drying conditions, the surface of the drying material attains the wet-bulb temperature if heat transfer is

Table 2.2 Definition of commonly encountered terms in psychrometry and drying (in alphabetical order)

Term/symbol	Meaning
Adiabatic saturation temperature, T_{as}	Equilibrium gas temperature reached by unsaturated gas and vaporizing liquid under adiabatic conditions. (Note: for the air–water system only, it is equal to the wet bulb temperature (T_{wb}).)
Bound moisture	Liquid physically and/or chemically bound to solid matrix so as to exert a vapour pressure lower than that of pure liquid at the same temperature.
Constant rate drying period	Under constant drying conditions, drying period when evaporation rate per unit drying area is constant (when surface moisture is removed).
Dew point	Temperature at which a given unsaturated air–vapour mixture becomes saturated.
Dry bulb temperature	Temperature measured by a (dry) thermometer immersed in vapour–gas mixture.
Equilibrium moisture content, X^*	At a given temperature and pressure, the moisture content of moist solid in equilibrium with the gas–vapour mixture (zero for non-hygroscopic solids).
Critical moisture content, X_c	Moisture content at which the constant drying rate first begins to drop (under constant drying conditions).
Falling-rate period	Drying period (under constant drying conditions) during which the rate falls continuously with time.
Free moisture, X_f; $X_f = X - X^*$	Moisture content in excess of the equilibrium moisture content (hence free to be removed) at given air humidity and temperature.
Humid heat	Heat required to raise the temperature of unit mass of dry air and its associated vapour through 1 degree (kJ kg^{-1} K^{-1} or Btu lb^{-1} °F^{-1}).
Humidity, absolute	Mass of water vapour per unit mass of dry gas (kg kg^{-1} or lb lb^{-1}).
Humidity, relative	Ratio of partial pressure of water vapour in gas–vapour mixture to equilibrium vapour pressure at the same temperature.
Unbound moisture	Moisture in solid which exerts vapour pressure equal to that of pure liquid at the same temperature.
Water activity, a_w	Ratio of vapour pressure exerted by water in solid to that of pure water at the same temperature.
Wet bulb temperature, T_{wb}	Liquid temperature attained when large amounts of air–vapour mixture are contacted with the surface. In purely convective drying, the drying surface reaches T_{wb} during the constant-rate period.

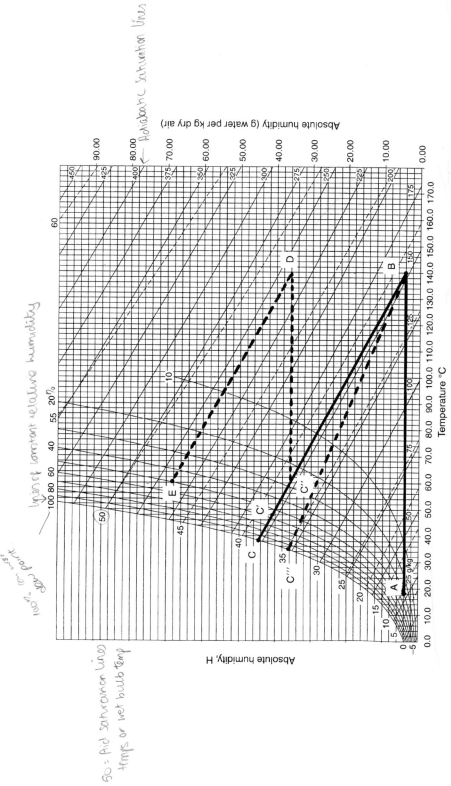

Figure 2.3 Psychrometric chart for air–water vapour system.

Handwritten annotations:

Adiabatic saturation lines

lines of constant relative humidity

50 = Adiabatic saturation lines
temps or wet bulb temp

100% dew point

Table 2.3 Psychrometric equations for an air–water vapour system (Geankopolis, 1983)

Parameter	Equation
Absolute humidity: kg H_2O per kg dry air	$Y = \dfrac{18.02}{29.97}\dfrac{p}{p_a - p}$
Saturation humidity	$Y_s = \dfrac{18.02}{29.97}\dfrac{p_w}{p_a - p_w}$
Percent humidity	$Y_p = 100\,\dfrac{Y}{Y_s}$
Relative humidity	$\psi = 100\,\dfrac{p}{p_w}$
Humid heat: kJ kg^{-1} dry air	$c_s = 1.005 + 1.88Y$
Humid volume: m^3 kg^{-1} dry air	$v_H = \dfrac{22.41}{273}\,T\left(\dfrac{1}{29.97} + \dfrac{Y}{18.02}\right)$ (T in K)
Total enthalpy: kJ kg^{-1} dry air	$H = (1.005 + 1.88Y)(T - T_r) + Y\lambda_r$ T_r = Reference temperature, K
Latent heat of vaporization λ, kJ kg^{-1}	$\lambda = a_1(a_2 T)^{a_3}$ $a_1 = 267.155,\ a_2 = 374.2,\ a_3 = 0.38$ T in °C

by pure convection. The wet-bulb temperature is independent of surface geometry as a result of the analogy between heat and mass transfer.

A simple example of the use of a psychrometric chart in dryer calculations is illustrated below.

Example: A food product is to be dried in an adiabatic cocurrent dryer. The inlet and outlet moisture contents are 0.28 and 0.12 kg H_2O per kg dry product respectively. Ambient air at 20°C and 30% relative humidity is heated indirectly by steam to the specified dryer inlet air temperature of 140°C. The difference between the dry-bulb temperature of the exhaust air and its dewpoint should be at least 10°C in order to avoid the possibility of condensation in the downstream ductwork and air cleaning devices. Calculate the mass flowrate of air required (kg h^{-1}, dry basis) per kg h^{-1} of bone-dry solids.

Consider the ambient air. Its condition can be represented on the psychrometric chart (Figure 2.3) by point A at which the 20°C and 30% relative humidity lines intersect. Note that the corresponding absolute humidity can be read from the graph as 4 g H_2O per kg dry air.

The ambient air is heated indirectly at constant absolute humidity to 140°C. Its condition is now represented by point B in Figure 2.3. The ideal path to saturation followed by the air in the dryer is represented by the adiabatic saturation line passing through point B, BC. This may be drawn slightly above and parallel to the 38°C line.

The condition of the exhaust air has to be determined by trial and error. Assume that its dry-bulb temperature is 60°C. The dewpoint (temperature corresponding to a relative humidity of 100% at the absolute humidity of the exhaust air) is approximately 35°C, and the difference between the two is $60-35$ or 25°C, which exceeds 10°C by a wide margin. Clearly, therefore, further drying can be permitted to take place. Air with a dry-bulb temperature of 48°C, denoted by C' on the psychrometric chart, has a corresponding wet-bulb temperature of around 37°C and therefore satisfies the criterion approximately. Its absolute humidity is 42 g H_2O per kg dry air.

The required air flowrate can now be determined from a material balance. The moisture lost per kg dry solids is $1 \times (0.28 - 0.12) \times 1000 = 160$ g H_2O h^{-1}. This moisture, which is gained by the air, is also equal to $G \times (42 - 4) = 38G$ g H_2O h^{-1}, where G is the mass flowrate of dry air in kg h^{-1}. Therefore, $G = 160/38 = 4.21$ kg h^{-1}.

Note that considerations other than psychrometry (e.g. equilibrium moisture content and drying kinetics, sections 2.3.2 and 2.4) must be taken into account when considering possible dryer designs. There is often a direct working relationship between the exhaust air temperature and the outlet moisture content of the solids. Let us assume that, in the present case, an exhaust air temperature of 60°C is required. This may be achieved by allowing the air to exhaust at this temperature (point C'' in Figure 2.3). In this case, the required air flow will rise to 5.0 kg h^{-1}. Alternatively, in some types of dryer, it is possible to reheat the air at an intermediate point in the drying cycle. This is illustrated by path $C''DE$. Here, when the temperature of the air drops to 60°C, it is reheated to 140°C, and drying continues until it has again dropped to 60°C. In this case, an air flow of 2.46 kg h^{-1} is required. Note that the 10°C difference between the temperature of the exhaust air and its dewpoint is still maintained under these conditions. In practice, the condition of the air may not exactly follow an adiabatic saturation line. This may be because of heat losses from the dryer and/or a significant heat of wetting. Path BC''' illustrates an example of such a situation.

2.3.2 Equilibrium moisture content

The moisture content of a wet solid in equilibrium with air of given humidity and temperature is termed the equilibrium moisture content (EMC). A plot of EMC at a given temperature versus the dry-basis moisture content of the solid is the so-called sorption isotherm. An isotherm obtained by exposing the wet solid to air of increasing humidity is termed the adsorption isotherm. That obtained by exposing the solid to air of decreasing humidity is known as the desorption isotherm. Clearly, the latter is of interest in drying as the moisture content of the solids progressively decreases. Most food materials display 'hysteresis' in that the two isotherms are not identical.

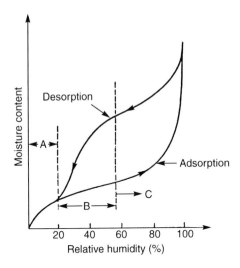

Figure 2.4 Typical sorption isotherms showing hysteresis.

Figure 2.4 shows the general shape of the sorption isotherms for a food material. They are characterized by three distinct zones, A, B and C, which are indicative of different water binding mechanisms at individual sites on the solid matrix. In region A, the water is tightly bound to the sites and is unavailable for reaction. In this region, there is essentially monolayer adsorption of water vapour and no distinction exists between the adsorption and desorption isotherms. In region B, the water is more loosely bound. The vapour pressure depression below the equilibrium vapour pressure of water at the same temperature is due to its confinement in smaller capillaries. Water in region C is even more loosely held in larger capillaries. It is available for participation in reactions and as a solvent.

Numerous hypotheses have been proposed to explain the hysteresis (Bruin and Luyben, 1980; Fortes and Okos, 1980; Bruin, 1988). For example, once a food material undergoes drying (desorption), the material shrinks and has fewer polar sites for water binding in a subsequent adsorption cycle.

Figure 2.5 shows schematically the shapes of the equilibrium moisture curves for various types of solids. Figure 2.6 show the various types of moisture (already defined in Table 2.2). The desorption isotherms are also dependent on external pressure. However, in all practical cases of interest, this effect may be neglected. According to Keey (1978) the dependence of the equilibrium moisture content on temperature can be correlated by

$$\left[\frac{\Delta X^*}{\Delta T}\right]_{\psi=\text{const}} = -\alpha X^* \tag{2.3}$$

where X^* is the dry-basis equlibrium moisture content, T is the temperature and ψ is the relative humidity of air. The parameter α ranges from 0.005 to 0.01

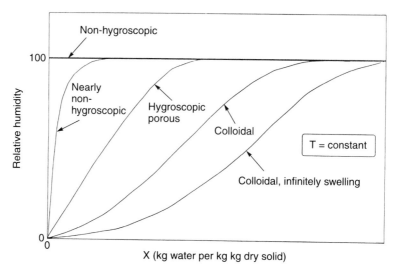

Figure 2.5 Equilibrium moisture curves for various types of solids.

K^{-1}. This correlation may be used to estimate the temperature dependence of X^* if no data are available.

For hygroscopic solids, the enthalpy of the attached moisture is less than that of pure liquid by an amount equal to this binding energy, which is also termed the enthalpy of wetting, ΔH_w (Keey, 1978). It includes the heats of sorption, hydration and solution and may be estimated from the following equation:

$$\frac{d(\ln \psi)}{d(1/T)}\bigg|_{X=\text{const}} = -\frac{\Delta H_w}{R_g T} \tag{2.4}$$

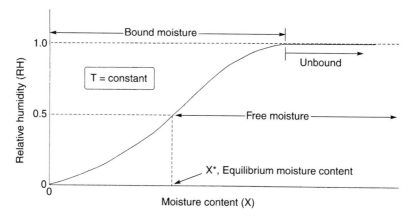

Figure 2.6 Equilibrium moisture curves showing different types of moisture (see Table 2.2 for definitions).

A plot of $\ln(\psi)$ against $1/T$ is linear with a slope of $\Delta H_w/R_g$ where R_g is the universal gas constant ($R_g = 8.314 \times 10^3$ kg kmol^{-1} K^{-1}). Note that the total heat required to evaporate bound water is the sum of the latent heat of vaporization and the heat of wetting; the latter is a function of the moisture content X. The heat of wetting is zero for unbound water and increases with decreasing X. Since ΔH_w is responsible for lowering the vapour pressure of bound water, at the same relative humidity, ΔH_w is almost the same for all materials (Keey, 1978).

For most foods, the moisture binding energy is positive; generally, it is a monotonically decreasing function of the moisture content, with a value of zero for unbound moisture. For hydrophobic materials (e.g. peanut oil, starches at lower temperatures), the binding energy can be negative.

In general, water sorption data must be determined experimentally. Some 80 correlations, ranging from those based on theory to those that are purely empirical, have appeared in the literature. Two of the most extensive compilations are due to Wolf et al. (1985) and Iglesias and Chirife (1982). Aside from temperature, water sorption is affected by the physical structure as well as the composition of the food. Sorption behaviour of regular and puffed pasta is different, for example. The pore structure and size, as well as gelatinization during processing, can cause significant variations in the moisture binding ability of the solid.

Composition related effects are harder to quantify and predict. Studies have been reported on the interaction of water and binding energy for proteins, lipids and polysaccharides, which are the key constituents of foods. Crapiste and Rotstein (1982) predicted reasonably well the isotherms for potatoes, peas, beans, corn and white rice based on a knowledge of the composition and sorption data of the basic components of the food. More general models need to be developed and validated, however.

2.3.3 Water activity

In a food material the availability of water for the growth of micro-organisms, the germination of spores, and participation in several types of chemical reaction depends on its relative vapour pressure, or water activity, a_w. This is defined as the ratio of the partial pressure, p, of water over the wet food system to the equilibrium vapour pressure, p_w, of water at the same temperature. Thus, a_w, which is also equal to the relative humidity of the surrounding humid air, is defined as

$$a_w = p/p_w \qquad (2.5)$$

Different shapes for the X versus a_w curves are observed, depending on the type of material (e.g. high, medium or low hygroscopicity solids).

Table 2.4 lists the measured minimum a_w values for microbial growth or spore germination. If a_w is reduced below this value by dehydration or by adding

Table 2.4 Minimum water activity, a_w, for microbial growth and spore germination (adapted from Brockmann, 1973)

Micro-organism	a_w
Organisms producing slime on meat	0.98
Pseudomonas, Bacillus cereus spores	0.97
B. subtilis, C. botulinum spores	0.95
C. botulinum, Salmonella	0.93
Most bacteria	0.91
Most yeasts	0.88
Aspergillus niger	0.85
Most moulds	0.80
Halophilic bacteria	0.75
Xerophilic fungi	0.65
Osmophilic yeast	0.62

water binding agents such as sugars, glycerol, or salt, microbial growth is inhibited. Such additives should not affect the flavour, taste or other quality criteria, however. Since the amount of soluble additives needed to depress a_w even by 0.1 is quite large, dehydration becomes particularly attractive for high moisture foods as a way of reducing a_w. Figure 2.7 shows schematically the water activity versus moisture content curve for different types of food. Rockland and Benchat (1987) provide an extensive compilation of results on water activity and its applications.

Figure 2.8 shows the general nature of the deterioration reaction rates as a function of a_w for food systems. Aside from microbial damage, which typically occurs for $a_w > 0.7$, oxidation, non-enzymatic browning (Maillard reactions) and enzymatic reactions can occur even at very low a_w levels during drying.

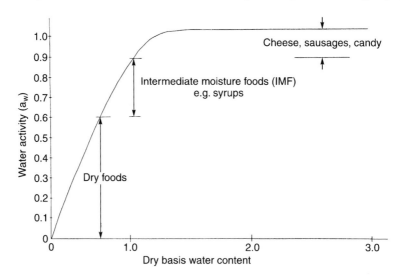

Figure 2.7 Water activity versus moisture content for different food types.

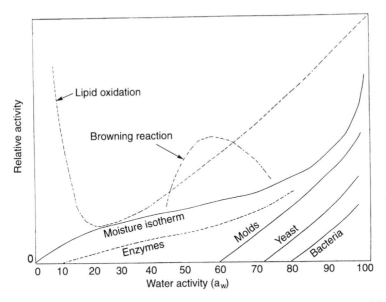

Figure 2.8 General behaviour of the determination reaction rates as functions of water activity for food systems.

Laboratory or pilot testing is essential to ascertain that no damage occurs during the selected process since this cannot in general be predicted.

2.4 Drying Kinetics

Consider the drying of a wet solid under fixed drying conditions. In the most general case, after an initial period of adjustment, the dry-basis moisture content, X, decreases linearly with time, t, following the start of the evaporation. This is followed by a non-linear decrease in X with t until, after a very long time, the solid reaches its equilbrium moisture content, X^* and drying stops. In terms of free moisture content, defined as

$$X_f = (X - X^*) \tag{2.6}$$

the drying rate drop to zero at $X_f = 0$.

By convention, the drying rate, N, is defined as:

$$N = -\frac{M_s}{A}\frac{\mathrm{d}X}{\mathrm{d}t} \quad \text{or} \quad -\frac{M_s}{A}\frac{\mathrm{d}X_f}{\mathrm{d}t} \tag{2.7}$$

under constant drying conditions. Here N (kg m^{-2} h^{-1}) is the rate of evaporation of water, A is the evaporation area (this may be different from the heat transfer area) and M_s is the mass of bone dry solid. If A is not known, then the drying rate may be expressed in kg water evaporated per h.

A plot of N versus X (or X_f) is called the drying rate curve and is always obtained under constant drying conditions. Note that, in actual dryers, the drying material is generally exposed to varying drying conditions (e.g. different relative gas–solid velocities, different temperatures and humidities, different flow orientations). Thus, it is necessary to develop a methodology to interpolate or extrapolate limited drying rate data over a range of operating conditions.

Figure 2.9 shows a typical 'textbook' drying rate curve displaying an initial constant rate period where $N = N_c$ = constant. At the so-called critical moisture content, X_c, N begins to fall with further decreases in X. The mechanism underlying this phenomenon depends both on the material and the drying conditions. Many foods and agricultural products do not display a constant rate period at all since internal heat and mass transfer rates determine the rate at which water becomes available at the exposed evaporating surface. The drying rate in the constant rate period is governed fully by the rates of external heat and mass transfer, since a film of free water is always available at the evaporating surface. The drying rate in this period is essentially independent of the material being dried. The rate N begins to drop at $X = X_c$ since water cannot migrate to the surface at the rate N_c because of internal transport limitations. Under these conditions, the drying surface becomes first partially unsaturated and then fully unsaturated until it reaches the equilibrium moisture content X^*. Detailed discussions of drying rate curves are given by Keey (1991), Mujumdar and Menon (1995), and Perry *et al.* (1996).

Note that a material may display more than one critical moisture content at which the drying rate curve shows a sharp change of shape. This is generally associated with changes in the underlying mechanisms of drying due to structural or chemical changes. Further, X_c is not a material property. It depends on

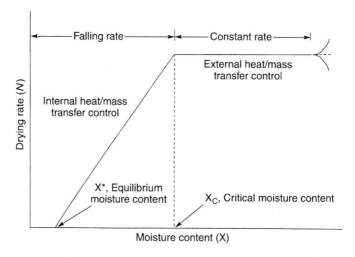

Figure 2.9 A textbook drying rate curve under constant drying conditions.

the drying rate under otherwise similar conditions. It must be determined experimentally.

It is easy to see that N_c can be calculated quite easily using empirical or analytical techniques to estimate the external heat/mass transfer rates (Keey, 1978; Geankoplis, 1983). Thus,

$$N_c = \sum q/\lambda_s \qquad (2.8)$$

where $\sum q$ represents the sum of heat fluxes due to convection, conduction and/ or radiation and λ_s is the latent heat vaporization at the solid temperature. In the case of purely convective drying, the drying surface is always saturated with water in the constant rate period and the liquid film attains the wet-bulb temperature. Because of the analogy between heat and mass transfer, the wet-bulb temperature is independent of the geometry of the drying object.

The drying rate in the falling rate period(s) is a function of X (or X_f) and must be determined experimentally for a given material being dried in a given type of dryer. If N versus X is known, the drying time required to reduce the solid moisture content from X_1 to X_2 is simply, by definition:

$$t_f = - \int_{X_1}^{X_2} \frac{M_s}{A} \frac{\mathrm{d}X}{N} \qquad (2.9)$$

Table 2.5 lists expressions for the drying time for constant rate, linear falling rates, and a falling rate controlled by liquid water diffusion in a thin slab. The subscripts c and f refer to the constant rate and falling rate periods, respectively. Different analytical expressions are obtained for the drying times, t_f, depending on the functional form of N or the model used to describe the falling rate e.g.

Table 2.5 Drying times for various drying rate models (constant drying conditions)

Model	Drying time
Kinetic model $N = -\dfrac{M_s}{A}\dfrac{\mathrm{d}X}{\mathrm{d}t_d}$	t_d = drying time to reach final moisture content X_2 from initial moisture content X_1
$N = N(X)$ (General)	$t_d = \dfrac{M_s}{A} \displaystyle\int_{X_2}^{X_1} \dfrac{\mathrm{d}X}{N}$
$N = N_c$ Constant rate	$t_c = -\dfrac{M_s}{A} \dfrac{(X_2 - X_1)}{N_c}$
$N = aX + b$ Falling rate	$t_f = \dfrac{M_s}{A} \dfrac{(X_1 - X_2)}{(N_1 - N_2)} \ln \dfrac{N_1}{N_2}$
$N = aX$ $X^* \le X_2 \le X_c$	$t_f = \dfrac{M_s X_c}{A N_c} \ln \dfrac{X_c}{X_2}$
Liquid diffusion model D_L = constant $X_2 = X_c$ Slab; one-dimensional diffusion, evaporating surface at X^*	$t_f = \dfrac{4a^2}{\pi^2 D_L} \ln \dfrac{8X_1}{\pi^2 X_2}$ X = average free moisture content a = half-thickness of slab dried uniformly from both sides

diffusion of liquid or vapour, capillarity, evaporation–condensation. For some solids, a receding front model (wherein the evaporating surface recedes into the drying solid) yields good agreement with observations. The principal goal of all models of falling rate drying is to allow reliable extrapolation of drying kinetic data over various operating conditions and product geometries.

The expression for t_f in Table 2.5 using the liquid diffusion model (Fick's second law form applied to diffusion in solids with no real fundamental basis) is obtained by solving analytically the partial differential equation

$$\frac{\partial X_f}{\partial t} = D_L \frac{\partial^2 X_f}{\partial x^2} \tag{2.10}$$

subject to　　　$X_f = X_i,$　　everywhere in the slab, at $t = 0$
　　　　　　　　$X_f = 0$　　　at $x = a$ (top, evaporating surface), and
　　　　　　$\partial X_f / \partial x = 0$　　at $x = 0$ (bottom, non-evaporating surface).

The model assumes one-dimensional diffusion with D_L = constant and no heat effects. X_2 is the average free moisture at $t = t_f$ obtained by integrating the analytical solution $X_f(x, t_f)$ over the thickness of the slab, a. The expression in Table 2.5 is applicable only for long drying times since it is obtained by retaining only the first term in the infinate series solution of the partial differential equation.

The diffusivity of moisture in solids is a function of both temperature and moisture content. For strongly shrinking materials, the mathematical model used to define D_L must account for the changes in diffusion path as well. The temperature dependence of diffusivity is adequately described by the Arrhenius equation as follows:

$$D_L = D_{LO} \exp[-E_a/R_g T] \tag{2.11}$$

where D_L is the diffusivity, E_a is the activation energy and T is the absolute temperature. Okos *et al.* (1992) have given an extensive compilation of D_{LO} and E_a values for various food materials. Zogzas *et al.* (1994) describe methods of moisture diffusivity measurement and provide an extensive bibliography on the topic.

Empirical models for D_L as a function of both moisture content and temperature are available in the literature for numerous food materials. Care should be taken in applying diffusivity correlations obtained with simple geometric shapes (e.g. slab, cylinder or sphere) to the more complex shapes actually encountered in practice.

Keey (1978) and Geankopolis (1983), among others, have provided analytical expressions for the diffusion and capillary models of falling rate drying. It is noteworthy that the diffusivity, D_L, is a strong function of X_f as well as temperature and must be determined experimentally. Thus, the diffusion model should be regarded purely as an empirical representation of falling rate drying. More advanced models are available but their widespread use in the design of dryers is hampered by the need for extensive empirical information required to solve the governing equations. Turner and Mujumdar (1996) provide a wide assortment of mathematical models of drying and dryers, and also discuss the application of different techniques for the numerical solution of the complex governing equations.

One simple approach to interpolating a given falling rate curve over a relatively narrow window of operating conditions is that first proposed by van Meel (1958). He found that the plot of normalized drying rate $v = N/N_c$ versus normalized fee moisture content $\eta = (X - X^*)/(X_c - X^*)$ was nearly independent of the drying conditions. This plot, called the characteristic drying rate curve, is illustrated in Figure 2.10. Thus, if the constant rate drying rate, N_c, can be estimated and equilibrium moisture content data are available, then the falling rate curve can be estimated using this highly simplified approach. Extrapolation over wide ranges is not recommended.

Waananen et al. (1993) have provided an extensive bibliography of over 200 references dealing with models of drying for porous solids. Basically, such models are useful to describe drying processes for the purposes of engineering design, analysis and optimization. A mathematical description of the process is based on the physical mechanisms of internal heat and mass transfer that control the process resistances, as well as the structural and thermodynamic assumptions made to formulate the model. In the constant rate period, the overall drying rate is determined by the heat and mass transfer conditions external to the material being dried, such as the temperature, gas velocity, total pressure and partial pressure of the vapour. In the falling rate period, the rate of internal mass transfer determines the drying rate. Modelling of drying is complicated by the fact that more than one mechanism may contribute to the total mass transfer rate and that the contributions from different mechanisms may change during the drying process.

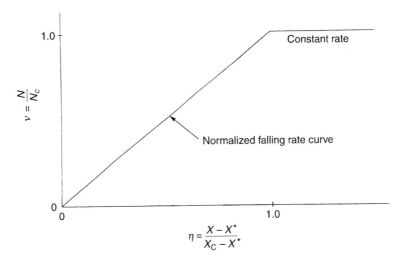

Figure 2.10 A typical characteristic drying rate curve.

Diffusional mass transfer in the liquid phase, as discussed earlier, is the most commonly assumed mechanism of moisture transfer used in modelling drying that takes place at temperatures below the boiling point of the liquid under locally applied pressure. At higher temperatures, the pore pressure may rise substantially and cause a hydrodynamically driven flow of vapour, which, in turn, may cause a pressure driven flow of liquid in the porous material.

For solids with continuous pores, driven by surface tension flow (capillary flow) may occur as a result of capillary forces caused by the interfacial tension between the water and the solid particles. In the simplest model, a modified form of Poiseuille flow can be used in conjunction with the capillary forces equation to estimate the rate of drying. Geankoplis (1983) has shown that such a model predicts the drying rate in the falling rate period to be proportional to the free moisture content in the solid. At low solid moisture contents, however, the diffusion model may be more appropriate.

The moisture flux due to capillarity can be expressed in terms of the product of a liquid conductivity parameter and the moisture gradient. In this case, the governing equation has, in fact, the same form as the diffusion equation.

For certain materials and under conditions such as those encountered in freeze drying, a 'receding-front' model involving a moving boundary between 'dry' and 'wet' zones often describes the mechanism of drying much more realistically than the simple diffusion or capillary model. Examination of the freeze drying of a thin slab of frozen material indicates that the rate of drying is dependent on the rate of heat transfer to the 'dry–wet' interface, and the mass transfer resistance offered by the porous dry layer to permeation of the vapour which sublimes from the interface. Because of the low pressures encountered in freeze drying, Knudsen diffusion effects may be significant. Liapis and Marchello (1984) have discussed models of freeze drying involving both unbound and bound moisture.

2.5 Food quality parameters

Numerous quality parameters can be defined for dried foods. Not all are applicable for a given product or for a given application. These quality parameters may be categorized as thermal, structural, optical, sensory, textural, biological and microbiological, as well as appearance. Giese (1995) has summarized the measurement techniques for selected quality properties of foods. Karathanos *et al.* (1996) have discussed some of the key quality properties of drying materials along with the method of measuring them. Table 2.6 summarizes quality properties that are appropriate to dehydrated foods. Most are not readily quantified and only a few can be measured on-line for control purposes.

Quality degradation during drying is a major concern in the selection, design and operation of a dryer for foods (Bimbenet and Lebert, 1992; Adamiec *et al.*,

Table 2.6 Quality changes in foods during drying

Physical	Chemical	Biochemical
Shrinkage	Loss of chemical activity	Degradation of cellular
Loss of density	Decomposition of some	structures and biomolecules
Alteration of shape, size,	chemical constituents	Oxidation of liquids
porosity		Denaturation of proteins
Crystallization		Enzymatic browning
Change of solubility		Maillard reaction
Reduced rehydration		Loss of vitamins
Loss of aroma, organoleptic		
properties		

1995; Bluestein and Labuza, 1975). The principal factors affecting the quality parameters are summarized in Table 2.7 and include:

- chemical changes e.g. browning reactions, lipid oxidation, discoloration;
- physical changes e.g. reconstitutability, cracking, texture, aroma loss;
- nutritional changes e.g. loss of vitamins and proteins, microbial growth.

Table 2.7 also lists the techniques employed to control the deterioration of foods by microbiological, chemical or physical factors. Detailed discussions of the quality parameters are given by Karel and Flink (1983), Karel (1992) and Heldman and Lund (1992).

Note that, in addition to moisture content, temperature and drying kinetics, food quality is also affected by other non-drying parameters. Thus, pH, composition of the food, feed pre-treatments and the presence of salts, solvents and oils, can all have a profound effect on the quality of the dried product. Factors that affect texture during dehydration include: crystallization of cellulose, degradation of pectins, gelatinization of starches, possible glass transitions, and case-hardening.

Phase transitions in food processing have been discussed in detail by Roos (1994). Glass transitions in foods may occur during drying and can have profound effects on product quality, both physical and chemical. The glass

Table 2.7 Deterioration of foods and control strategies

Microbiological	Chemical	Physical
Pathogenic micro-organisms leading to infection	1. Enzymatic reactions	1. Crystallization and collapse
	2. Lipid oxidation	
Control by:	3. Nonenzymatic browning	2. Wrinkling, attrition, abrasion
• lowering temperature to slow growth	4. Thermal destruction of vitamins	
• remove or bind water to prevent growth	Control by:	Control by:
• lower pH	• pH, O_2, CO_2, water activity modification	• careful handling/storage
• Control O_2, CO_2	• temperature control	• temperature/humidity control
• remove nutrients by composition modification	• additives, e.g. anti-oxidants	

Table 2.8 Transformations occurring during drying

Biochemical	Enzymatic	Chemical	Physical
Cell destruction as a result of loss of cellular water	Loss of activity of enzymes Loss of vitamins	Decrease in nutritive value and activity Aroma loss Discoloration	Changes in solubility, rehydration, shrinkage Textural changes

transition temperature, at which the material changes from a 'glassy' to a 'rubbery' state, is a function of moisture content as well as the concentration of solutes, such as sugar, if present. Studies have shown that glass transitions can cause changes in the diffusivity of components, resulting in aroma loss, as well as affecting the diffusion of reactants that affect non-enzymatic browning, and hydrolysis, for example, or the mobility of polymer chains, such as starch and proteins, which can cause textural softening. In general, the glass transition temperature is a monotonically decreasing function of moisture content and must be determined experimentally. Further, it is a rate dependent parameter so the data must be interpreted carefully. The stickiness of material being dried is related to the glass transition temperature; in the design and operation of dryers it is desirable to avoid contact between the material and the dryer walls at or near the 'sticky point', a temperature close to the endset.

Table 2.8 summarizes the key changes in physical, chemical and biochemical quality properties that can occur during food drying. Not all changes are relevant to the drying of a given food product. Most of these changes are temperature dependent. Thus, the highest quality parameters are generally obtained by vacuum or freeze drying. Table 2.9 lists the various quality properties of dried foods. Due consideration of these is of paramount importance in selecting a dryer and drying conditions in practice. Not all properties may be relevant to a given food product.

Rahman (1995) has provided a very comprehensive compilation of various properties of foods relevant to food processing in general, including drying. Methods for the measurement and prediction of water activity, sorption isotherms, phase transitions (e.g. crystallization, gelatinization, glass transition) as

Table 2.9 Quality properties of dried foods

Type	Key properties
Thermal	Glass transition temperature; melting point; thermo-mechanical properties
Structural	Bulk density; porosity; pore size distribution, specific surface area; crystal structure
Textural	Viscoelasticity; creep; compressive strength at failure; maximum stress at fracture; chewiness; stickiness; hardness
Optical	Colour; spectrophotometric measurement of browning
Sensory	Aroma, taste and flavour (using expert panels)
Nutritional	Vitamin losses; lipid oxidation; protein denaturation
Microbial	Contamination by various micro-organisms
Physical	Rehydration; wetting properties
Appearance	Surface morphology; roughness; flowability

well as extensive data and correlations are presented for convenient use in process calculations.

2.6 Summary

This chapter gives a brief overview of the basic aspects of drying with special reference of food materials. It is noted that heat and mass transfer considerations are important in the selection, design and operation of dryers but quality constraints have the controlling role in the selection and operation of food dryers.

Acknowledgements

The author wishes to acknowledge with appreciation the prompt word processing of the manuscript by Louise Miller and preparation of the artwork by Sakamon Devahastin. Dr Z. Pakowski of Lodz Technical University provided Figure 2.3. This psychrometric chart was generated using the commercial software entitled dryPAK written by Dr Pakowski and distributed by TKP Omnikon Ltd, 90-950 Lodz, P.O. Box 281, Poland.

References

Adamiec, J., Kaminski, W., Markowski, A.S. and Strumillo, C. (1995) Drying of biotechnological products, in *Handbook of Industrial Drying* (ed. A.S. Mujumdar), Marcel Dekker, NY, pp. 775–808.

Bimbenet, J.J. and A. Lebert (1992) Food drying and quality interactions, in *Drying '92*, Pt. A (ed. A.S. Mujumdar), Elsevier, Amsterdam, pp. 42–57.

Bluestein, P.M. and Labuza (1975) Effects of moisture removal on nutrients in nutritional evaluation of food processing, in *Nutritional Evaluation of Food Processing*, 2nd edn (eds R.S. Harris and E. Karmas), The AVI Publishing Co., Westport, CT, USA, pp. 289–323.

Brockmann, M.C. (1978) Intermediate moisture foods, in *Food Dehydration* (eds W.B. van Arsdel, J.H. Copley and A.I. Morgan), The AVI Publishing Co., Westport, CT, USA.

Bruin, S. (ed.) (1988) *Preconcentration and Drying of Food Materials*, Elsevier, Amsterdam.

Bruin, S. and Luyben, K.Ch.A.M. (1980) Drying of food materials: a review of recent developments, in *Advances in Drying*, vol. 1 (ed. A.S. Mujumdar), Hemisphere, NY, pp. 115–216.

Crapiste, G.H. and Rotstein, E. (1982) Prediction of sorptional equilibrium data for starch-containing foodstuffs. *Journal of Food Science*, **47**, 1501–7.

Fortes, M. and Okos, M.R. (1980) Drying theories: their bases and limitations as applied to foods and grains, in *Advances in Drying*, vol. 3 (ed. A.S. Mujumdar), Hemisphere, NY, pp. 119–154.

Geankopolis, C.J. (1983) *Transport Processes and Unit Operations*, Allyn & Bacon, Boston.

Giese, J. (1995) Measuring physical properties of foods. *Food Technology*, **54**(2), 54–63.

Heldman, D.R. and Lund, D.B. (eds) (1992) *Handbook of Food Engineering*, 1992, Marcel Dekker, NY.

Iglesias, H.A. and Chirife, J. (1982) *Handbook of Food Isotherms*, Audemi Press, NY.

Karathanos, V.T., Maroulis, Z.B., Marinos-Kouris, D. and Saravacos, D.G. (1996) Hygrothermal and quality properties applicable to drying, data sources and measurement techniques. *Drying Technology*, **14**(6) 1403–18.

Karel, M. (1992) Optimization of quality of dehydrated foods and biomaterials, in *Drying '92*, Pt. A, Elsevier, Amsterdam, pp. 3–16.

Karel, M. and Flink, J.M. (1983) Some recent developments in food dehydration research, in *Advances in Drying*, vol. 2, (ed. A.S. Mujumdar), Hemisphere, NY, pp. 103–49.

Keey, R.B. (1978) *Introduction to Industrial Drying Operations*, Pergamon Press, Oxford.

Keey, R.B. (1991) *Drying of Loose and Particulate Materials*, Hemisphere, NY.

King, C.J. (1971) *Freeze Drying of Foods*, CRC Press, Cleveland.

Liapis, A. and Marchello, J.M. (1984) Advances in modeling and control of freeze drying, in *Advances in Drying*, vol. 3 (ed. A.S. Mujumdar), Hemisphere, NY, pp. 217–44.

Mujumdar, A.S. (ed.) (1995) *Handbook of Industrial Drying*, 2nd revised and expanded edition, Marcel Dekker, NY. (Also Chapter 1, pp. 1–40.)

Mujumdar, A.S. and Menon, A.S. (1995) Drying of solids: principles, classification and selection of dryers, in *Handbook of Industrial Drying*, 2nd edn (ed. A.S. Mujumdar), Marcel Dekker, NY, pp. 1–46.

Okos, M., Narsimhan, G., Singh, R.K. and Weitnauer, A.C. (1992) Food dehydration, in *Handbook of Food Engineering* (eds D.R. Heldman and D.B. Lund), Marcel Dekker, NY, pp. 437–562.

Pakowski, Z., Bartczak, Z., Strumillo, C. and Stenstrom, S. (1991) Evaluation of equations approximating thermodynamic and transport properties of water, steam and air for use in CAD of drying processes. *Drying Technology*, **9**(3), 753–73.

Pakowski, Z. (1996) *dryPAK Software Package*, Lodz Technical University, Lodz.

Perry, R.H., Chilton, D.W. and Maloney, J.V. (1996) *Perry's Chemical Engineering Handbook*, McGraw-Hill, NY.

Rahman, S. (1995) *Food Properties Handbook*, CRC Press, Boca Raton, USA.

Rockland, L.B. and Beuchat, L.R. (eds) (1987) *Water Activity, Theory and Applications to Food*, Marcel Dekker, NY.

Roos, Y.H. (1994) Phase transitions in food systems, in *Handbook of Food Engineering* (eds D.R. Heldman and D.B. Lund), Marcel Dekker, NY, pp. 145–98.

Schmidt, E. (1969) *Properties of Water and Steam in SI Units*, Springer Verlag, Berlin.

Turner, I.W. and Mujumdar, A.S. (eds) (1996) *Mathematical Modeling and Numerical Techniques in Drying*, Marcel Dekker, NY.

van Meel, D.A. (1958) Adiabatic convection batch drying with recirculation of air, *Chemical Engineering Science*, **9**, 36–44.

Waananen, K.M., Lichfield, J.-B. and Okos, M.R. (1993) Classification of drying models for porous solids, *Drying Technology*, **11**(1), 1–40.

Wolf, W., Spies, W.E.L. and Jung, G. (1985) *Sorption Isotherms and Water Activity of Food Materials*, Science and Technology Publishers, London.

Zogzas, N.P., Maroulis, Z.B. and Marinos-Kouris, D. (1994) Moisture diffusivity: methods of experimental determination, *Drying Technology*, **12**(3), 483–515.

3 Through-flow dryers for agricultural crops

S. SOKHANSANJ

3.1 Introduction

In through-flow dryers (Figure 3.1), the drying air passes through a permeable bed of product, which is usually in the form of granular, stemmy or leafy-solid particles. The transfer of moisture takes place at the surface of these particles.

Through-flow dryers are available in a variety of forms, and may be characterized by the following features:

- operating mode (batch, continuous or combination);
- bed type (fixed or fluidized, thick or thin);
- product flow (stationary, motion induced by gravity, or by a moving band);
- relative direction of flows of product and air in continuous dryers (cocurrent flow, countercurrent flow, cross-flow, and mixed-flow).

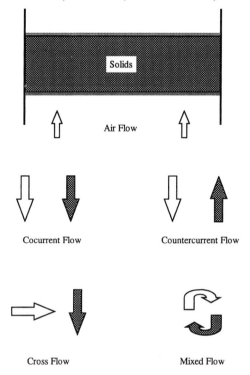

Figure 3.1 Definition of a through-flow dryer.

Agricultural crop dryers, in which a moving bed of solids flows downwards under gravity, are discussed in this chapter. Other through-flow dryers, such as band dryers and fluidized bed dryers, are considered elsewhere in the book.

Following a general description of the principal types of crop dryers, equations to aid in their design and to characterize their performance are presented. Quality aspects of the dried product are then considered, and the chapter concludes with a discussion of drying costs.

Through-flow agricultural dryers can be loosely divided into commercial and on-farm types. Commercial dryers are generally of high throughput, typically more than $10 \, t \, h^{-1}$. They are normally owned by business enterprises, including cooperatives. In Canada, the larger dryers are often located at the inland or port terminals. On-farm dryers have throughputs of less than $10 \, t \, h^{-1}$ and are owned and operated by farmers. Commercial dryers are generally stationary of a type; on-farm dryers are mobile and can be batch, continuous or a combination of these.

3.2 On-farm dryers

Most of the larger farmsteads in the industrialized nations are equipped with a grain drying system. Such equipment allows the grain to be harvested at a higher moisture content and thus to reduce uncertainties in crop quality that might be brought about by adverse weather conditions. The safe moisture content varies from grain to grain, ranging from 8 to 10% for oilseeds, 12 to 14% for starchy seeds, and 13 to 15% for legumes. The lower moisture contents quoted apply for storage periods of more than 1 year.

There are two types of on-farm grain drying systems: hot-air dryers and in-bin, unheated-air dryers. In hot-air dryers, air is heated to a temperature close to the maximum safe drying temperature of the grain, which is dried in one or two passes and within a relatively short time period. The grain is then cooled and moved into the storage area. High temperature grain dryers are common to the humid regions and low temperature dryers adapt well to drier, cooler areas.

3.2.1 Heated-air dryers

In a typical on-farm, heated-air dryer, the grain flows by gravity from the top to the bottom, and hot air flows through the grain. The maximum temperatures in this type of dryer are set by the grain quality. Usually, grain for use as seeds is not exposed to temperatures exceeding 50°C, and grain for baking or for oil extraction to not more than 80°C (Brooker et al., 1974). Grain for feeding has been dried at roasting temperatures (more than 140°C (Brooker et al., 1992)). High drying temperatures result in degradation of some essential amino acids (for example, lysine). Also, grain dryers are prone to fire hazards at high temperatures. Heated-air dryers feature a variety of configurations and operational characteristics. Some of the more popular designs are discussed in the following sections.

(a) Intermittent dryers.   In these dryers, which are generally of the batch type, the grain is dried with a pulsating flow of hot air. The particular dryer illustrated in Figure 3.2 has a holding tank on the top and 16 vertical drying chambers, which are of rectangular cross-section, 305 mm wide by 711 mm long, and 1450 mm high. After the chambers are filled, the flow of grain stops. Heated (or cold) air is forced through the grain drying chambers on one side of the dryer to dry (or cool) the grain.

After 15 s, exhaust dampers stop the flow of air through the chambers on one side and, at the same time, hot (or cold) air is directed through the grain drying chambers on the opposite side. This damper action is repeated until the grain is dried completely. The cooling cycle follows the drying cycle. After completion of both cycles, the grain is dropped through the unloading gates into the dry-grain holding chambers, where it is discharged by an auger while the dryer chambers are being refilled for the next drying cycle.

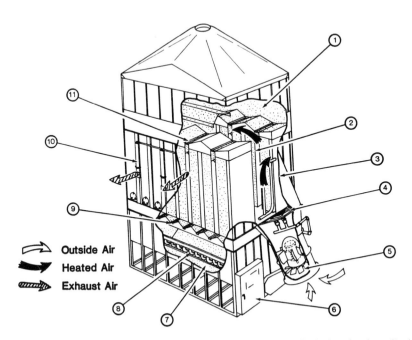

Figure 3.2 Intermittent dryer. (1) Wet grain holding chamber; (2) grain drying chamber; (3) air plenum; (4) burner, (5) control panel; (6) fan; (7) dry grain holding chamber; (8) discharge auger; (9) unload gate; (10) exhaust air damper; and (11) load gates. (Courtesy, PAMI, Humboldt, Saskatchewan, Canada.)

(b) Batch internal-recirculating dryers. Figure 3.3 shows a batch internal-recirculating dryer. Air and grain move in cross-flow directions. The annular grain drying chamber has internal and external diameters of 2.00 and 2.45 m, respectively, and is cone-shaped on the top and open at the bottom. The top section of the dryer holds wet grain. Outside air is heated by a burner and is forced into the central plenum and through the grain column. The grain is fed continuously into the bottom of a vertical auger by gravity and is lifted to the top of the drying chamber for recirculation. An optional cleaner attachment removes fines and small weed seeds in the process. The length of the drying cycle is determined by a pre-set grain temperature probe, which automatically shuts off the burner to start the cooling cycle. Dry, cooled grain is discharged at the top of the dryer.

(c) Mixed-flow dryers. The mixed-flow dryer is a continuous dryer, as shown in Figure 3.4. It consists of one or two rectangular columns, a typical column

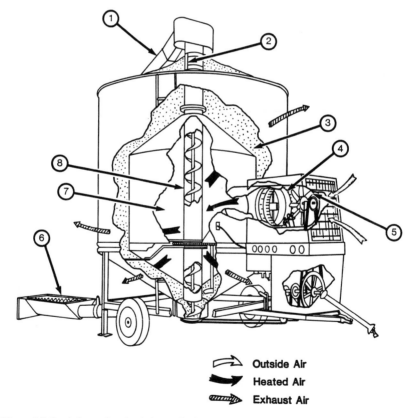

➩ **Outside Air**

➤ **Heated Air**

➩ **Exhaust Air**

Figure 3.3 Batch internal-recirculating grain dryer. (1) Unloading chute; (2) grain cleaner; (3) grain chamber; (4) burner; (5) fan; (6) loading auger; (7) air plenum; and (8) vertical auger. (Courtesy, PAMI, Humboldt, Saskatchewan, Canada.)

being 762 mm thick. A series of inverted V-shaped ducts are staggered trans-
versally within the column. Alternate rows of these ducts are open to the inside
plenum of the dryer and to the outside. Grain flows from the top to the bottom
around these ducts. In this design, grain is mixed somewhat as it flows
downward.

The central plenum is divided into a heating plenum and a cooling plenum.
Grain is dried in the top section of the column, which is about 3 m in length and
cooled in the bottom section, which is about 1.2 m in length. A single blower
forces the outside air through the grain columns. Air on the top is heated with a
burner located in the dryer plenum. Part of the air from the blower enters the
lower section for cooling. An auger levels the grain on the top of the dryer to
ensure even grain flow.

(d) Cross-flow dryers.   Figure 3.5 shows a cross-flow dryer. Both batch and
continuous versions are available; that shown in the figure is a batch version.
The grain drying section is a hexagonal chamber enclosing the grain plenum.

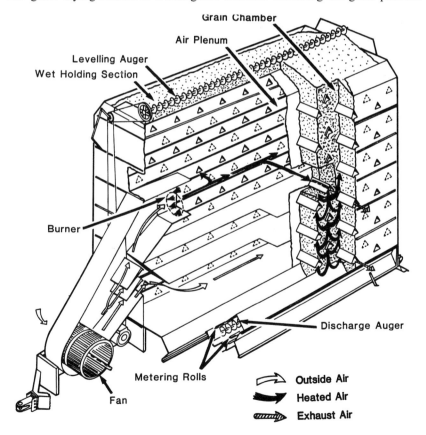

Figure 3.4 Mixed-flow dryer. (Courtesy, PAMI, Humboldt, Saskatchewan, Canada.)

The wet grain enters the top-rear centre of the dryer and is carried along the length of the dryer by a leveling auger. Once the grain chamber is filled, outside air is forced by the fan past the burner into the air plenum and through the grain chamber to dry or cool the grain. In the cooling cycle, the burner is naturally inoperative.

In the continuous version, unheated ambient air is drawn into the dryer from the lower section of the grain column. The air cools the grain, passes through a burner in which it is heated, and is then blown through the upper section of the grain column. Dry, cool grain is discharged at the bottom rear center of the dryer by an auger. The effective thickness of the drying column is 305 mm.

(e) Characterization and performance of heated air dryers. Grain dryers are characterized by drying temperature, drying and cooling times, and the overall throughput. The Prairie Agricultural Machinery Institute (PAMI) (1982, 1983,

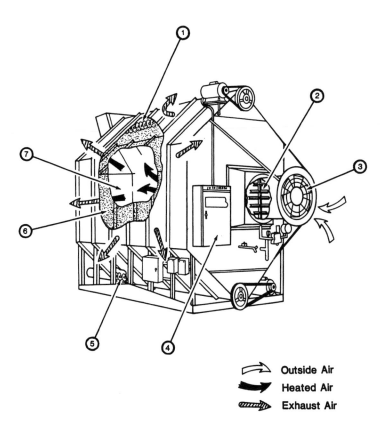

Figure 3.5 Cross-flow dryer. (1) Levelling auger; (2) burner; (3) fan; (4) control panel; (5) discharge auger; (6) grain chamber; and (7) air plenum. (Courtesy, PAMI, Humboldt, Saskatchewan, Canada.)

Table 3.1 Results of on-farm dryer trials conducted by PAMI. Drying temperature and specific energy consumption (kJ of heat energy per kg of water removed)

Dryer configuration	Wheat		Barley		Canola		Corn	
	Temperature (°C)	kJ kg^{-1}	Temperature (°C)	kJ kg^{-1}	Temperature (°C)	kJ kg^{-1}	Temperature (°C)	kJ kg^{-1}
Mixed flow	90	4700	93	5100	72	4600	110	4400
Cross flow	82	4200	71	4700	66	3900	104	3600
Intermittent	104	4900	71	5400	66	4700	110	4200
Internal recirculating	82	4300	93	4100	71	4100	110	4100

1984, 1985) tested a number of on-farm heated-air dryers in which wheat, feed barley, oilseed canola, and feed corn were processed. The results of these tests are summarized in Tables 3.1, 3.2 and 3.3.

Table 3.1 gives the drying temperature and energy (kJ) used to evaporate 1 kg of water from grain. Drying temperature is often specific to a particular dryer design. The maximum temperature is controlled by the desired quality of the dried grain. In mixed-flow dryers, it is higher than in other (e.g. cross-flow) types, as the grain and air flow cocurrently and the time of exposure of the grain to high temperatures is shorter.

The specific-energy values listed in Table 3.1 range from 3600 to 4900 kJ kg^{-1}. Considering that the theoretical value is 2250 kJ kg^{-1} for free water, these values show that to extract moisture from grain takes around twice as much energy. Part of this additional requirement is used to heat the grain, and

Table 3.2 Results of on-farm dryer trials conducted by PAMI. Drying and cooling times

Dryer configuration	Wheat		Barley		Canola		Corn	
	Dry (h)	Cool (h)	Dry (h)	Cool (h)	Dry (h)	Cool (h)	Dry (h)	Cool (h)
Mixed flow	Grain flows continuously							
Cross flow	0.8	0.4	0.9	0.2	0.9	0.3	1.0	0.4
Intermittent	0.7	0.4	0.9	0.2	1.3	0.3	1.1	0.3
Recirculating	2.0	0.6	1.1	0.9	1.9	0.5	2.2	0.9

Table 3.3 Results of on-farm dryer trials conducted by PAMI. Holding capacity and the throughput

Dryer configuration	Holding capacity (m^3)	Wheat (t h^{-1})	Barley (t h^{-1})	Canola (t h^{-1})	Corn (t h^{-1})
Mixed flow	19.1	9.1	8.1	3.7	5.3
Cross flow	8.2	3.9	3.9	3.2	3.2
Intermittent	22.4	3.7	3.3	2.3	2.6
Recirculating	12.6	2.7	2.7	2.5	1.9

part to free the bound moisture within the grain. The remaining energy is wasted in the exhaust air.

Table 3.2 lists drying and cooling times for the batch dryers tested. The values were for a moisture reduction of 5%, except for corn, which was 10%. The batch internal-recirculating dryer exhibited the longest drying and cooling times, mainly as a result of the thick drying column. It can be noted that, for the dryers tested, the cooling time was about half to one third of the drying time.

Table 3.3 lists the capacity and the throughput of the dryers tested by PAMI. The throughput depended upon the grain being processed and the type of dryer. The throughputs for wheat and barley were the highest; those for canola and corn were lower. The batch internal-recirculating dryer exhibited the lowest drying capacity.

3.2.2 Unheated-air dryers

In these systems, grain is dried while stored in a bin, which can be either flat-bottomed (Figure 3.6) or hopper-bottomed (Figure 3.7). Unheated air is drawn from the outside into a plenum under the grain pile. The air is then introduced into the stationary grain through a perforated floor. The moist air exits from the top or, in some designs, from the side of the bin. The air is heated by 3–5°C when it passes through the fan. In some instances, a small auxiliary heater is fitted to increase the temperature of the grain. The air should not be heated by more than 5–10°C, as higher temperatures cause large moisture gradients within the pile, which may give rise to condensation on the upper layers of grain. The volumetric flows of air used in these systems vary from 1 to 3 $m^3\,min^{-1}$ per tonne of grain.

(a) Flat-bottom bins. In flat-bottom bins (Figure 3.6), air enters the grain either at the centre of the grain mass, through ducts placed on the floor of the

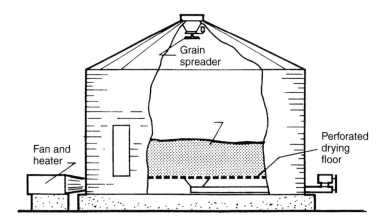

Figure 3.6 Typical flat-bottom grain bin with ventilation system for aeration and drying.

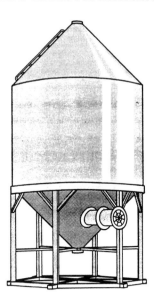

Figure 3.7 Hopper-bottom bin with natural drying fan. (Courtesy, PAMI, Humboldt, Saskatchewan, Canada.)

bin, or through a perforated floor. The floor ducts are either circular or semi-circular in cross-section, with recommended diameters of about 400 mm and 500–700 mm, respectively. They are constructed from galvanized steel and are often corrugated for strength. The perforated floor consists of planks, which are usually 175 mm wide and of an appropriate length. The planks are made from 20 gauge steel (0.95 mm thick), and perforated to provide a 19% open area. They are ribbed for extra strength and are slightly crowned so that the weight of the stored grain forces the seams of the planks together to form a positive seal. The completely smooth surface facilitates easy cleaning of the bin.

Flat-bottomed bins are also ventilated with a combination of horizontal and vertical perforated ducts. The horizontal ducts are positioned radially and are connected to an upright central duct. They are either laid on the floor or are held about 500 mm above it. In this design a series of air-exhaust ports is located in the sides of the bin.

One of the fundamental problems with flat-bottom bins is the difficulty experienced during unloading, since any above-floor ventilation system tends to impede the flow of the grain. It is often necessary for an operator to enter the bin and sweep the grain out, especially from underneath the ducts. There is also a tendency to unload these bins from the side door. This causes off-centre unloading, which may cause them to collapse.

(b) Hopper-bottom bins. Hopper- or conical-bottom bins are becoming popular as they provide a safe and convenient way to load and discharge grain

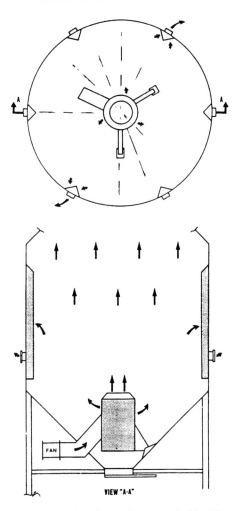

Figure 3.8 Ventilation duct systems in a hopper bottom grain bin. (Courtesy, PAMI, Humboldt, Saskatchewan, Canada.)

(PAMI, 1989). Grain drying in these bins, however, is a challenge as the air should be uniformly distributed across the bin. Two of the more popular systems are shown in Figure 3.8, where outside air is blown through a pipe that extends into the hopper section of the bin. Moist air that has passed through the grain exits from the top or, in some designs, from the side of the bin, or from both top and side. In a system similar to the mixed-flow design (not shown), the air enters the dryer through an inverted V embedded horizontally in the grain.

Figure 3.8 also shows a design in which air enters the grain from a vertical perforated pipe at the centre of the bin. The air exits the bin from the side and from the top hatch. In some designs, air enters the grain through a false perforated floor installed inside the hopper. In others, the grain ducts are installed on the inside wall of the hopper.

One of the main problems with these systems is non-uniformity of the final moisture content within the bin. Gu *et al.* (1996), for instance, found that the grain in the centre of the bin remained at a moisture content of 20% while the average bin moisture content was 14.5%. Installation of a vertical central ventilation duct alleviated this problem somewhat. PAMI tested a number of these systems and found extreme variations in the drying time.

(c) Characterization and performance of unheated-air dryers. A number of control strategies have been developed which optimize the operation of heater and fan in low temperature drying systems (Brook, 1987; Morey *et al.*, 1981; Johnson and Otten, 1979; Fraser and Muir, 1981; Nellist and Bartlet, 1988). Some of these control strategies are in the form of management practices and some in the form of semi- or fully automated controllers. The success of the control schemes, manual or automated, depends on the specific crop and climatic conditions for which these schemes have been developed. Nellist and Barlet (1988) concluded that continuous fan operation without the use of heat is not adequate to dry grain to 14% m.c.

Sokhansanj *et al.* (1991) conducted a detailed analysis of unheated low-temperature drying systems on the Canadian Prairies. (Table 3.4 shows typical harvest weather conditions in this region.) Sokhansanj and Lischynski (1991) and Arinze *et al.* (1994) simulated the management and control of unheated-air in-bin drying of wheat and barley. The following conclusions were drawn from these studies.

1. Continuous fan operation with at least $1 \, m^3 \, min^{-1} \, t^{-1}$ is adequate for successfully drying grain having a moisture content less than or equal to 19% in August. The air flow must be increased to $1.5 \, m^3 \, min^{-1} \, t^{-1}$ or $2 \, m^3 \, min^{-1} \, t^{-1}$ when drying conditions are typical of those encountered in September or October, respectively.
2. Small air flow rates of $0.5 \, m^3 \, min^{-1} \, t^{-1}$ give rise to lower direct drying costs but the risk costs associated with these air flow rates are higher (section 3.7).
3. Continuous fan operation with supplementary heat to raise the plenum temperature by 5 to 10°C results in the lowest drying costs.

Table 3.4 Climatic conditions on the Canadian Prairies during harvest

Month	Dry bulb temperature (°C)	Wet bulb temperature (°C)
August	15±3	12±3
September	10±2	8±4
October	6±2	4±5
November	−5±2	−6±7

Source: ASAE (1995)

4. Drying grain at moisture contents of 19% or higher in October requires supplementary heat.
5. The drying time almost doubles when the drying front is allowed to sweep the entire grain bed (through-drying) as compared to drying grain to a bin-average moisture content of 14.5%. Through-drying results in 1 to 2 t of over-drying in a 47-t bin under Saskatoon climate conditions.
6. Through-drying costs from 30% to 100% more than average drying.

3.2.3 Combination heated- and unheated-air dryers

Heated-air dryers are characterized by efficient drying in terms of processing time and throughput. Low temperature dryers, on the other hand, dry grain slowly, but its quality is often superior to that of heated-air dryers. Moreover, low temperature dryers use much less energy than their heated-air counterparts. Attempts have therefore been made to combine these two processes in order to utilize the advantages of each.

In combination drying, grain is pre-dried in a conventional heated-air dryer without cooling. The hot grain is then transferred into a bin where it is cooled and dried to the target moisture content. Combination drying has been credited with increased throughput, improved energy efficiency, and improved grain quality over conventional heated-air drying.

The idea of transferring partially dried grain from a heated-air dryer to temporary storage for further drying and cooling was first introduced by Foster (1964). The main feature of his technique was an 8 to 12 h steeping process prior to final drying and cooling. Calling the method 'dryeration', Foster demonstrated that corn dried by this technique developed less stress cracks than corn dried by the conventional heated-air drying/cooling method. Dryeration was further tested and developed by Muhlbauer and Kuppinger (1975) and by Lasseran (1977) for high moisture corn. In spite of the superior features of the process, it did not become popular, mainly because of the inherent delay in drying during the steeping process and the extra management requirements.

Field tests were performed during the 1984 and 1985 harvest seasons with Canada No. 1 and Canada No. 3 Red Spring Wheat at the PAMI grain drying facility in Humboldt, Saskatchewan. The details of these tests and the results of grain quality evaluations have been reported by Lischynski et al. (1986). Table 3.5 summarizes the experimental results, where the quantities of grain, the initial, transfer and final moisture contents, along with the drying time and specific energy consumption recorded in each test, are given. The measured air flow rates in the unheated-air drying experiments ranged from 1.6 to 2.0 $m^3 min^{-1} t^{-1}$.

Combination drying lowered the drying time (81 h versus 346 h for unheated-air drying). However, the energy efficiency of combination drying was less than half that of conventional unheated-air drying but somewhat greater than that of heated-air drying (3430 kJ kg^{-1} versus 3950 kJ kg^{-1}).

Table 3.5 Test results for combination drying

Drying method	Grain mass (t)	Moisture content (%)			Drying time (h)	Energy (kJ kg^{-1})
		Initial	Transfer	Final		
PH/PL 1	50.9	26.2	14.9	14.1	53.5	3890
PH/PL 2	53.3	22.0	16.1	14.5	81.3	3430
PH/PL 3	50.4	25.9	17.3	14.5	125.4	3030
TH	–	23.0	21.6	–	14.3	3950
TL	51.5	22.2	–	14.5	346.0	1750

Key: P = partial drying; T = total drying; H = high temperature; L = low temperature

3.3 Commercial dryers

3.3.1 Commercial grain dryers

Most commercial grain dryers are of a cross-flow design, but other configurations, such as cocurrent- and mixed-flow, are also employed. In cross-flow designs (Figure 3.9), the grain column is usually rectangular in cross-section. Its thickness varies from 305 mm at the top to 430 mm at the bottom. This varying cross-section allows the grain to flow at different speeds in different parts of the dryer. The residence time of the grain is controlled by varying its feed rate. A specially configured inverter redirects the grain from the hot side to the cold side and vice versa. This inversion is important since, in cross-flow dryers, the grain closest to the air plenum tends to over-dry. Another feature of the dryer is the use of wire mesh to construct the grain column wall in order to minimize the pressure drop experienced by the air. Louvres in the upper section of the dryer prevent back-pressure caused by wind.

Commercial grain dryers operate for longer periods of the year than on-farm dryers. Their throughput ranges from 10 t h^{-1} to more than 40 t h^{-1} of dried grain. To save energy, part of the exhaust air from the cooler is recycled into the top section of the dryer after re-heating. The fraction of the exhaust air that can be recirculated is limited by its lower drying capacity; Sokhansanj and Bakker-Arkema (1981) showed that the maximum possible exhaust air recirculation in an experimental cocurrent dryer was 15%.

Differential grain-speed cross-flow drying, multi-stage cocurrent-flow drying, multi-stage mixed-flow drying and cascading rotary drying are evolving techniques in agricultural crop drying. Accurate prediction of grain quality for seed, food and feed use has become the most pressing issue in the optimum design and operation of processing equipment, including dryers. Similarly, the understanding and modelling of the mechanics of moisture diffusion within the cellular structure of a grain kernel are fundamental to the advancement of commercial drying technology.

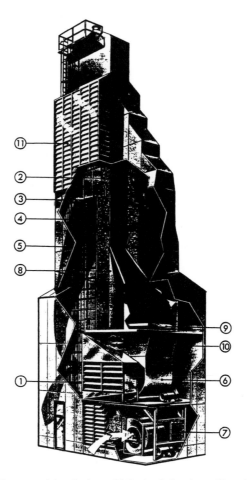

Figure 3.9 A typical commercial grain dryer. (1) Recirculating ducts; (2) grain mixer; (3) tapered column design; (4) panel; (5) galvanized enclosure; (6) air duct; (7) control panel; (8) wire-mesh grain panel; (9) air flow straighteners; (10) air dampers; and (11) exhaust-air louvres.

3.3.2 Malt dryers

Malt drying (kilning) is an important part of beer production. In this process, batches of wet germinated barley, known as green malt, are dried in a hot, forced-air flow. The objectives of kilning are to arrest botanical growth and internal modification, to reduce moisture for safe storage, to ensure the survival of enzymes, where appropriate, and to develop colour and flavour compounds in the malt.

Most modern kilns are fixed beds, because this design is simpler and cheaper than the traditional multi-storey kilns. Figure 3.10 depicts a typical modern double-deck kiln. These are generally circular in shape as the air flow distribution can be controlled better than in rectangular kilns. The walls and floor of the kiln

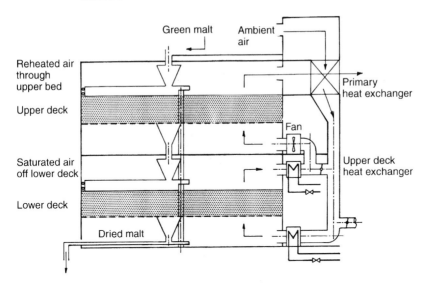

Figure 3.10 A double-deck malt drying kiln. (Adapted from Bamforth and Barclay, 1993.)

were formerly of brick construction, but are now made from a high-grade steel capable of resisting corrosion. The kiln requires loading and unloading conveyors, levelling devices, and temperature monitoring and control systems. The optimum loading of the kiln is 350–500 kg of malt per m^2 of floor area (Bamforth and Barclay, 1993). This gives an effective bed depth of 0.85 to 1.2 m.

The germinated barley, initially at 45% m.c., is dried to approximately 4% moisture content. The inlet air temperature follows a defined schedule in which it increases progressively from 50°C to 80°C. The characteristics of the finished malt depend on the cycle times, temperature, relative humidity, and volume of air drawn through the kiln bed (Sumner *et al.*, 1989). High volumes of warm air will produce lightly coloured, enzymatic malt, while reduced flow of hot, wet, recirculated kiln air will 'stew' the malt and produce high colour and more toasted-malt flavours.

A basic principle of kilning is that drying should commence at a relatively low temperature to ensure survival of the most heat sensitive enzymes. This should be followed by a progressively increasing temperature to effect flavour and colour changes and, subsequently, to complete the drying.

In kilning, the drying process is divided into four phases.

1. Free drying down to approximately 23% moisture content. This is the initial drying stage of kilning. It is designed to remove most of the free and bound moisture while protecting against enzyme denaturation with a lower temperature. During free drying, the flow of air (5000–6000 m^3 min^{-1} t^{-1}) at an air temperature of 45–65°C ensures unrestricted removal of water.
2. An intermediate stage, from 23% to 12% moisture content. Once the moisture content approaches 23%, the decreased ability of water to reach the

surface of the grain and the fact that the residual moisture is bound, both restrict evaporation. The air-off temperature above the bed suddenly starts to increase; 'break point' has been reached. The relative humidity of the air emerging from the bed starts to drop. Therefore, the air-on temperature can be increased, and air flow may be reduced to increase the relative humidity.

3. The bound-water stage, from 12% to 6% moisture content. When the moisture content reaches approximately 12%, all the water in the kernel is said to be bound. The air-on temperature is increased further and the fan speed is again reduced. As the relative humidity of the air declines, some of the air can be recirculated.

4. Curing, in which the moisture content is typically reduced from 6% to 2–3%. The air-on temperature is increased to between 80 and 120°C, depending on the type of malt required.

Kilning produces a low moisture content malt that is microbiologically and enzymatically stable. It also facilitates rootlet removal, enhances milling characteristics and generates the characteristic malt flavour and colour. Variations in curing temperature result in malts of different colour and flavour.

3.4 Dryer control

Dryer controls have the following functions: efficient dryer operation, safety of the operator, safety of the machine, desired uniform final moisture content, and easy maintenance. Three types of controls should form part of a dryer: (1) gas-flow and burner controls; (2) electrical controls for fans, gas manifolds, the burner system, filling and discharge mechanisms; and (3) mechanical controls for filling, profiling the burner, directing air, recirculating air, and inverting, diverting, metering and discharging the grain.

The controls for gas flow and burner operation include those on the gas supply line. These are provided by the gas supplier and do not usually form part of the dryer. They include a manual gas shut-off valve, a gas pressure regulator, a gas meter, a pressure gauge test point, and any other safety controls that are required to meet the local codes. The gas flow controls on the dryer include: a high-pressure gas safety switch, an interlocked gas safety shut-off, a gas control manifold, gas plumbing, and burner operation. The gas-train parts supply fuel to the dryer. The ASAE Standard S248.3 (ASAE, 1995) specifies the amount of gas that may be supplied to the burner pilot light. This should not exceed 3% of the overall heat capacity of the burner, or a maximum of 21.1 MJ h^{-1}, whichever is the higher.

In dryers that utilize large amounts of heat, a modulating fuel supply system is preferred to an on–off system. The sensing device for the modulating valves is a gas-filled capillary bulb mounted in the plenum. The gas pressure in the bulb is transmitted to a bellows in a modulating valve to allow the flow of a certain amount of fuel to the burners. A spring in the valve is adjusted so as to give specified maximum and minimum gas flows. A pressure regulating valve must be installed ahead of the modulating valve. When temperatures in the dryer are excessive, a master solenoid valve shuts off the flow of gas to the dryer.

Operational safety depends on safe temperatures being maintained at several points in the dryer. The critical values are the temperatures at the burner point, in the plenum, in the dryer chambers and in the exhaust. These temperatures also signal that the grain has been dried to its final moisture content. In continuous-flow dryers, the exhaust air temperature controls the rate of grain discharge.

3.5 Design of through-circulation dryers

Drying involves the transfer of moisture and heat between the material and the air. Since dryers exhibit a variety of air–material configurations, a general approach to modelling is followed. A dryer may be modelled under: (1) steady-state conditions; (2) well-mixed transient conditions; and (3) unmixed transient conditions. Condition (1) is used for sizing the dryer and for calculating the overall energy consumption and performance of the system. Conditions (2) and (3) are used to develop operational controls and for detailed dryer design and analysis, respectively. The calculations become more complex on moving progressively from condition (1) to condition (3).

3.5.1 Design calculations

(a) Steady-state continuous drying.   In a continuous-flow dryer, wet material enters at one end of the dryer and the dried material exits from the other end. Meanwhile, heated air flows through the dryer, removing moisture that is evaporating from the drying product. The relative movement of air and product can be counter-flow, concurrent-flow, cross-flow, or mixed-flow. A steady-state counter-flow dryer is shown schematically in Figure 3.11.

The calculation procedures that follow enable the time for a given moisture content reduction to be estimated. As the grain feed rate is specified, the dryer size can then be determined. To start the development of the calculation procedure, we can write simple heat and mass balance equations for air and product over the dryer chamber.

Figure 3.11 Straws entering and leaving a continuous counter-flow dryer.

Moisture balance:

$$G'Y_i + S'X_i = G'Y_o + S'X_o \qquad (3.1)$$

Heat balance:

$$G'H_i + S'h_i = G'H_o + S'h_o + Q_1 \qquad (3.2)$$

In the above equations, G' and S' are the dry-basis flowrates of air and product per unit area, and Y and X are the humidity of the air and the dry-basis moisture content of the solids. Subscripts i and o denote inlet and outlet conditions, respectively, and Q_1 is the heat loss (or gain) from the dryer to its immediate surroundings. For a dryer with good insulation, it is reasonable to assume that $Q_1 = 0$.

Definitions of the enthalpies H and h are as follows.

For air mixed with water vapour:

$$H = c_a(T_a - T_r) + Y[c_v(T_a - T_r) + \lambda_r] \qquad (3.3)$$

and for a product that contains water and dry solids

$$h = c_d(T_s - T_r) + X[c_l(T_s - T_r) + \Delta H_w]. \qquad (3.4)$$

Here c_a, c_d, c_l and c_v are the heat capacities of dry air, dry solids, liquid water, and water vapour, respectively. T_r is a reference temperature, which is often chosen as 0°C. λ_r is the latent heat of evaporation and ΔH_w is the heat of wetting. Both λ_r and ΔH_w are evaluated at T_r.

Equations 3.1 and 3.2 describe mass and heat balances between the air and the grain as they pass through the dryer. With these two equations, any two unknowns can be calculated. These can be air temperatures, product moisture contents, grain or air flow rates. It is important to note that the drying parameters in equations 3.1 and 3.2 do not change with time or with the size of the dryer.

Table 3.6 shows the solution to a typical problem involving steady-state operation of a continuous dryer. Locations 1 and 2 refer to the two end points of the dryer. For the particular example shown, the SOLVER function in the EXCEL spreadsheet was used to calculate air exit humidity and temperature with other conditions specified.

Table 3.6 Steady-state simulation results

Parameter	Air in (Location 1)	Air out (Location 2)	Parameter	Grain in (Location 2)	Grain out (Location 1)
G' (kg m^{-2} s^{-1})	11	11	S' (kg m^{-2} s^{-1})	1.80	1.80
T (°C)	80.0	25.0	T (°C)	25.00	75.00
Y (kg kg^{-1})	0.007	0.023	X (kg kg^{-1})	0.20	0.10
c_a (kJ kg^{-1} K^{-1})	1.01816	1.049	C_d (kJ kg^{-1} K^{-1})	2.00	2.00
c_v (kJ kg^{-1} K^{-1})			C_l (kJ kg^{-1} K^{-1})	4.20	4.20
λ (kJ kg^{-1})	2250	2250	h (kJ kg^{-1})	71.00	181.50
H (kJ kg^{-1})	97.2028	79.014			

(b) Sizing the dryer.   The dryer is also a heat exchange device. The sensible heat exchange between the product and air is described by the following equation

$$G'(H_i - H_o)A_x = UA_h\Delta T_{lm} \tag{3.5}$$

where A_x is the cross-sectional area available for air flow, U denotes the overall heat transfer coefficient, A_h the heat exchange area, and ΔT_{lm} the log mean temperature difference between the air and grain. This is defined as

$$\Delta T_{lm} = \frac{(T_a - T_s)_1 - (T_a - T_s)_2}{\ln[(T_a - T_s)_1/(T_a - T_s)_2]} \tag{3.6}$$

Equations 3.5 and 3.6 along with 3.1 and 3.2 can be used to estimate the physical size of the dryer.

Well-mixed transient dryer.   In this stage of the calculations, the assumption is that air and the product are well mixed. Air enters the dryer at an initial condtion specified by subscript *i*. The air and the product are in contact for a period of time Δt after which each may exit from the dryer. The product and the air may or may not be in equilibrium after time Δt. The most important concept here is that there are no variations in either temperature or moisture content within the dryer. The parameters vary only with time.

The model development begins by rewriting equations 3.1 and 3.2 for a short time step Δt:

$$G'Y_i\,\Delta t + sX_i = G'Y_o\,\Delta t + sX_o \tag{3.7}$$

$$G'H_i\,\Delta t + sh_i = G'H_0\,\Delta t + sh_o \tag{3.8}$$

Here s is the grain loading (mass per unit area). Equations 3.7 and 3.8 can be re-arranged to yield

$$G'(Y_i - Y_o)\,\Delta t = -s\,\Delta X \tag{3.9}$$

$$G'(H_i - H_o)\,\Delta t = -s\,\Delta h \tag{3.10}$$

where $\Delta X = (X_i - X_o)$ and $\Delta h = (h_i - h_o)$.

Differentiating equation 3.4 with respect to time, noting that the air temperature and humidity are almost independent of time, we obtain:

$$\Delta h = [c_l(T_s - T_r) + \Delta H_w]\,\Delta X + (c_d + Xc_l)\,\Delta T_s. \tag{3.11}$$

Combining equations 3.10 and 3.11 yields:

$$G'(H_i - H_o)\,\Delta t = -s\{[c_l(T_s - T_r) + \Delta H_w]\,\Delta X + (c_d + Xc_l)\,\Delta T_s\} \tag{3.12}$$

An additional equation relating to the sensible heat exchange between air and the product over time Δt can be written as:

$$G'(H_i - H_o)A_x\,\Delta t = -U(T_s - T_a)A_h\,\Delta t. \tag{3.13}$$

Note that equation 3.13 is time dependent because the enthalpies and temperatures are time dependent. In practice, we may assume that the heat transfer coefficient U is large and therefore $(T_s - T_a)$ is small. As a result, equation 3.13 is not employed further in the calculations.

To complete the calculations, we need to know the rate of loss of moisture from the particles to the drying medium. An empirical equation obtained from experiments on the drying of individual grain kernels might have the following first-order form:

$$-\frac{dX}{dt} = k(X - X^*) \tag{3.14}$$

where X^* is the equilibrium moisture content.

Experience has shown that the solution to equations 3.7–3.12 is difficult because heat and mass transfer are involved at the same time, especially in the early stages of drying. Thompson et al. (1968) developed a two-step procedure to compute the grain and air conditions during the time interval Δt.

As a first step in the solution procedure, Thompson et al. (1968) assumed an initial zero mass transfer ($\Delta X = 0$) when the air and grain come into contact in the dryer. This simplified the solution as equations 3.9 and 3.10 were reduced to solving equation 3.10 only for the grain temperature at the end of the time period Δt. Using the new grain temperature and the old air relative humidity, a new moisture content was calculated. This was then used in equations 3.9 and 3.12 to calculate new values for relative humidity and temperature. The solution may require a few iterations.

Following Thompson's procedure, $X_o = X_i$ and $Y_o = Y_i$; equation 3.8 simplifies to:

$$G'[(c_a + c_v Y_i)(T_{ai} - T_{ao})] \Delta t = -s(c_d + c_l X_i)(T_{si} - T_{so}). \tag{3.15}$$

This may be written in the form:

$$R(T_{ao} - T_{ai}) = (T_{si} - T_{so}) \tag{3.16}$$

where

$$R = \frac{(c_a + c_v Y_o)G' \Delta t}{(c_d + c_l X_i)s}. \tag{3.17}$$

By assuming $T_{ao} = T_{so}$, equation 3.16 can be rearranged to solve for the exit air temperature:

$$T_{ao} = \frac{RT_{ai} + T_{si}}{R + 1}. \tag{3.18}$$

The exit air humidity Y_o is calculated from the change in the moisture content of the grain in the layer:

$$Y_o = \frac{s(X_i - X_o)}{G' \Delta t} + Y_i. \tag{3.19}$$

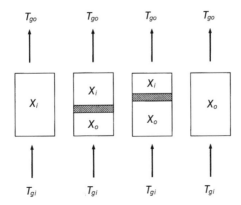

Figure 3.12 Progress of drying in a fixed bed through-drying situation.

At the conclusion of this calculation, the new X and H values are used in equations 3.9 and 3.12 to arrive at a new equilibrium temperature. The calculations are then repeated for a new layer.

Time and space dependent drying. This situation arises when the grain bed is more than, say, 5 cm deep and also remains unagitated during drying. In these circumstances, a drying front is formed within the bed and this front moves from the hot side to the cold side. Figure 3.12 depicts this situation.

The method of modelling time and space dependent drying is similar to that developed in the previous section for time dependent drying. Here we replace s that represented the whole mass of dry grain with $\rho s \, \Delta x$, where Δx is the thickness of a thin layer of the grain. We assume that grain and air are well mixed within this thin layer. Equations 3.7 and 3.8 are recast, as follows.

For the mass balance:

$$G'Y_i \, \Delta t + \rho_s \, \Delta x X_i = G'Y_o \, \Delta t + \rho_s \, \Delta x X_o \qquad (3.20)$$

For the heat balance:

$$G'H_i \, \Delta t + \rho_s \, \Delta x h_i = G'H_o \, \Delta t + \rho_s \, \Delta x h_o \qquad (3.21)$$

and for the heat exchange between the grain and air within the layer

$$G'(H_i - H_o) = -U_x a_x (T_s - T_a)_x \, \Delta x \qquad (3.22)$$

To proceed with the calculations, the grain bed is divided into thin layers of depth Δx. For each layer, a procedure similar to that outlined in the previous section is followed. The inlet air condition to each layer is the same as the output condition of the previous layer. Calculations are undertaken for all layers. The drying time is then progressed forward by Δt and a new round of calculations over the layer is repeated.

In a two-pass solution scheme, the initial grain condition at $t = 0$ for all locations, x, in the bed is specified. The air temperature and humidity are known at the entrance to the bed ($x = 0$). In the first pass, X and T_s are evaluated from equations 3.20 and 3.21 for all x at $t = \Delta t$. In the second pass, these new values are used to compute air temperature T and humidity Y at all x locations after $t = 2\Delta t$. This procedure is repeated until the simulation is completed over the specified time period. The subscripts on the dependent variables show the points of evaluation. Note that if T_a, T_s, Y, and X are known at (x, t), the drying-product conditions can be computed at $(x, t + \Delta t)$ and the air conditions at $(x + \Delta x, t)$.

The fixed-bed drying equations, 3.20 and 3.21, may be modified to simulate a steady-state cross-flow drying process by substituting $\Delta t = \Delta y/v_s$ (Cenkowski and Sokhansanj, 1989). Here, Δy represents a short distance in the direction of grain motion and v_s is the volumetric flowrate of grain per unit area. For the numerical solution, a time equivalent to $\Delta t_m = \Delta y/v_s$ is computed. Here, Δt_m is equal to $j\Delta t$, where j is the number of iterations with a time step Δt required to move a layer of product forward by a distance Δy. After a lapse of Δt_m, a layer of dry product exits the dryer while a new layer of moist product enters the dryer. This scheme is outlined in Figure 3.13.

It should be pointed out that two major assumptions were made in the development of the drying equations: (1) moisture loss from the solid is by diffusion but the moisture distribution is uniform; and (2) heat transfer is by convection from the surface of the solid (no conduction) and the temperature within the solid is uniform. Sokhansanj and Bruce (1987) demonstrated that these assumptions do not hold in relation to an accurate prediction of grain moisture content and temperature, which are important in estimating the crop quality during drying.

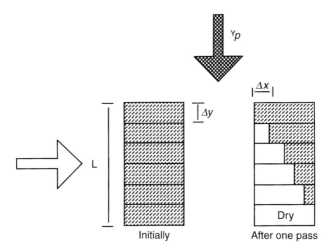

Figure 3.13 Simulation of cross-flow drying using fixed-bed model.

3.5.2 Grain properties

A knowledge of both the physical and thermal properties of air and grain is required in dryer design and performance calculations. Equations describing the properties of moist air are given in Chapter 2. Those relating to the equilibrium moisture content and drying rates of grain are discussed in the following paragraphs. Values of other physical properties of grain can be found in the literature. Examples include the pressure drop across grain beds (Jayas and Sokhansanj, 1989; Li and Sokhansanj, 1994), and bulk and kernel densities (Lang et al., 1993, 1994; Sokhansanj and Lang, 1996).

(a) Equilibrium moisture content data. The equilibrium moisture content of grain, X^* (dry basis, fraction), is correlated with temperature (T, °C) and relative humidity (ψ, fraction) by the equation:

$$X^* = 0.01 \left[\frac{-\ln(1 - \psi)}{-a_4(T + a_5)} \right]^{1/a_6} \tag{3.23}$$

Table 3.7 lists the values of a_4, a_5 and a_6 for a number of products (ASAE, 1995).

(b) Drying rate data. Experimentally determined drying kinetic data are required to calculate the $-\Delta X/\Delta t$ term in dryer calculations. The following semi-empirical equation has been found to describe the relationship between X and t:

$$\eta' = \frac{X - X^*}{X_i - X^*} = \exp(-kt^n) \tag{3.24}$$

Equation 3.24 can be differentiated with respect to time to yield

$$N = -\frac{dX}{dt} = (X_i - X^*) \exp(-kt^n)(nkt^{n-1}) \tag{3.25}$$

Table 3.8 lists values of k and n for a number of crops (ASAE, 1995).

Table 3.7 Constants in the equilibrium moisture content correlation (equation 3.23)

Product	a_4	a_5	a_6
Alfalfa	9.171	2.609	6.894
Barley	2.2919 E−5	195.267	2.0123
Canola	50.0 E−5	40.1204	1.5702
Corn	8.6541 E−5	49.81	1.8634
Rice	1.9187 E−5	51.161	2.4451
Soybean	30.5327 E−5	134.136	1.2164
Wheat	2.3007 E−5	55.815	2.2857

Table 3.8 Values of constants in the drying-rate equation (equation 3.25)

Product	k	n
Corn	$\exp[-7.1735 + 1.2793 \ln (T) + 0.1378 \, v] \; (\mathrm{h}^{-1})$	$0.0811 \ln (\psi) + 0.78X_i$
Canola	$1.3552 - 0.00301X_i - 0.00751T$	$0.5068 - 0.0015X_i + 0.0103T$
	$- 0.5112v \; (\mathrm{h}^{-1})$	$- 0.244v$
Lentils	$0.182626 + 0.0043T \; (\mathrm{h}^{-1})$	0.527
Wheat	$139.3 \exp[-4426/(T + 273)] \; (\mathrm{s}^{-1})$	1

Key: T, Temperature (°C); v, air velocity (m s^{-1}); ψ, relative humidity (%); X, moisture content (dry basis, decimal).

3.5.3 Verification of design calculation

(a) Hot-air dryer. A computer program (Cenkowski and Sokhansanj, 1989) was used to simulate an internal recirculating batch dryer (GT380, Gilmore & Tatge Mfg. Co., Clay Center, Kansas), rated at 2.7 t h^{-1} for wheat undergoing a five point moisture removal (19.5 to 14.5%). Experimental data obtained in the course of the performance tests on the dryer (PAMI, 1983) were used to verify the simulation model.

The fixed-bed drying equations 3.1–3.4 were modified to simulate a steady-state cross-flow drying process. In the numerical solution, a time interval equivalent to $\Delta t_m = \Delta y/v_s$ was computed (p. 52).

The manufacturer's specified capacity for the vertical grain recirculating auger (55 t h^{-1}) was used to calculate the residence time of grain (9 min) in the drying column. Following each re-circulation, the grain was assumed to be completely mixed. The process was repeated until a target moisture content was reached.

Canada Red Spring No.1 wheat, with initial moisture contents ranging from 16.9% to 24.2%, was dried in the dryer. The drying temperature varied from 82°C to 84°C. The ambient air temperature during the test ranged from 3°C to 22°C and the relative humidity from 90% down to 50%. The quantity of grain dried in each test ranged from 7.5 to 8.6 t per batch.

Table 3.9 lists the experimental and calculated results for eight individual test runs. The simulated and experimental final moisture contents were within 0.5% except for test runs 5 and 8 for which there was almost a 1% difference. The simulated grain temperatures were consistently higher than the experimental values by as much as 6°C, which may have been caused by the fact that heat losses were neglected in the simulation. The differences between the simulated and experimental residence times, ranging from 2 to 6 min, represented less than 5% of the total drying time.

Recording accurate grain temperatures and moisture contents during drying is difficult, as is simulation of the internal recirculating batch dryer. Considering these factors, the agreement between the experimental and simulated results was considered satisfactory.

Table 3.9 Experimental and simulated results of wheat drying in a batch internal-recirculation grain dryer

Value measured	Type of run	Test runs							
		1	2	3	4	5	6	7	8
Initial moisture content (%)		16.9	17.1	19.1	19.5	22.1	22.3	23.9	24.2
Final moisture content (%)	Exp.	15.0	15.7	13.9	15.1	15.4	15.5	15.2	15.4
	Sim.	14.5	15.3	13.8	15.1	16.3	15.8	15.2	14.7
Final grain temperature (°C)	Exp.	36	34	38	37	38	36	36	37
	Sim.	42	39	44	41	38	39	41	41
Drying time (min)	Exp.	77	59	141	115	139	154	208	221
	Sim.	80	64	145	121	145	161	210	226

Key: Exp. = experimental; Sim. = simulated.

(b) Unheated-air dryer. Verification trials were conducted on a 4.3 m diameter bin equipped with a fully perforated floor and an in-line centrifugal fan. No auxiliary heaters were attached. Drying started on August 17, 1984 when freshly harvested wheat was pre-dried in the batch internal recirculating dryer and was transferred to the bin at approximately 2-h intervals. The temperature and moisture content of the grain at the time of transfer were recorded and the fan was turned on after loading the first batch. A total of seven batches was needed to fill the bin to a height of 4.3 m. When the bin was full, the static pressure was measured on the pressure side of the fan. The performance curve of the fan supplied by its manufacturer was used to estimate the air flow rate, 87 m^3 min^{-1}, equivalent to 1.85 m^3 min^{-1} t^{-1}.

Ambient temperature, relative humidity and the plenum temperature were recorded hourly. The grain at four levels in the bin was sampled daily for moisture content determination. The fan operated continuously until the grain was considered to be dry at 12.5% on August 26.

To simulate layer drying, the total airflow for the full bin (87 m^3 min^{-1}) was divided by the cumulative mass of grain in the bin as new layers were added. The temperature (37°C, typically) and moisture content of each layer at the time of transfer were taken as the initial condition for that layer, and the recorded hourly plenum temperature and relative humidity (measured by a dewpoint cell) used as input to the simulation program.

Typical values of the experimental and simulated moisture contents for the unheated-air layer drying of wheat are given in Table 3.10. In this particular test, the first layer of grain was transferred from the heated-air dryer into the bin at 1.50 p.m. on August 17, when its average moisture content was 15.7%. The moisture content of the fourth layer transferred at 8.50 p.m. (the same day) was 17.2%; that of the last (seventh) layer was 16.6% at the time of transfer, 4.00 a.m. on August 18.

As Table 3.10 indicates, the actual moisture contents were generally somewhat lower than the values calculated in the simulation program. The actual

Table 3.10 Experimental and simulated moisture contents for the unheated drying of layers of partially dried wheat transferred from the heated-air dryer

Grain depth (m)	Type of run	Moisture content (% wet basis)				
		Aug 17 1.50 pm	Aug 17 8.50 pm	Aug 18 4.00 am	Aug 22	Aug 26
4.3	Exp.			16.6	15.6	14.8
	Sim.			16.2	15.8	15.8
3.0	Exp.		17.2	16.2	15.0	13.1
	Sim.		17.2	16.0	15.9	14.7
1.2	Exp.		15.6	15.6	12.3	10.8
	Sim.		15.5	15.5	12.5	11.0
0.3	Exp.	15.7	13.0	12.4	12.1	10.5
	Sim.	15.7	14.7	14.3	11.5	10.8
Avg.	Exp.	15.7	15.6	15.6	14.0	12.5
	Sim.	15.7	16.0	15.8	14.2	12.8

Key: Exp. = experimental; Sim. = simulated. Depth is from the bottom of the bin.

airflow might have been more than that assumed in the simulation, thereby contributing to the discrepancy. Once the bin was full, the experimental and simulated moisture contents of the individual layers differed by less than 1%. The average simulated and actual bin moisture contents were within 0.5%.

3.6 Crop quality and drying

The crop quality is affected by species and variety, and by the growing conditions. Harvesting, handling, drying and storage also affect quality. In near-ambient drying, quality changes are similar to those occurring during storage. Thus, for present purposes, 'drying' refers to moisture removal above 30–35°C, during which a kernel of grain is subjected to an ever-changing temperature and relative humidity environment. Moreover, dried hot grain is often cooled after drying, thereby exposing it to an immediate temperature change.

It is known that high temperature drying causes increased breakage, stress cracking and discoloration, and decreased millability, oil recovery, protein quality, baking quality and germination (Sokhansanj and Jayas, 1982). The maximum allowable drying temperature depends on the moisture content and the type of grain, in addition to the use to which the grain will be put. Some critical crop (as opposed to air) temperatures are listed in Table 3.11. In certain dryer configurations, such as cocurrent flow, the product temperature is significantly lower than the air temperature. In cross-flow dryers, however, it approaches the air temperature.

The allowable temperatures decrease with the moisture content of the product (Table 3.12). Likewise, as the relative humidity of the drying air increases, the critical kernel temperature decreases. In other words, grain is more susceptible to heat when moisture is present in the air.

Table 3.11 Maximum safe temperatures for seeds and forages

Crop	Temperature (°C)		
	Seed	Commercial	Feed
Starchy seeds	50	65.5	93
Oilseed	50	71.1	93
Pulses	43	60	93
Green alfalfa chops	–	140	–

3.6.1 Quality models

Grain and forage are particularly susceptible to high temperatures. The overall quality, however, depends upon a combination of temperature, moisture content, and time of exposure. Several equations have been developed to predict the quality of the product during drying. Some of these equations are presented below.

(a) Exponential model. Schreiber *et al.* (1981) proposed the use of a kinetic model to predict changes in quality:

$$-\frac{dq}{dt} = K(t)q^m \tag{3.26}$$

where q is the quality characteristic, $k(t)$ is a rate constant and m is the order of the reaction. $k(t)$ depends on temperature and moisture content of the product as follows:

$$K(t) = K(X(t),\ T(t)) \tag{3.27}$$

Equation 3.26 can be written:

$$-\frac{dq}{dt} = K(x(t),\ T(t))q^m \tag{3.28}$$

or

$$\ln\left(-\frac{dq}{dt}\right) = \ln K + m \ln q \tag{3.29}$$

Table 3.12 Effect of moisture content on the allowable drying temperature for wheat

Moisture content (%)	Temperature (°C)	
	Seed	Baking
23	53	55
19	57	60
15	63	67

Table 3.13 Quality constants for wheat (equation 3.32)

Germination	$\ln (K_0)$	a'	b'
Germination	-27.28	0.2328	0.3505
Gluten	-32.30	0.3117	0.2455
Bread volume	-24.08	0.1867	0.3360

K can be found experimentally using the following linear relationship:

$$\ln(K) = \ln(K_o) + a'T + b'X \tag{3.30}$$

Equation 3.29 yields

$$\ln\left(-\frac{dq}{dt}\right) = \ln(K_o)a'T + b'X + m\ln(q) \tag{3.31}$$

On the basis of experimental data; Schreiber *et al.* (1981) concluded that the value of m was small; therefore, equation 3.31 reduces to:

$$\ln\left(-\frac{dq}{dt}\right) = \ln(K_o) + a'T + b'X \tag{3.32}$$

The quality characteristic q may represent, for example, either the germination ratio, the relative volume of bread, or relative gluten. Table 3.13 lists values of the constants K_o, a' and b' for wheat.

(b) Linear model. The time in days, t_g, before wheat in the store loses its germination capacity by 5% has been related to a constant moisture content and constant temperature as follows (Fraser and Muir, 1981):

$$\log(t_g) = 6.234 - 0.2118\,X - 0.053\,T \qquad 12\% \leq X \leq 19\% \tag{3.33}$$

$$\log(t_g) = 4.129 - 0.0100\,X - 0.058\,T \qquad 19\% < X \leq 24\% \tag{3.34}$$

To simulate the dynamic loss in germination during drying, a value of t_g was computed for each time interval Δt. Theoretically, grain loses 5% of its germination capability when the sum of computed $\Delta t/t_g$ values for each layer over the simulated drying period equals unity:

$$S_i = \sum \Delta t/t_g \simeq 1 \tag{3.35}$$

S_i is a storage index and its instantaneous value represents the progress of grain spoilage; it varies between 0 and 1.

(c) Probit model. Finney (1971) and Nellist and Bruce (1987) showed that seed viability decreased exponentially at a given temperature and could be modelled using the normal density function. Therefore variation in the germination of seed was characterized by two parameters, mean and standard deviation(s), in the domain of the normalized standard deviate. Analysis in this domain is known as probit analysis (Finney, 1971). Tang and Sokhansanj (1993)

used this method to predict lentil seed viability as a function of drying parameters.

Sokhansanj and Patil (1996) also used probit analysis to calculate the degradation of the green colour of alfalfa leaf and stem during drying at different temperatures. The expression to predict the greenness of alfalfa was as given below.

$$\gamma = \frac{1}{\sigma_\gamma (2\pi)^{1/2}} \int_{-\infty}^{t} \exp[-(t - t')^2/2\sigma_\gamma^2] \, dt \qquad (3.36)$$

where γ is percent greenness of alfalfa compared to the original colour, σ_γ is the standard deviation of the distribution of greenness with time, and t' is the mean period during which greenness drops to half its original value. The authors derived correlations that enabled σ_γ and t' to be calculated for both leaf and stem.

3.6.2 Stress cracks and broken kernels

Drying cereal grains with hot air induces stress cracks which eventually lead to broken kernels. Stress cracks occur primarily in grains whose main constituent is starch, for example, corn, rice and pulses such as peas and beans. Wheat and barley are also damaged by stress cracking but not as extensively.

The breakage susceptibility of most grains is measured by using a standard Stein Breakage Tester. The tester imparts shear and impact forces on a 100 g grain sample for about 3 min. The sample is then passed through a 4.7 mm round-hole sieve and the quantity of broken kernels and foreign material is measured. The following equation, relating the percentage of broken corn kernel in the Stein Breakage Tester to the drying rate, has been proposed:

$$\% \text{ broken} = 2.9 \, \Delta X/\Delta t + 20 \qquad (3.37)$$

where $\Delta X/\Delta t$ is the drying rate of the kernels in decimal moisture loss per hour.

3.7 Crop drying costs

Two types of costs should be considered in drying agricultural crops: (1) fixed and variable costs; and (2) risk costs based on potential losses. Both of these can be important in comparing drying policies. Details of how to calculate these costs are given in the following paragraphs.

3.7.1 Fixed and variable costs

These costs are associated with using the drying equipment. The components are capital costs, labour costs, energy costs, and costs of over-drying.

(a) Capital costs.   The annual fixed cost C_1 is given by:

$$C_1 = C_0(R_f + R_m) \tag{3.38}$$

where C_0 is the initial cost of the drying system, and R_m is the ratio of the maintenance costs to the initial cost of the drying system (say, 0.01). Typical components of C_0 are as follows: bin and floor \$700, fan \$1000$(G_v)^{0.6}$, heater \$1200, controls (humidistat and thermostat) \$175.

The ratio of the annual capital cost to the initial cost of the drying system, R_f, can be expressed as follows (Audsley and Boyce, 1974):

$$R_f = i\left(\frac{1+r}{2} + \frac{1-r}{n_l}\right) + \left(\frac{1-r}{n_l}\right) \tag{3.39}$$

where i is the annual interest rate, r is the ratio of the salvage value to the original value (typically 0.10), and n_l is the life of the drying system. For a perforated floor, $n_l = 20$ years. For the heater, fan, and controllers n_l was estimated from:

$$n_l = \frac{5040}{n_o} \tag{3.40}$$

where 5040 is the expected life in hours and n_o is the number of operating hours in a year.

(b) Labour costs.   Costs associated with regular inspection of the bin (C_2) can be considered equal for all types of drying system at \$25, as the same amount of labour input is required regardless of the drying time. Thus,

$$C_2 = 25 \tag{3.41}$$

(c) Energy costs.   Both electrical energy (E_e) and fuel energy (E_f) are consumed in the drying process. The total energy cost C_3 is calculated as follows:

$$C_3 = E_e C_e + E_f C_f \tag{3.42}$$

where C_e and C_f are the unit costs of electricity and fuel, respectively.

(d) Cost of over-drying.   Over-drying, although not a visible quality loss, can represent an economic loss at the time of grain sale. No. 1 Canadian wheat can be sold on a mass basis at a maximum moisture content of 14.5% without penalty. The following equation can therefore be used to estimate the cost of over-drying, C_4:

$$C_4 = w(1 - X_{wi}/100)\left(\frac{14.5}{85.5} - \frac{X_{wo}}{100 - X_{wo}}\right)C_g \tag{3.43}$$

In the above equation, w is the mass of grain, X_{wi} and X_{wo} are the average percentage moisture contents of the wet and dried grain, respectively, and C_g is the grain price.

3.7.2 Risk costs

These are costs that cannot be attributed to capital or operating costs. Nevertheless, they have to be considered as part of the cost/benefit analysis of agricultural drying systems. Their methods of calculation are outlined in the following sections.

(a) Inventory cost.   The inventory cost C_5 is derived from the lost interest on the value of the grain during the time that it remains unsold (t_u, h). Taking w as the grain mass, and C_g as the price of No. 1 CWRS wheat (currently \$127 t^{-1}), the inventory cost can be estimated by:

$$C_5 = \frac{wC_g it_u}{8760} \qquad (3.44)$$

The constant 8760 is the number of hours in a year.

(b) Cost of quality loss.   The risks associated with the loss of quality are higher as long as the temperature and moisture content of the grain remain high. Let us assume that, if the wheat experiences a 5% loss in germination $(S_i = 1)$, its selling price decreases from that of No. 1 food grain to feed grain. The potential cost of quality loss C_6 may be calculated as

$$C_6 = wS_i \, \Delta C \qquad (3.45)$$

where ΔC is the price difference (currently \$37 t^{-1}) between No. 1 food grain and feed grain.

(c) Late harvest cost.   There is a risk in grain being downgraded as a result of rain or frost damage when it is left in the field unharvested. The risk cost C_7 associated with the late harvest is estimated by:

$$C_7 = f_p w \, \Delta C \left(\frac{X_{ws} - X_{wi}}{X_{ws} - 14.5} \right) \qquad (3.46)$$

where X_{ws} is a grain moisture content at which, if grain is harvested, it would not be exposed to bad weather. Let us assume that $X_{ws} = 22\%$. X_{wi} is the moisture content of the harvested wet grain to be dried, and f_p is a probability factor representing the incidence of bad weather. Based upon personal field experiences, we assign $f_p = 0.20$ for August, $f_p = 0.25$ for September, and $f_p = 0.33$ for October.

3.7.3 Total costs

The drying cost is the sum of the costs associated with capital C_1, labour C_2, energy C_3, and loss of weight due to over-drying, C_4. The risk cost is the sum

INDUSTRIAL DRYING OF FOODS

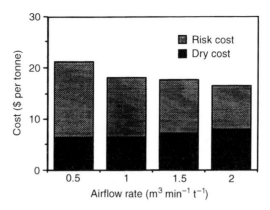

Figure 3.14 Drying costs and risk costs associated with airflow rates from 0.5 to 2 m³ min⁻¹ t⁻¹.

of the potential losses associated with unsold grain C_5, grain deterioration C_6, and late harvest C_7.

Figure 3.14 shows the effect of increasing the air flowrate on the drying cost for 19% moisture content grain in August on the Canadian prairies. As may be expected, the increased air flowrate increased the drying costs slightly, from \$6.2 t⁻¹ for 0.5 m³ min⁻¹ t⁻¹ to \$7.5 t⁻¹ for 2 m³ min⁻¹ t⁻¹. The risk costs, on the other hand, decreased as the potential costs associated with spoilage decreased with air flow, from \$15 t⁻¹ to \$8 t⁻¹. The sum of the two costs also showed a general decrease with the increasing airflow rate.

3.8 Summary

Through-flow drying systems are used for drying agricultural grains. Batch or continuous, fixed or moving bed dryers are available. The most popular and perhaps the simplest configuration is the cross-flow dryer. Grain dryers are divided into on-farm and commercial systems. On-farm dryers are of low capacity (less than 10 t h⁻¹) and are often mobile. Commercial dryers may have a capacity as high as 50 t h⁻¹ and are usually stationary and located at the main grain collection depots.

Steady-state dryer calculations are described. These can be used to estimate the size of the dryer and to calculate energy requirements. Unsteady-state calculations are more complex and are generally used for dryer control and detailed analysis. Grain quality during drying is important. Generally a maximum exposure temperature of 80°C is recommended for grain used for commercial purposes. Grain for seed should not be exposed to temperatures greater than 50°C.

References

Arinze, E.A., Sokhansanj, S., Schoenau, G.J. and Sumner, A.K. (1994) Control strategies for low temperature in-bin drying of barley for feed and malt. *J. Agric. Eng. Res.*, **58**, 73–88.

ASAE Standards 1995. American Society of Agricultural Engineers, St Joseph, MI, USA.

Audsley, E. and Boyce, D.S. (1974) A method of minimizing the costs of combine-harvesting and high temperature grain drying. *J. Agric. Eng. Res.*, **19**(1), 173–88.

Bamforth, C.W. and Barclay, A.H. (1993) Malting technology and uses of malt. In: *Barley: Chemistry and Technology* (eds A.W. MacGregor and R.S. Bhatty), American Association of Cereal Chemists, Inc., St Paul, Minnesota, USA.

Brook, R.C. (1987) *A Look at Control in Bin Grain Dryuing Equipment*. ASAE Paper No. 87-6037, American Society of Agricultural Engineers, St Joseph, MI.

Brooker, D.B., Bakker-Arkema, F.W. and Hall, C.W. (1974) *Drying Cereal Grains*. AVI Publishing Company, Westport, CT, USA.

Brooker, D.B., Bakker-Arkema, F.W. and Hall, C.W. (1992) *Drying and Storage of Grains and Oilseeds*. AVI Publishing, Van Nostrand Reinhold, New York, USA.

Cenkowski, S. and Sokhansanj, S. (1989) Mathematical modelling of radial continuous cross-flow agricultural dryers. *J. Food Process Eng.*, **10**, 165–81.

Finney, D.J. (1971) *Probit Analysis*. 3rd edn, Cambridge University Press, Cambridge, UK.

Foster, G.H. (1964) Dryeration – a corn drying process. *USDA Agricultural Marketing Services Bull. 532*, Washington, DC, USA.

Fraser, B.M. and Muir, W.E. (1981) Airflow requirements predicted for drying grain with ambient air and solar-heated air in Canada. *Trans. ASAE*, **24**(1), 208–10.

Gu, D., Sokhansanj, S. and Norum, D.I. (1996) Intergranular air movement and grain drying in silos with fully, partially and slanted perforated floor. *Drying Technol.*, **14**(3&4), 615–45.

Jayas, D.S. and Sokhansanj, S. (1989) Design data on the airflow resistance to canola (rapeseed). *Trans. ASAE*, **32**(1), 295–6.

Johnson, P.D.A. and Otten, L. (1979) *Solar Assisted Low Temperature Corn Drying in Canada*. ASAE Paper No. 79-3019, American Society of Agricultural Engineers, St Joseph, MI.

Lang, W., Sokhansanj, S. and Sosulski, F.W. (1993) Bulk volume shrinkage during drying of wheat and canola. *J. Food Process Eng.*, **16**, 305–14.

Lang, W., Sokhansanj, S. and Rohani, S. (1994) Dynamic shrinkage and variable parameters in Bakker-Arkema's mathematical simulation of wheat and canola drying. *Drying Technol.*, **12**(7), 1687–708.

Lasseran, J.C. (1977) Dryeration or intermittent drying. *Perspective Agricoles*, **6**, 59–66.

Li, W. and Sokhansanj, S. (1994) Generalized equations for airflow resistance of bulk grains with variable density, moisture content and fines. *Drying Technol.*, **12**(3), 649–67.

Lischynski, D.E., Wasserman, J.D., Frehlick, G.E. and Sokhansanj, S. (1986) *Combination Drying of Wheat in the Prairie Provinces*. Final Report ERDAF 01SG.01916-4-FC69. Engineering and Statistical Research Institute Research Branch, Agriculture Canada, Ottawa, Canada.

Morey, R.K., Cloud, H.A. and Hansen, D.J. (1981) Ambient air wheat drying. *Trans. ASAE*, 24(1), 1312–16.

Muhlbauer, W. and Kuppinger, H. (1975) Better dryer performance, lower energy requirement and improved grain quality as a result of dryeration. *Mais*, **3**(3), 17–20.

Nellist, M.E. and Bruce, D.M. (1987) Drying and cereal quality. *Aspects Appl. Biol.*, **15**, 439–56.

Nellist, M.E. and Bartlet, D.I. (1988) *A Comparison of Fan and Heater Control Policies for Near-ambient Drying*. Report No. 54. AFRC Institute of Engineering Research, Wrest Park, Silsoe, Bedford, UK.

PAMI (1982) *Evaluation Report 289*. Prairie Agricultural Machinery Institute, Humboldt, Saskatchewan, Canada.

PAMI (1983) *Evaluation Report 308*. Prairie Agricultural Machinery Institute, Humboldt, Saskatchewan, Canada.

PAMI (1984) *Evaluation Report 352*. Prairie Agricultural Machinery Institute, Humboldt, Saskatchewan, Canada.

PAMI (1985) *Evaluation Report 424*. Prairie Agricultural Machinery Institute, Humboldt, Saskatchewan, Canada.

PAMI (1989) *Evaluation Report on Hopper Bin Natural Air Drying Systems*. Prairie Agricultural Machinery Institute, Humboldt, Saskatchewan, Canada.

Schreiber, H., Muhlbauer, W., Wassermann, L. and Kuppinger, H. (1981) *Influence of Drying on Wheat Quality - Study of the Reaction Kinetics*. NIA Paper No. 517. National Institute of Agricultural Engineering, Wrest Park, Silsoe, Bedford, UK.

Sokhansanj, S. (1982) Grain drying simulation with respect to energy conservation and grain quality, in *Advances in Drying*, **3** (ed. A.S. Mujumdar), Hemisphere, McGraw–Hill, New York, pp. 121–80.

Sokhansanj, S. and Bakker-Arkema, F.W. (1981) Waste heat recovery in grain dryers. *Trans. ASAE*, **24**(5), 1321 and 1325.

Sokhansanj, S. and Bruce, D.M. (1987) A conduction model to predict grain drying simulation. *Trans. ASAE*, **30**(4), 1181–4.

Sokhansanj, S., Lang, W.G. and Lischynski, D. (1991) Combination drying of wheat. *Canadian Agric. Eng.*, **33**(2), 265–72.

Sokhansanj, S. and Lischynski, D. (1991) Low temperature drying of wheat with supplement heat. *Canadian Agric. Eng.*, **33**(2), 273–8.

Sokhansanj, S. and Jayas, D.S. (1995) Drying of foodstuff, in *Industrial Drying Handbook* (ed. A.S. Mujumdar), Marcel Dekker, New York, pp. 589–625.

Sokhansanj, S. and Lang, W. (1996) Kernel and bulk volume of wheat and canola exposed to humid and dry air – experimental data and theoretical analysis. *J. Agric. Eng. Res.*, **63**, 129–36.

Sokhansanj, S. and Patil, R.T. (1996) Kinetics of dehydration of green alfalfa. *Drying Tech.*, **14**(5), 1197–234.

Sumner, A.K., Crowle, W.L., Spurr, D.T., Sokhansanj, S. and Kernan, J.A. (1989) The effect of drying temperature on the properties and quality of immature barley. *J. Plant Sci.*, **69**, 1083–92.

Tang, J. and Sokhansanj, S. (1993) Drying parameter effect on lentil seed viability. *Trans. ASAE*, **36**(3), 855–61.

Thompson, T.L., Peart, R.M. and Foster, G.M. (1968) Mathematical simulation of corn drying – a new model. *Trans. ASAE*, **11**(4), 588–96.

4 Fluidized bed dryers

R.E. BAHU

4.1 Introduction

Fluidized bed dryers are one of the most common types of dryer used in industry to produce dry particulate products. They are particularly strong candidates for drying powders in the size range 50–2000 μm. Their advantages are simple construction and low capital and maintenance costs. There is a particularly wide variety of equipment configurations and options available, which reflects the versatility of fluidization technology. The main operational problems can arise from poor fluidization and powder carry-over, both of which can be dealt with by appropriate equipment selection and design.

Typical operating characteristics of fluidized bed dryers are their ability to handle:

- particulate feed and product in the size range 20 μm–5 mm;
- slurry or solution feeds in fluidized bed granulators;
- throughput in the range 10 kg h^{-1}–100 t h^{-1} solids;
- drying times from 1 min to 2 h.

4.2 Basics of fluidization

An understanding of fluidization is the key to successful fluidized bed drying. Many of the problems experienced when fluidized beds were first introduced into drying practice were simply due to a lack of appreciation of the underlying principles. A concise, practical guide to fluidization has been given by Pell (1990).

Figure 4.1a illustrates schematically a typical fluidized bed. An idealized plot of the pressure drop across the bed (ΔP_b) versus gas velocity (u) is shown in Figure 4.1b. Consider a bed of particles initially at rest, supported by, say, a simple perforated plate (the 'distributor'). As the gas velocity is progressively increased, ΔP_b normally increases linearly with u. (For large and dense particles, however, the relationship is parabolic.) At a certain gas velocity, termed the minimum fluidizing velocity, u_{mf}, the pressure drop across the bed becomes equal to the weight of the bed per unit area, ΔP_{mf}. At this point, the particles become freely suspended in the air stream and the bed begins to expand. Although further increases in u result in added expansion of the bed, because of the presence of gas bubbles, ΔP_b remains constant at ΔP_{mf}. If the gas velocity is now decreased, the point of minimum fluidization is more clearly defined.

The above description is an idealization for a bed of uniformly sized particles. Figure 4.2 shows four variants on the ideal plot of pressure drop versus gas velocity. If there is a wide particle size distribution, the smaller particles will try to become fluidized at a gas velocity at which it is not possible to fluidize the larger particles. Panel A shows an erratic pressure drop as the gas velocity increases but a smooth defluidization curve on decreasing it. Note that, in this case, there is no single minimum fluidization velocity, rather a range of gas velocities. In deep beds, slugging can occur if bubbles coalesce, and the pressure drop will increase after fluidization, as shown in Panel B. Channelling occurs if the gas can find an easy route through the bed. This is usually due to poor gas distribution in beds of fine particles. This leads to areas which are defluidized and hence, as part of the bed is unsupported, the situation depicted in Panel C results. A development of fluidized beds based on exploiting channelling has a

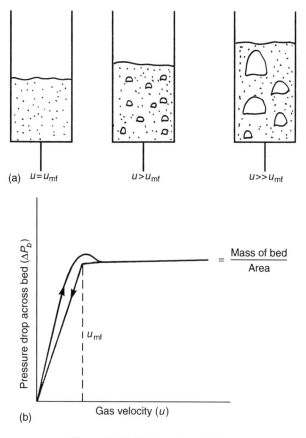

Figure 4.1 Fluidization characteristics.

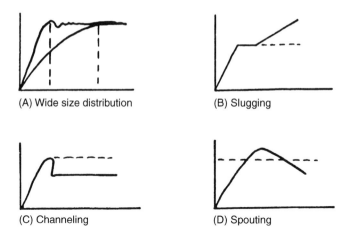

(A) Wide size distribution   (B) Slugging

(C) Channeling   (D) Spouting

Figure 4.2 Pressure drop plots for special cases of fluidization.

central gas jet which entrains particles in a fountain and is called a spouted bed (Panel D).

Even after the minimum fluidization velocity has been reached, not all solids will be well fluidized. Usually, the operating gas velocity is set at between 2–4 times the minimum fluidization velocity. Geldart (1973) has classified materials into four categories according to their fluidization behaviour:

- Type A – aeratable particles, fluidize smoothly (typically 20–200 μm);
- Type B – sand-like particles, bubbly with most gas passing through as discrete bubbles (typically 200–800 μm);
- Type C – cohesive, difficult to fluidize (typically <20 μm);
- Type D – spoutable, difficult to fluidize as very high gas velocities are needed (typically >1000 μm). These solids may best be handled in spouted beds.

Note that the above values of typical particle sizes are for a particle fluid density difference of some 1000 kg m^{-3}. At significantly lower or higher density differences, the particle sizes will be higher or lower, respectively.

Clearly, there is an upper limit to the operating gas velocity set by the acceptable fraction of the bed which is carried out with the exhaust gas. When the bed is well fluidized, a significant number of particles of various sizes will be ejected, primarily as a result of bubbles bursting on the surface. Most of these particles will fall back into the bed after some 10–100 mm but some will be entrained for a much greater height. The limiting height by which all such solids have disentrained is called the transport disengagement height (TDH) (Guguoni and Zenz correlation, 1980). The so-called 'freeboard' above the bed will generally be designed to be somewhat greater than the TDH. In some cases, carryover of particles may be encouraged to achieve some fines removal but particle size classification is rather coarse. If the particle size distribution is wide or if the freeboard height is limited, then an expanding section can be used to reduce the superficial gas velocity or a vibro-fluidized bed can be considered (section 4.4.6).

Another effect of fluidization to consider is attrition. This can occur either through particle collisions with other particles or solid surfaces, such as walls, or by abrasion. In general, attrition is not a serious problem in fluidized beds unless the particles are quite friable. In many cases, a useful rule of thumb is that the extent of attrition is proportional to the cube of the gas velocity.

It is important to consider the effect on the equipment of the vigorous fluidization in deep beds, especially for Type B materials. Large bubbles bursting at the surface of the bed can cause vibration in the dryer and supporting structure. Indeed, if the bubble production rate is close to the natural frequency of the unit, resonance could occur and cause catastrophic damage.

The effect of wet particles on fluidization can be dramatic. If the feed or starting material has surface moisture then it could be much more cohesive and difficult to fluidize than the equivalent dry powder. Also, internal moisture will raise the actual density of the particles. A fluidized bed dryer has therefore to cope with the changes in fluidization characteristics of the material as its moisture content decreases from the feed to product. In food applications, care must be taken not to expose the particles to excessive temperatures, which generally will be highest in the region within 20 mm of the distributor. Also, any larger particles which are not well fluidized may suffer case-hardening and not dry down to the required moisture content.

4.3 Features of construction

A fluidized bed dryer consists of three important elements:

- distributor
- plenum chamber
- freeboard region and gas-cleaning systems.

The correct specification and design of each of these components are necessary to ensure that the dryer operates in a satisfactory and controlled manner. The principle features are discussed in the following sections.

4.3.1 Distributor

The distributor plate is vital to the success of the overall design. The requirements are as follows.

1. It must distribute the fluidization gas around the bed sufficiently evenly to avoid excessive local gas velocities, which may cause channelling; or low velocity regions, where defluidization may occur.
2. The pressure drop through the holes should not be too high.
3. The holes must not be so large that bed material can fall through them into the plenum chamber.

4. The distributor must be strong enough to provide support for the solids in the bed and to withstand thermal stresses from the heating and cooling of the bed and the gas.

To achieve 1 and 3, a large number of small holes is desirable, but this makes the distributor plate expensive to construct. Fewer, larger holes reduce the pressure drop, but give poorer fluidization. In practice, distributors rarely have holes of less than 1 mm in diameter because they are so costly to bore. The hole diameter does not normally exceed five times the particle diameter; larger holes would allow excessive drainage of solids. Conversely, the holes must not be so small that the orifice velocity is high enough to cause particle attrition. A reasonable rule-of-thumb is that the velocity should never exceed 90 m s^{-1}. Holes are normally drilled on a triangular pitch.

With regard to point 2, as a rough guide, for shallow beds up to about 0.2 m, a distributor pressure drop equal to the pressure drop across the bed (bed weight per unit area) will usually be adequate. This is sufficient to promote uniform fluidization without being excessive.

In order to satisfy point 4, plate thickness is usually considerably greater than hole diameter, but plates as thin as 0.75 mm have been manufactured.

A considerable number of distributor plate designs have been tried; the most common are illustrated in Figure 4.3. The simplest and most widely used form is the perforated plate as shown in A. This is also the design most likely to allow

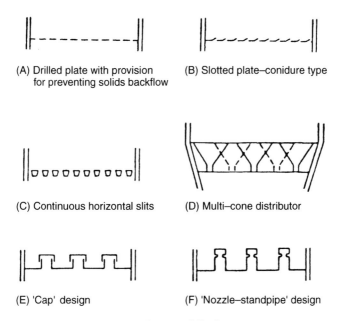

(A) Drilled plate with provision for preventing solids backflow

(B) Slotted plate–conidure type

(C) Continuous horizontal slits

(D) Multi–cone distributor

(E) 'Cap' design

(F) 'Nozzle–standpipe' design

Figure 4.3 Types of distributor.

solids to fall back into the plenum chamber, especially at shutdown or if the gas flow is temporarily interrupted; special provision to prevent this is usually required. The use of a slotted 'conidure' type of plate as in B inhibits backflow but increases pressure drop. Narrower continuous slits too small for particles to fall through are employed in the design shown in C. In D the air passages are conical; this spreads the airflow more evenly and reduces pressure drop while ensuring that the bottom of the plate is still strong and rigid.

On some perforated plate distributors, nozzles are fitted over each hole. These are designed so that, at a pre-determined pressure drop, the air flows in one direction and prevents particles back-sifting into the plenum chamber. In the last two types, a lid is mounted over the inlet nozzle and the gas emerges around the side of it. The 'cap' design in E is similar to a bubble-cap tray in a distillation column. Another variant is the 'nozzle-standpipe' design shown in F.

A further important variant is where the distributor is constructed of a woven wire mesh. This is particularly common on batch units and pilot plants. It is less robust than a plate and is therefore less suitable for heavy duty continuous applications.

A mesh or cloth may also be used to increase the pressure drop across the distributor, if required. This is usually mounted at the top of the plenum chamber, immediately below the distributor plate. If the product characteristics change or a different bed depth is used, the pressure drop can be adjusted simply by replacing the cloth with one of a different thickness.

4.3.2 Plenum chamber

The hot gas entering the dryer is fed into a plenum chamber below the fluidized bed distributor. The chamber is generally the same diameter as the bed, and much larger than the supply pipe. Its function is to provide a homogeneous region of gas below the distributor plate, with little swirl or localized flows. If this is not achieved, 'channelling' of gas will tend to occur, giving rise to high velocities in certain regions of the bed. This, in turn, leads to irregular fluidization, with severe fines entrainment in some parts of the bed or slumping (loss of fluidization) in other regions, or both.

Conical shaped plenum chambers and inverted central supply lines have been used successfully in well mixed beds (section 4.4.3), whereas round or spherical shaped chambers have given rise to problems. In plug flow beds (section 4.4.4), distribution can be aided by narrowing the plenum chamber from the feed end to the discharge end. Alternatively, deliberate maldistribution can be used to yield higher gas flows where fluidization is most difficult, such as the feed end of a plug flow bed.

The gas supply duct into the plenum chamber should be of sufficient cross-section not to create a strong jet, and hence potential distribution problems. Kunii and Levenspiel (1991) recommend that the pressure drop on entering the chamber should be less than 1% of the pressure drop through the distributor. The wide availability of computational fluid dynamics tools will aid the design of such chambers.

4.3.3 Freeboard region and gas cleaning systems

The region above the fluidized bed is known as the freeboard. It must be sufficiently high to allow disentrainment of particles which have been thrown up from the bed by transient local high flows of gas. Even so, the gas leaving the dryer will contain a significant proportion of entrained fine particles which must be removed. Usually the gas discharges into a cyclone, which separates out most of the particles, and any remaining dust is removed by a back-up bag filter or similar unit.

The recovered fines will already be dry because of their high surface area-to-volume ratio, so if they are sufficiently valuable and their presence is not a problem, they can be mixed in directly with the final product. Otherwise they may be agglomerated using a suitable binder, or even used as seeds in a fluidized bed granulator (section 4.4.8).

4.3.4 Exhaust air recycle

In recycle systems, a proportion of the exhaust gas is bled off after the fines have been removed, and is then recycled to the entrance of the dryer. Energy savings result because the heat in the warm exhaust gas is recovered. The circulating air contains more water vapour than a once-through system; this may affect combustion and an indirect heater is therefore used. The increase in inlet air humidity will reduce the driving forces for drying. Therefore, in order to maintain the same drying rate, the dryer has to operate at a higher exhaust temperature. This may limit the amount of recirculation possible with heat-sensitive products.

A once-through system with heat exchange between exhaust and inlet air may then be preferred, although its capital cost will be higher. Recycling can also be beneficial if there is an odour problem with the product. The reduced volume of exhaust air concentrates the odour, thereby making it easier to remove.

An alternative flowsheet configuration utilizes direct firing in the recycle loop. A high ratio of recycle to purge is used (typically 5:1). The circulating gas then contains a significant proportion of combustion products (up to 15%) but the oxygen concentration is correspondingly reduced to 3–8% from its normal value of 21%. For mildly combustible dusts, this gives a safe margin below the oxygen concentration which would allow combustion of the product and a dust explosion. This is called a 'self-inerting' dryer. If the water vapour concentration becomes too high to give an adequate humidity driving force in the dryer, a condenser can be added, although there is a corresponding loss in energy efficiency. The condenser may also be used to remove fines and soot if they would contaminate the product to an unacceptable extent. Again, odour removal is made easier.

4.4 Types of fluidized bed dryer

4.4.1 Introduction

There is a variety of fluidized bed designs to cater for widely different applications, namely:

- batch fluidized bed dryers;
- well-mixed continuous fluidized bed dryers;
- plug-flow continuous fluidized bed dryers;
- multi-stage fluidized bed dryers;
- internally heated fluidized bed dryers;
- vibrated fluidized bed dryers;
- fluidized bed granulators/coaters;
- mechanically agitated fluidized bed dryers;
- centrifugal fluidized bed dryers;
- Jetzone dryer.

The basic features of each type are described in the following sections.

4.4.2 Batch fluidized bed dryers

Batch fluidized bed dryers (Figure 4.4) have to a certain extent superseded tray ovens as the most economic method of drying powders. They are used for low throughput (normally $<50\,\mathrm{kg\,h^{-1}}$ and good for $<1000\,\mathrm{kg\,h^{-1}}$), multi-product applications. The wet feed is loaded into a tub with a distributor plate as its base. The capacity of the product container ranges from $0.02\,\mathrm{m^3}$ for pilot plant units to $0.5\,\mathrm{m^3}$ for the larger units. In situations where materials are particularly difficult to fluidize, the product container can be equipped with an agitator to enhance fluidization. The tub is wheeled into the drying cabinet and locked in position.

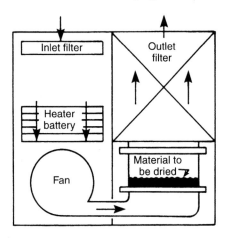

Figure 4.4 Batch fluidized bed dryer.

The fan is switched on to start the drying process with the hot air being heated either indirectly by steam or hot water, or directly by natural gas. The exhaust air passes through filter bags located in the top of the cabinet to trap any entrained solids. The major reason for siting the fan upstream of the bed and the solids feedpoint is one of safety. In this case there is no danger of a dust explosion within the fan and it may not be necessary to fit explosion relief panels.

If ambient air is being drawn in from the environment, it will normally be passed through disposable glass fibre pre-filters to remove any dust. If a particularly clean air supply is required HEPA filters may be used.

A vibrated batch unit is available for materials which are difficult to fluidize. Alternatively, a static fluidized bed may be fitted with a suitable agitator.

A development in distributor design for a batch fluidized bed of cylindrical section involves the use of directional slits. This patented distributor is particularly useful for rapidly discharging the dried materials from the bed. In modern batch fluidized bed dryers, the feeding, drying and discharging operations are automatically initiated and controlled.

4.4.3 Well-mixed fluidized bed dryers

These continuous dryers have a circular or square cross-section which produces near perfect mixing of the solids (Figure 4.5). The result is a uniform bed temperature equal to the product and exhaust gas temperatures. However, the

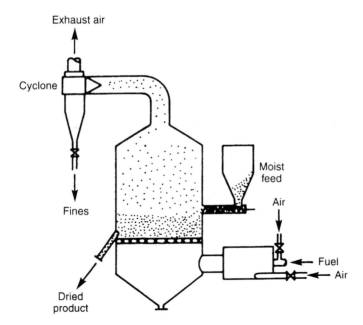

Figure 4.5 Well-mixed fluidized bed dryer.

perfect mixing means that the product moisture content distribution will span the range from inlet moisture content to zero, which may be unacceptable for many food applications. Also for this reason, it is impractical to use these devices to dry down to low moisture contents. From a product quality viewpoint, it may be undesirable for some particles to remain in the dryer for many times the average residence time. One advantage of the perfect mixing is that the feed falls into a bed of relative dry material and so is easy to fluidize.

The height of the layer of particles in a well-mixed fluidized bed is normally in the range 250–450 mm. The dryer may also have an underflow weir which allows coarse material to flow across the distributor plate into a discharge chute. By adjustment of the air velocity in the freeboard above the bed to the elutriation velocity of the smaller particles, coarse classification of the material can be effected simultaneously with the drying operation, if so desired. The entrained particles are separated from the exhaust gas in cyclones or other gas cleaning equipment.

4.4.4 Plug-flow fluidized bed dryers

In plug-flow fluidized bed dryers, the bed usually has a length-to-width ratio in the range 4:1 to 30:1 and the solids flow continuously along it like a river. A simple straight channel may be 1–2 m wide and up to 20 m in length. The greater the length to width ratio the more uniform will be the product moisture content. A number of geometries are available as shown in Figure 4.6. However, care must be taken to minimize corners, where dead zones may appear and cause deviations from plug flow. The main operational problems occur at the feed end where wet feedstock must be fluidized directly rather than mixed with drier material as in a well mixed unit. In some installations, mechanical stirrers are used to overcome initial fluidization problems; in others, a higher local gas velocity is used. It is also worth noting that, as the air at the discharge end of the dryer performs little drying, the thermal efficiency of plug flow devices is lower than that of the well mixed type, unless the plenum chamber is split and lower gas temperatures are used in the later stages of drying. Thermally sensitive materials may also suffer from being exposed to hot gas in this region, where they are essentially dry.

Drying rates in the initial section may be so high that the air leaving the bed is close to its dewpoint; it can then condense, particularly on the metal surfaces above the bed. Corrosion may result but, more seriously, the water runs back down the walls and rewets the solids, lowering efficiency and encouraging agglomeration. Further along the bed, drying rates are much lower so the heat available in the air is not fully utilized and energy efficiency falls.

4.4.5 Multi-stage fluidized bed dryers

There are often advantages in using two or more fluidized beds in stages. A typical example is a well mixed unit followed by a plug flow one (Figure 4.7),

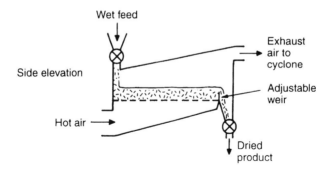

(a) Straight path

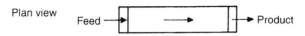

(b) Reversing path

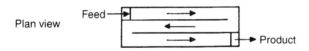

(c) Spiral path

Figure 4.6 Plug-flow fluidized bed dryer.

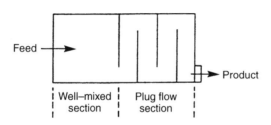

Figure 4.7 Integrated well-mixed and plug-flow fluidized bed dryer.

allowing wet and sticky feedstocks to be handled effectively without getting too wide a spread of final moisture content. In some cases, the distributor is constructed so that the well mixed and plug flow sections are part of the same bed. It is also common to have a plug flow dryer in which different sections are supplied with gas at different temperatures and/or velocities, or to have a fluidized cooler as the final stage, which can be beneficial for food applications (Figure 4.8).

The first stage of a two-stage system need not, of course, be another fluidized bed. Since surface moisture tends to cause stickiness problems, a dryer with a short residence time, which removes surface moisture rapidly, is particularly effective. For a solid feed, a pneumatic conveying (flash) dryer (Chapter 9) can be used as the first stage (Figure 4.9). If the initial feed is a liquid or slurry, a spray dryer may be used as the first stage (Chapter 5). In both cases, the fluidized bed is the second stage and provides the long residence time needed to remove internal moisture. Note that hybrid dryers are now available with an integral fluidized bed located within the base of the spray dryer. This eliminates transfer problems between the two stages. Also, by operating the fluidized bed at a higher velocity than normal, an agglomerated product can be produced.

4.4.6 Vibro-fluidized bed dryers

In order to overcome the fluidization problem of traditional plug flow devices, designs are available in which the distributor is vibrated (Figure 4.10). Materials handled in this type of dryer include those having a wide size distribution with a significant number of oversize particles and those which are cohesive, sticky or fragile. With a vibro-fluidized bed, the gas velocity can be set low enough to avoid excessive elutriation, while the large particles are kept moving by the vibrations. For the same reason, vibro-fluidized beds are often used for beds consisting entirely of large particles (1 mm diameter or greater) so that the air velocity need only be slightly above the minimum fluidization velocity.

The current designs are normally limited to a maximum inlet air temperature of 250°C by materials of construction and to a maximum length of 10 m owing to flexing of the distributor. However, improved distributor design is allowing static fluidized beds to handle many of the feedstocks previously processed in vibrofluidized units.

4.4.7 Internally heated fluidized bed dryers

In this type of dryer, part of the heat is supplied by heating elements immersed in the fluidized bed of solids. The main application is the drying of fine, heat sensitive materials. With such materials, both the air velocity and inlet gas temperature are limited to low values. If all the heat were supplied by the air, this would lead to a very large distributor area. Therefore, the use of submerged

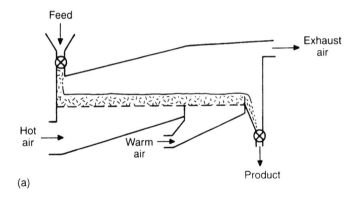

(a)

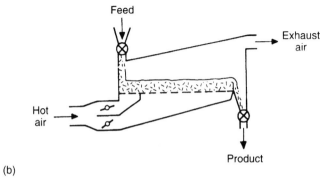

(b)

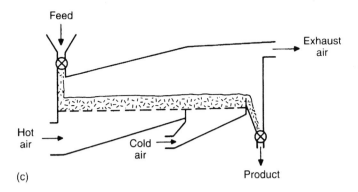

(c)

Figure 4.8 Sectioned fluidized bed dryers: (a) multi-temperature, (b) multi-velocity, (c) dryer/
cooler.

heating elements can lead to significant capital and operating cost savings.
However, heat transfer considerations make this option most attractive for fine
particles of the order of 100 μm in size.

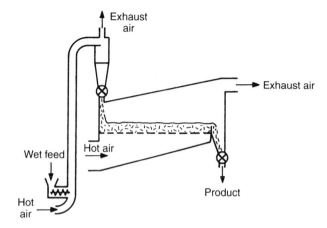

Figure 4.9 Pneumatic conveying/fluidized bed dryers in series.

4.4.8 Fluidized bed granulators/coaters

Fluidized bed granulators produce solid or agglomerated powders from slurry and solution feedstocks which are sprayed onto or into the fluidized bed. Batch and continuous units are available (Figure 4.11). Two growth mechanisms are possible. Firstly, 'onion ring' growth, in which layers are built up on the particles, yields a hard dense product. This mechanism is also utilized when a coating is to be applied to particles. Secondly, when the sprayed liquid acts as a binder, the fine particles are agglomerated into 'raspberry' clusters. The dominant mechanism will depend on the nature of the material and the operating conditions. The main operating problem arises if the liquid to solid ratio in the bed is too high and 'wet choking' occurs.

4.4.9 Mechanically agitated fluidized bed dryers

With some difficult to handle feedstocks, one option is to use a fluidized bed with a mechanical agitator mounted in the base and air entering tangentially via

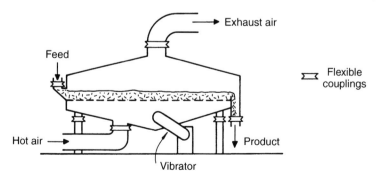

Figure 4.10 Vibro-fluidized bed dryer.

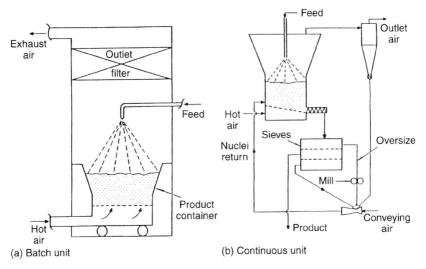

Figure 4.11 Fluidized bed granulators: batch and continuous.

a distributor in the wall (Figure 4.12). The dry product is carried out of the dryer by the gas stream, and can undergo further drying in the duct above, which acts as a pneumatic conveying (flash) dryer. For this reason, this device is sometimes called a spin flash dryer. The agitator rotates at some 100 rpm, which should be

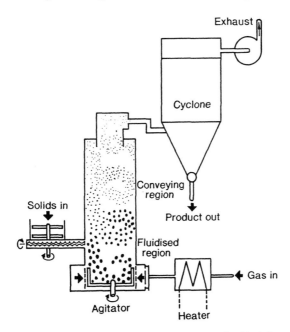

Figure 4.12 Mechanically agitated (spin-flash) fluidized bed dryer.

sufficient to break up most filter cakes and lumps and mix them into the dry powder. A very similar unit without the agitator is marketed as the swirl fluidizer. More reliance is placed on the tangential air jets to break up agglomerates and other oversized particles before they are swept away.

4.4.10 Centrifugal fluidized bed dryers

One example of process intensification in the field of drying is the centrifugal fluidized bed. High relative gas velocities are achieved by placing the solids in a rotating perforated drum and balancing the centrifugal and drag forces (Figure 4.13). The main applications are pre-drying and removal of surface moisture, although the technique has been applied to segmented foods such as diced vegetables. The hold-up is relatively low at 10–20%, but the drying rates are correspondingly higher than in a traditional fluidized bed, leading to a more compact dryer for a given throughput.

4.4.11 The Jetzone dryer

This unit has a solid base instead of a perforated distributor. Air is blown downwards at high velocity from an array of pipes on to this surface, lifting the bed of particles away from it and creating an air cushion. An advantage is that markedly non-spherical objects, such as flakes, can be processed if their density is reasonably low.

4.5 Operating considerations

The various aspects of dryer operation in general are discussed in Chapter 12. Specific aspects relating to fluidized bed dryers are highlighted in the following paragraphs.

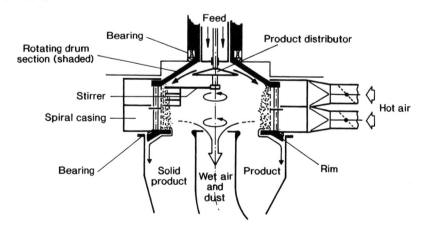

Figure 4.13 Centrifugal fluidized bed dryer.

4.5.1 Control

Williams-Gardner (1971) recommends the following control scheme for fluidized bed dryers in particular:

- maintain a constant fluidizing gas velocity and temperature;
- control solids feedrate to maintain a constant fluidized bed temperature;
- control dry product discharge rate to maintain a constant bed depth.

In addition to the control of drying, any scheme must take account of how changes in operating conditions will affect fluidization.

4.5.2 Fire and explosion hazards

In a fluidized bed dryer, the most likely location for a dust explosion is in the freeboard above the bed, where fines are being elutriated. Other, less likely positions are in the cyclone or, if solid has fallen back through the distributor plate, the plenum chamber. The air space above storage hoppers can also contain a significant dust concentration; it should be remembered that everyday materials like grain, flour, custard powder and dried milk powder are combustible, and all of these have been responsible for major dust explosions. Finally, a dust layer, if disturbed, will create a high local airborne concentration which can be ignited. Calculation of explosive limits is discussed in the Institution of Chemical Engineers Guide (Abbott, 1990). It is also possible to carry out combustion tests and the Guide lists firms who have the necessary specialized expertise.

4.5.3 Energy conservation

As with any other type of dryer, good housekeeping measures are essential in order to maintain fluidized bed dryers in prime operating condition and to minimize energy consumption.

Exhaust air recycle, heat recovery, and self-inerting systems are all possible with fluidized bed dryers, but at the expense of additional capital cost.

4.6 Applications

Fluidized bed drying can be used to dry a wide variety of particles. The principal requirement is that the feed must be fluidizable at a reasonable gas velocity; this rules out very small and light particles (below about 20 μm), large and heavy particles (above about 1 mm) or solids which are very sticky across the entire range from feed to product moisture contents. Even where the feed material is too wet to fluidize properly, a fluidized bed can be used as a second-stage dryer following a pneumatic conveying dryer or similar unit which removes mainly surface moisture. The dryer system can be adapted to handle almost any throughput and naturally to deal with thermally sensitive materials. It is capable

Table 4.1 Selective list of published food applications of fluidized bed dryers

Baby foods	Instant coffee
Baker's yeast	Mashed potato
Casein	Milk sugar
Cheese	Nuts
Citric acid	Powdered milk
Dairy products	Preserves
Desiccated coconut	Rice
Fish meal	Sauce mixtures
Flavours	Seeds
Flour	Tea
Grain	Vitamins
Herbs	

of drying particles down to low moistures close to their equilibrium moisture content. It can also be used for fragile and friable materials, as it handles the solids more gently than any other dispersion dryer in which a solid feed is processed.

As with most major types of dryer, a very large range of materials has been successfully dried in standard or modified types of fluidized bed dryer. Table 4.1 is a selected list of food applications taken from manufacturers' literature. In several cases, the products are granulated or coated in the fluidized bed. Some of the larger products are processed in vibro-fluidized beds.

Operating conditions for various food applications can range up to gas velocities of 15 m s^{-1} and residence times up to 30 min (Okos *et al.*, 1992). Fluidized beds can also be used to create puffed products (Shilton and Niranjan, 1994).

4.7 Test procedures

There are three sets of measurements which must be made on a material in order to provide the data necessary to design a new fluidized bed dryer or analyse the performance of an existing unit, namely:

- fluidization characteristics;
- equilibrium moisture contents;
- batch drying curves (dry basis moisture content as a function of time).

Clearly, the tests must be performed on samples which represent as closely as possible the actual material from the full-scale drying system. This can be a problem with natural food products which are inhomogeneous and subject to variability depending on source and season. In particular, for food products, it is always advisable to use feed material rather than re-wet previously dried material. Often the best approach is to locate the test equipment in the food factory, rather than transport samples, which may involve significant time

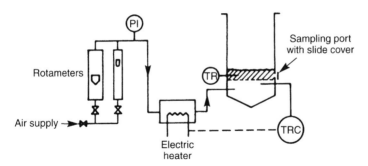

Figure 4.14 Apparatus for fluidization and drying tests.

intervals during which there is a risk that the material will degrade or that moisture will re-distribute.

A simple test apparatus is shown in Figure 4.14. This can be used for the fluidization and drying tests. The test fluidized bed should be a glass section of some 10 cm or more in diameter (the larger the better), with a perforated or porous plate distributor. Table 4.2 summarizes the fluidization measurements required for the various types of fluidized bed dryer. In addition, it is useful to note:

• bed height as a function of gas velocity and bed mass (dry basis) per unit area (needed to set overflow weir height);
• particulate ejection height as a function of gas velocity and bed height (needed to set freeboard height).

The bed pressure drop can be measured using a manometer. One arm is connected to a capillary tube having one end resting on the distributor plate and the other arm is open to atmosphere. Note that these tests are performed at ambient conditions and the data will need to be extrapolated to process conditions (Wen and Yu, 1966).

The desorption equilibrium isotherm (equilibrium moisture as a function of absolute temperature and relative humidity) is also needed. The simplest measurement method involves placing a sample of the test material in a sealed

Table 4.2 Required fluidization measurements

Gas mass velocity	Batch	Well-mixed	Plug-flow
Minimum fluidization			
Product	–	–	Yes
Minimum uniform fluidization			
Feed	Yes	–	Yes
Product	*	Yes	*
Minimum significant elutriation			
Feed	*	–	*
Product	Yes	Yes	Yes

*For rare cases where product is stickier than feed

container together with a saturated salt solution, which is maintained at constant temperature. By using a series of containers with different salt solutions, a wide range of relative humidities can be covered (ASTM, 1985).

It is also recommended that at least three batch drying curves are measured. These should be obtained at two gas mass velocities and two bed masses per unit area. In addition, curves at a different gas inlet humidity and gas inlet temperature may be measured but are not usually required. The conditions should be close to those envisaged for the full scale unit. The selection of optimum conditions has been discussed by Reay and Allen (1982) and Reay (1989).

The simplest method of measuring the batch drying curve is to take samples of material periodically from the bed either via the sampling port or by a scoop. It is important that less than 5% of the initial bed weight is removed in total, otherwise the bed mass per unit area will change significantly. For fast-drying materials, or where sample sizes are limited, an exhaust gas monitoring system utilizing an infra-red gas analyser linked to a computer can generate drying curves directly. Such a system is operated by the Separation Processes Service at AEA Technology, Harwell, UK.

4.8 Design methods

4.8.1 Scoping design

It is possible to calculate roughly the size or throughput of a continuous fluidized bed dryer using psychrometry. There are two basic assumptions, namely that the inlet gas is indirectly heated and that there are no heat losses. The sizing calculation is given as follows:

1. Calculate the water evaporation rate, W_{ev}

$$W_{ev} = S(X_i - X_{o,avr}) \qquad (4.1)$$

where S (kg s^{-1}) is the dry basis solids feedrate, X_i (kg kg^{-1}) the dry basis inlet moisture content and $X_{o,avr}$ (kg kg^{-1}) the dry basis outlet moisture content.

2. Locate inlet conditions on a psychrometric chart (Figure 4.15) using inlet gas temperature T_{ai} (°C) and inlet gas humidity Y_i (kg kg^{-1}) (default value 0.005 kg kg^{-1}). Draw a line of constant enthalpy from the inlet condition to the saturation line (100%RH). Read off the exhaust gas humidity Y_o (kg kg^{-1}) at the intersection with 20%RH curve. This rule of thumb will yield a conservative estimate of the dryer size in most cases.

3. Calculate required dry basis gas mass flowrate:

$$G = W_{ev}/(Y_o - Y_i) \qquad (4.2)$$

4. Calculate approximate gas density at inlet conditions at atmospheric pressure p_a (kPa):

$$\rho_a = 353 \, p_a / [(273 + T_{ai}) \, 101.4] \qquad (4.3)$$

5. Calculate bed area A_b (m^2) from the specified gas velocity u (m s^{-1}) (perform fluidization test or previous experience or default value 1 m s^{-1}):

$$A_b = G/(\rho_a u) \qquad (4.4)$$

The throughput can be calculated for a given bed area by using the above procedure in reverse.

Worked example: Tea is to be dried from 30% to 5% at a throughput of 3000 kg h^{-1} dry basis. A simple fluidization test indicates that a gas velocity of 1 m s^{-1} is sufficient for good fluidization. An inlet temperature of 120°C can be used.

1. $S = 3000$ kg h^{-1} = 0.833 kg s^{-1}. Hence

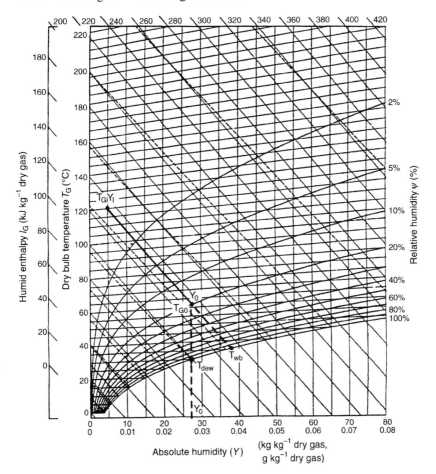

Figure 4.15 Psychrometric chart for scoping design calculation.

$W_{ev} = 0.833 \ (0.30 - 0.05) = 0.208 \ \text{kg s}^{-1}$

2. $Y_o = 0.027 \ \text{kg kg}^{-1}$ (Figure 4.15) (assumes $Y_i = 0.005 \ \text{kg kg}^{-1}$)

3. $G = 0.208/(0.027 - 0.005) = 9.45 \ \text{kg s}^{-1}$

4. $\rho_a = 353 \times 101.4/[(273 + 120) \times 101.4] = 0.898 \ \text{kg m}^{-3}$
   (assumes $p_a = 101.4 \ \text{kPa}$)

5. $A_b = 9.45/[0.898 \times 1] = 10.5 \ \text{m}^2$

4.8.2 Detailed design calculations

It is beyond the scope of this chapter to cover the detailed design of the wide variety of fluidized bed dryer types which are available. Instead, a scientific approach to modelling fluidized bed dryers for design and performance simulation will be presented. Reay (1989) has reviewed this approach in detail. It is based on the rationale that any overall model of a dryer is composed of two subsidiary models, the equipment model and the material model.

The equipment model describes the characteristics of the equipment used for drying. It deals with:

• how the material moves through the dryer;
• the gas flow patterns;
• the heat transfer from the gas to the material.

The material model describes the characteristics of the material being dried. It covers:

• drying kinetics – the rate at which the material dries;
• drying equilibria – the absolute limit to how far the material can dry.

The ultimate aim of this approach is to enable dryers to be designed from a small number of small-scale laboratory tests, such as those described in section 4.7. Its development into a procedure for the design and scale-up of fluidized bed dryers has been described by Bahu (1994).

The concept of a so-called 'integral' model of drying was first proposed by Vanacek et al. (1964) based on the following equation for calculating the average outlet moisture content $X_{o,avr}$ from a continuous dryer:

$$X_{o,avr} = \int_0^\infty E(t)X(t) \ dt \qquad (4.5)$$

Here $E(t)$ is a function describing the residence time distribution of the material in the dryer and represents the equipment model. $X(t)$ is a drying curve function describing how the moisture content varies as the material passes through the dryer and represents the material model.

In a continuous well-mixed fluidized bed dryer, where the residence time distribution is that of a perfectly mixed vessel, equation 4.5 becomes:

$$X_{o,\,avr} = \frac{1}{t_m} \int_0^\infty X(t) \exp\left(-\frac{t}{t_m}\right) dt \qquad (4.6)$$

where t_m is the mean residence time.

Ideally in a continuous well-mixed unit, a drying curve is required which accurately follows the drying of particles at both constant inlet gas and bed temperatures. This cannot be achieved in a batch test as either the inlet gas temperature is kept constant, in which case the bad temperature will rise during drying, or the bed temperature is kept constant by progressively reducing the inlet gas temperature. In practice, the isothermal bed batch drying curve (IBBDC) has been adopted. However, it is difficult to measure experimentally. Reay and Allen (1983) developed a simple procedure for converting the easy to determine isothermal inlet gas batch drying curve (IIGBDC) into an IBBDC.

McKenzie and Bahu (1991) extended the method to account for the effect of inlet gas humidity using the following equation on successive time intervals on the drying curve:

$$\Delta t_2 = \Delta t_1 \frac{(p_w - p_i)_1 (X_{av} - X^*)_1 (X_i - X^*)_2}{(p_w - p_i)_2 (X_{av} - X^*)_2 (X_i - X^*)_1} \qquad (4.7)$$

where Δt is a time interval on the batch drying curve, p is the water vapour pressure, and X is solids moisture content. Subscript 1 refers to IIGBDC, 2 to IBBDC, w to saturation, i to inlet and av to average and superscript * to equilibrium.

By solving equation 4.6 for a range of mean residence times, a so-called design curve can be constructed and used to extract the mean residence time required to achieve a specified outlet moisture content. The bed area can then be calculated for a given throughput S and bed mass per unit area m_b:

$$A_b = \frac{S}{m_b} \qquad (4.8)$$

The choice of the bed mass per unit area will be dictated by the need to break up the wet feed and to ensure rapid dispersion away from the feedpoint.

For a plug flow fluidized bed dryer, the residence time will deviate from idealized plug flow due to back mixing. This can be accounted for by employing the axial dispersion number, $B = \Gamma_p t/L^2$, where Γ_p is the particle diffusivity given by Reay's (1978) correlation and L is the bed length. Levenspiel (1972) gives a comprehensive account of dispersed plug flow. For small deviations $(B < 0.1)$, the residence time function is:

$$E(t) = \frac{1}{2(\pi B)^{1/2}} \exp\left(\frac{-(1 - t/t_m)^2}{4B}\right) \qquad (4.9)$$

In the plug flow unit, the inlet gas temperature remains constant, whereas the bed temperature rises as the solids progress along the dryer. This is directly analogous to the IIGBDC. Again, a design curve is constructed and equation 4.8

is used to calculate the bed area. In a plug flow unit, the bed depth should be shallow in order to minimize back mixing effects, which can become significant when the depth exceeds 150 mm. In order to calculate the ratio of the bed width to length, the fraction of product which has a moisture content in excess of a specified value must be fixed. Equation 4.9 can then be used to calculate the bed length, and hence, from the bed area, the width.

In the above method, the design curve applies to the design conditions. In order to explore how changes in operating conditions affect the design, dimensionless normalization factors have been proposed by Reay and Allen (1982) and McKenzie and Bahu (1991). These factors allow for changes in gas inlet temperature and humidity, solids inlet and equilibrium moisture contents, bed depth, gas velocity and in-bed heaters. They also vary between easy to dry materials and slow drying materials such as wheat grains, where drying is controlled by internal diffusion and is therefore independent of external conditions such as gas velocity and bed mass per unit area.

It is always advisable to run a pilot plant test to verify the final design calculation in case there are deviations. For example, Kerkhof (1994) has highlighted that, when drying foods and bioproducts with high internal mass transfer resistance under extreme conditions of very shallow beds and very high gas velocities, the above normalization procedures may break down. In such cases, it will simply be necessary to re-measure the batch drying curve at the new design conditions.

Acknowledgements

Much of the work described in this chapter was conducted at the Separation Processes Service (SPS), AEA Technology, Harwell, OX11 0RA, UK by myself and colleagues, particularly, David Reay, Ray Allen, Bob Huber, Ian Kemp and Ken McKenzie. I would like to acknowledge the permission of SPS to use diagrams and other material from the SPS Drying Manual Volume III on fluidized bed drying. The manual volume is accompanied by a PC computer program called FLUBED, which undertakes both the scoping and full design calculations described in this chapter.

References

Abbott, J.A. (ed.) (1990) *Prevention of Fires and Explosions in Dryers – A User Guide*, 2nd edn, Institution of Chemical Engineers, Rugby, UK.
ASTM (1985) *Standard Practice for Maintaining Constant Relative Humidity by Means of Aqueous Solutions*. Designation E, 104–85.
Bahu, R.E. (1994) Fluidised bed dryer scale-up, *Drying Technology*, **12**(1&2), 329–39.
Geldart, D. (1973) Types of gas fluidisation, *Powder Technology*, **7**, 285–92.
Guguoni, R.J. and Zenz, F.A. (1980) Correlation for theoretical disengagement height, in *Fluidization III* (eds J.R. Grace and J.M. Matsen), Plenum, New York, p. 501.

Kerkhof, J.A.M. (1994) A test of lumped-parameter methods for the drying rate in fluidized bed driers for bioproducts. Proceedings IDS '94 (9th International Drying Symposium), Gold Coast, Australia, 131–40.

Kuni, D. and Levenspiel, O. (1991) *Fluidization Engineering*, 2nd edn, Butterworth–Heinemann, Boston.

Levenspiel, O. (1972) *Chemical Reaction Engineering*, 2nd edn, J. Wiley, New York.

McKenzie, K.A. and Bahu, R.E. (1991) Material model for fluidised bed dryers, in *Drying '91* (ed. A.S. Mujumdar), Elsevier, Amsterdam. Paper originally presented at IDS '90, Prague.

Okos, M.R., Narsimhan, G., Singh, R.K. and Weitnauer, A.C. (1992) Food dehydration, *Handbook of Food Engineering* (ed. D.R. Heldman and D.B. Lund), Marcel Dekker, New York, pp. 437–562.

Pell, M. (1990) Gas fluidisation. *Handbook of Powder Technology*, Vol. 8, Elsevier, Amsterdam.

Reay, D. (1978) Particle residence time distributions in shallow rectangular fluidised beds. Proceedings of the First International Drying Symposium, Montreal, Canada, Science Press, Princeton, USA, pp. 136–44.

Reay, D. (1989) A scientific approach to the design of continuous flow dryers for particulate solids, *Multiphase Science and Technology*, **4**, 1–102.

Reay, D. and Allen, R.W.K. (1982) Predicting the performance of a continuous well-mixed fluid bed dryer from batch tests, in *Proceedings of the Third International Drying Symposium, Birmingham, UK* (ed. J. Ashworth), Drying Research Ltd, Vol. 2, 130–40.

Reay, D. and Allen, R.W.K. (1983) The effect of bed temperature on fluid bed batch drying curves. *J. Sep. Proc. Technology*, **3**(4),11.

Shilton, N.C. and Niranjan, K. (1994) Puff drying of foods in a fluidised bed , in *Developments in Food Engineering* (ed. T. Yano *et al.*). Chapman & Hall, London, pp. 337–9.

Vanacek, V., Picka, J. and Najmr, S. (1964) Some basic information on the drying of granulated NPK fertilizers, *Int. Chem. Eng.*, **4**(1) 93–9.

Wen, C.Y. and Tu, Y.H. (1966) A generalized method of predicting the minimum fluidization velocity, *AICHE Journal*, **12**, 610–2.

Williams-Gardner, A. (1971) *Industrial Drying*. Leonard Hill, London.

5 Spray dryers

K. MASTERS

5.1 Introduction

The demand for tailor-made agglomerated powdered foodstuffs throughout the industrialized world has never been so great as it is today. Of all the food processing techniques capable of producing such powders and fulfilling this market demand, spray drying offers the best solutions. By combining atomization, fluidization and agglomeration technologies in advanced spray dryer designs, it is possible to meet the end-product quality specifications that ensure nutritive advantage and consumer acceptance within a safe, hygienic and environmentally friendly process.

Spray drying as a concept can be traced back to a patent of the 1860s, and it is interesting to note that the invention related to the advantages of preserving fresh food (fruit, vegetables) if converted into powders. It took nearly 50 years for the first commercially successful spray dryer design to be developed and operated on so-called heat-sensitive products. It took another 30 years and a world war to establish spray drying as a suitable and recommendable industrial process for continuous, high production output of powdered food and dairy products, eggs and food ingredients. Powders produced in those days were, compared with today's standards, of poor flowability and solubility, difficult to handle, dusty, and prone to misflavour as a result of degradation in the drying process. However, it was the breakthrough, and in the subsequent years spray drying finally became established in the eyes of industry. Whatever the limits on quality compared with fresh raw materials, spray dried products had a clear advantage of still being superior to the same products produced by other available drying techniques.

What has been achieved in the last 50 years is quite remarkable. Developments have led to the following:

1. Control of particle structure during drying to specific agglomerated forms.
2. Ability to handle the most difficult drying applications involving high fat and sugar-containing foods and food ingredients.
3. Capability of designing extremely large dryers producing over 10 t of powder per hour in an operation that retains full industrial confidence in meeting quality specifications at all times. It is worth appreciating that, at these high capacities, even the smallest of design and operational faults can result in the production of tonnes of subgrade quality powder.

Food and food-rated products and ingredients that are spray dried are listed in Table 5.1.

Table 5.1 Typical spray dried foodstuffs including ingredients and beverages

Baby food	Lactose
Butter	Malt extract
Buttermilk	Maltodextrine
Carbohydrates	Milk: flavoured, mixed, skim,
Casein	sweetened condensed, whole
Caseinates	Milk replacer
Cheese	Mother liquor
Coconut milk	Permeate
Coffee	Sorbitol
Coffee whitener	Soup mixes
Corn syrup	Soy isolate
Cream	Soy sauce
Dextrose	Starch/sweeteners
Eggs	Tea
Fructose	Tomato paste
Fruit juices with filler	Total sugar
High fat powders	Vegetable purées with filler
Hydrolysed products	Vegetable protein
Ice-cream mix	Whey: acid, sweet
Ingredients/ready mixes	Whey permeate

By definition, spray drying involves the atomization of a liquid feedstock containing solids in solution, suspension or emulsion, and directing the resulting spray of droplets into a flow of hot drying air. Contact takes place in a drying chamber. The method of atomization and hot air introduction is critical with food-based feedstocks, since the most lenient drying conditions must be achieved whereby product temperatures are held low at all times, and drying is completed in as short a time as possible. One of the most important successes in recent years has been developments in spray dryer plant design that keep powder (particle) temperatures at very low levels, even though drying air enters the drying chamber at high temperatures that would certainly be detrimental to product quality in any other system.

Spray drying is a convective suspended-particle process. The atomization stage creates a very large wet surface area in the form of millions of droplets which, when exposed to the hot drying air, results in the occurrence of very high rates of heat and mass transfer. Drying times become short and it is possible for product heating to be restricted to levels at which thermal degradation does not occur. For every cubic metre of feed atomized in a typical food application, tens of thousands of square metres of surface are provided; this is a unique situation. The atomization stage gives the spray drying plant further unique features compared with other systems. It is not only a dryer, it is also a particle formation system. Liquid feedstocks are not only just dried, but are also converted into powders having specific particulate structures as determined by the atomization and drying conditions adopted, and also the drying chamber and plant layout selected. This is particularly important in food drying where some products are required as fine powders consisting of individual particles, some as loose, porous or compact agglomerates, and others as a more solid structure resembling a granule. Through the selection of design and mode of operation, various powder forms can be achieved (section 5.4). The basic flow sheet of spray dryer is shown in Figure 5.1.

5.2 Spray drying principles

Various process stages are involved in the spray drying of foodstuffs.

5.2.1 Atomization

Irrespective of the application, the first stage involves the pumping of a liquid feed (usually as a concentrate to minimize the amount of water to be evaporated during drying) to an atomizer. The atomizer is operated so that a spray of droplets of the desired size distribution is produced. There are two main types of atomizers based upon the use of centrifugal energy (rotating wheel atomizer) or pressure energy (pressure nozzle atomizer). For very high viscosity feeds, the third alternative, that of using kinetic energy (pneumatic or two-fluid nozzle atomizer) may be necessary, but this is the least used in actual production plants.

(a) Wheel atomization. The wheel atomizer features the feed being fed into the centre of a wheel rotating at high speed (peripheral velocities are in the range of 60–150 m s^{-1}). The wheel contains radial vanes which direct the feed to the periphery. This enables the liquid, now as a thin film on the vaned surface, to accelerate and acquire the peripheral speed of the wheel. Atomization occurs immediately as the thin film leaves the wheel edge. Spherical droplets are formed as the thin film disintegrates because of the influence of the imparted energy, the physical properties of the liquid, and air turbulence effects around the rotating vaned wheel. The rotating wheel also creates an air pumping effect and plays a major role, albeit negative, in widening the droplet size distribution and causing aeration to the product. Wheel speed is by far the most significant operational variable, with higher speeds at constant wheel diameter and feed rate creating smaller particles. Smaller particles are the easiest to dry. The majority of spray dryers for foodstuffs operate in the medium speed wheel range.

The design of the wheel also influences end-product particle properties and manufacturers offer a selection of designs for given applications. Straight vane wheels are considered to be the basic standard. Curved vanes produce higher bulk density powders. Combinations of vaned and vaneless surfaces are claimed to give powders of improved free flowability through higher densities and less aeration. There are also wheels that are flushed with steam to achieve a more compact particle structure. However, this technique is being replaced by air fluidization of powder within the plant in order to avoid the possible product

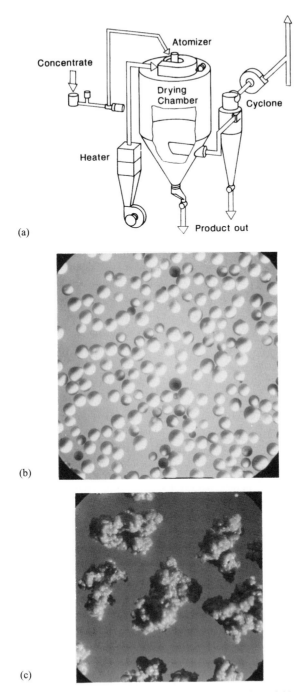

(a)

(b)

(c)

Figure 5.1 Spray dryer, basic layout. (a) Particle forming process; (b) and (c) particle types. (By courtesy of Niro A/S.)

overheating problems associated with live steam contact. A rotating wheel atomizer in operation is shown in Figure 5.2a.

(b) Pressure nozzle atomization. The pressure nozzle atomizer has two vital internal features: a device to create liquid rotation within the nozzle head, and an orifice through which the liquid is discharged as a conical spray. The size of these two components and the operating pressure determine the nozzle capacity and the size of the droplets in the resulting spray. The mechanism of liquid break-up at the orifice is influenced by the imparted energy and liquid physical properties. Local air turbulence plays a lesser role. The higher the pressure at a fixed feed rate, the smaller the resulting droplet size. The spray angle from this type of atomizer is 60–110°, much narrower than the 180° associated with wheel atomizers with their horizontal droplet trajectory. Pressure nozzles, with fixed size of swirl and orifice components, operate successfully only over a fairly narrow range of feed rates. Furthermore, the maximum capacity per nozzle is a limiting factor. A pressure nozzle in operation is shown in Figure 5.2b.

On comparing nozzle with wheel atomizers, wheels have an infinite turndown ratio and can be sized for feed rates well in excess of the maximum possible with an individual nozzle. High capacity spray dryers with nozzle atomizers therefore require multi-nozzle arrangements. This in itself does not create operational problems, but the nozzle atomizer assembly arrangement is more complex as it must be possible to remove each nozzle individually from the assembly during plant operation, should it malfunction or if cleaning of the nozzle is required. Due to the small size of the nozzle orifices, partial blockage or product pre-drying on the nozzle head can occur and create non-uniform sprays, and even squirts of liquid moving directly on to the chamber walls. Modern control systems can readily cope with multi-nozzle operations.

Both pressure nozzles and wheel atomizers have their own advantages and disadvantages as shown in Table 5.2, and selection is according to product application and operational experience. Wheel atomizers are usually preferred for high feed rates and where standard powders of small particle size (non- or partially agglomerated structure) are required and where powder handling in pneumatic conveying systems is adopted. Nozzle systems are often preferred with high fat feedstocks or where free flowability from a larger particle size is desired. Nozzles are also used in spray dryers incorporating fluid beds (wheel atomizers are also used in some cases, e.g. high capacity) or moving belts. Here, the nozzle spray angle is much better suited to the accompanying air flow patterns in the drying chamber.

5.2.2 Spray–air contact and flow

The second stage involves an air disperser to create the best conditions of contact between the spray and hot air entering the drying chamber. Contact and the following movement of droplets and air in the drying chamber can follow

(a)

(b)

Figure 5.2 Atomization. (a) Rotary wheel; and (b) pressure nozzle. (Both photographs by courtesy of Niro A/S.)

Table 5.2 Comparison of atomizer types

Rotary atomizer	Pressure nozzle
Easy control of droplet size through speed control	Less easy control of droplet size through pressure control
Single wheel irrespective of capacity	Nozzle duplication at higher capacities
Large flow areas not prone to block	Small flow areas prone to block
Coarse feed filtering	Fine feed filtering
Large feed turn-down capability	Small feed turn-down capability
Capacity independent of pressure	Capacity proportional to square root of pressure
Wide spray angle gives increased wall deposit tendencies	Narrower spray angle gives reduced deposit tendencies
Low pressure system	High pressure system

three distinct modes: cocurrent, countercurrent and mixed flow. Of the three, countercurrent is not applicable to food drying, as the driest particles will contact the hottest air, causing unacceptable heat damage in the product.

In the cocurrent mode, two designs of air dispersers are available. One causes air rotation within the drying chamber and is used with wheel or nozzle atomizers. The alternative design creates streamlined or non-rotating flow in the drying chamber and is used exclusively with nozzles. Air temperature profiles in these designs are illustrated in Figure 5.3. The main feature of cocurrent operation is the relative uniformity of temperature levels in the drying chamber. Uniformity is greater with the air dispersers creating air rotation. Temperatures throughout the majority of the drying chamber volume approach the outlet temperature level set to achieve the required powder moisture content. Product temperatures are thus also held low; this is essential for obtaining the lenient drying conditions characterized by cocurrent drying.

In the mixed-flow mode, there are two alternative designs of air disperser: one creates the rotary air flow suitable for operation with wheel atomizers, the other causes a more downward direct flow appropriate with nozzles. The mixed flow mode is associated with chamber designs incorporating integrated fluid beds. To utilize the full flexibility of integrated fluid bed operation in producing different particle structures (agglomerated powder characteristics), air dispersers have to operate over a much wider range of inlet air velocities than is the case with the more standard cocurrent mode air dispersers.

5.2.3 Evaporation, drying and particle formation

The combination of the first two stages creates the conditions necessary for droplet drying and particle formation. The passage of every droplet formed at the atomizer through the drying chamber, whether under cocurrent or mixed flow conditions, will depend upon the air flow profile present. These profiles are very complex and each droplet will be subjected to different local temperature and humidity conditions as it dries and moves according to the air flow pattern in the drying chamber. This, coupled with the fact that the swirling air contains droplets of different size, means that the rates of moisture removal from the droplets differ and that the droplets form particles having different structures. Droplets subjected to the hottest environments might shrink or become mis-shapen, whereas those in the cooler regions more easily retain their original atomized droplet shape. There is always the possibility of particles of different moisture content colliding to form a loose agglomerate structure although, in all cases, the bonding between the individual particles is very weak. Collision of particles cannot be avoided and therefore, for producing non-agglomerated powders, cocurrent designs are used with pneumatic conveying of powders so that the mechanical forces encountered in the conveying system will break up this weakly bonded structure.

Operational temperatures also play a leading role in the third stage, where hot inlet air temperatures are maximized to obtain the best thermal efficiencies without leading to product damage. Outlet temperature levels are selected to give the residual moisture content required in the powder discharged from the drying chamber. Both temperature and humidity influence the residual moisture

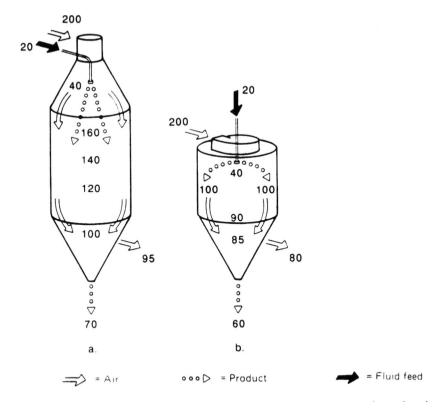

Figure 5.3 Temperature profiles in cocurrent spray dryers. (a) Cocurrent: non-air rotation air disperser; and (b) air rotation air disperser.

content level but, for convenience, normally only temperature is used as the control variable. However, in plant operations where absolute humidities can vary substantially, for example between winter and summer, the temperature required to achieve a constant residual moisture content may have to be reset according to the season.

Drying chambers with integrated fluid beds give a further control dimension, where the selection of outlet drying temperatures and inlet temperatures to the fluid bed can influence the particle structure and type of powder produced. Two possible temperature profiles are shown in Figure 5.4. In both cases, temperatures in the drying chamber approximate to the outlet drying temperature, giving excellent lenient drying conditions. However, if the higher outlet temperature is used, powder entering the fluid bed is in a drier state. The possibility for interparticle adhesion is reduced and a non-agglomerated powder is produced that requires a low inlet air temperature to the fluid bed, which then acts as a cooler (Figure 5.4a). If a lower outlet air temperature from the drying chamber is used, powder entering the fluid bed has a much higher moisture content. The possibility for particles to stick together is increased and, with this, a degree of agglomeration is achieved. Agglomerated powder is formed having a moisture content that requires further moisture removal in the fluid bed. Therefore warm air has to be passed through the fluid bed, which then acts as an after-dryer (Figure 5.4b).

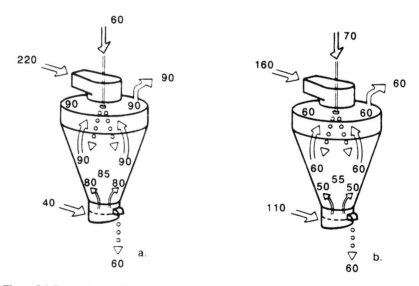

Figure 5.4 Temperature profiles in mixed flow, integrated fluid bed spray dryers. (a) Fluid bed as cooler; and (b) fluid bed as agglomerator.

5.2.4 Powder handling

The majority of particles formed during the above three stages fall onto the wall of the lower conical section of the drying chamber and slide down to the outlet at its base. A small amount of material, however, remains entrained in the drying air and passes out of the chamber. This is recovered in the externally mounted cyclones. Filters can also be used for fines separation and recovery. These are mounted either internally (metallic/cartridge) or externally (fabric) depending upon the spray dryer layout.

Subsequent powder handling depends upon whether single or multi-stage processing is being used. Single stage (Figure 5.5a) is associated with food drying applications where the powder produced in the drying chamber has the required properties of particle size, moisture content and bulk density, and no further drying or agglomeration is required. Such powder is usually passed into a pneumatic conveying system into which the fines fraction collected in the cyclone is also passed. All powder is then conveyed to a conveying cyclone and discharged into bags or to a silo system for bulk storage. The conveying system acts to cool the powder further, and cooled and dehumidified conveying air may well be necessary to reach the powder temperatures required for ideal packing and storage.

If the desired powder specification cannot be achieved in the single stage process, multi-stage operations are necessary. Fluid bed assemblies act as the second or third stages to conduct the agglomeration, after-drying and conditioning.

In conventional two-stage processing (Figure 5.5b), vibrating fluid beds are installed outside the drying chamber base. Special construction and air

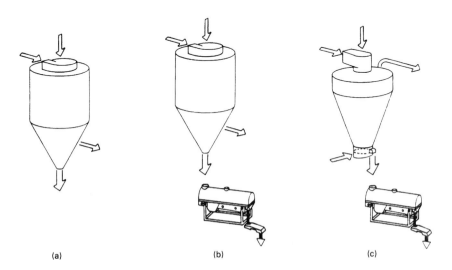

(a)   (b)   (c)

Figure 5.5 Spray dryer layouts. (a) Single stage; (b) two-stage; and (c) three-stage.

distributor plate design are necessary to meet the hygiene standards required for food applications. In the latest, three-stage processing involving drying chambers with integrated fluid beds, the second stage is a fixed bed mounted within the base of the drying chamber. The third stage is an externally mounted vibrating fluid bed (Figure 5.5c). The performance of the air distribution plates mounted in both the fixed and vibrating beds is critical for efficient multi-stage drying. The plate must not only create uniform fluidization conditions, but also control the direction of particle (powder) flow. In this way, greater control of powder residence time in the fluid bed stages, and also the degree of agglomeration, drying, cooling, attrition and size classification that takes place, is achieved.

Conventional perforated plates having vertical, circular holes do not have the characteristics required to achieve controlled particle flow; directional plates featuring punched openings (gills) are necessary. The gills can be arranged in any pattern to create a certain particle movement (FLEX PLATE^{TM}). In cases where powder penetration through the plate cannot be accepted during shutdown, a special non-sifting plate design is now available, again based upon the gill design principle (Figure 5.6).

Multi-stage drying gives rise to the possibilities of product agglomeration and close residual moisture content control, and has the added advantage of producing dried particulate foodstuffs in a process having a much higher thermal efficiency. This is shown in Table 5.3.

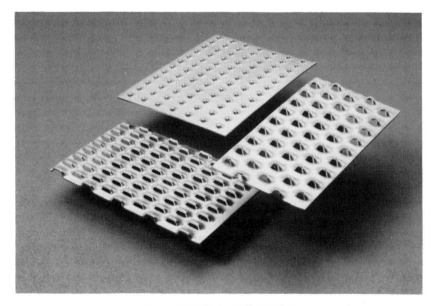

Figure 5.6 Gill air distributor plates.

Table 5.3 Energy requirements: comparison between layouts

Quantity measured	One-stage (Figure 5.5a)	Two-stage (Figure 5.5b)	Three-stage (Figure 5.5c)
Drying air to air disperser: rate (kg h^{-1})/temperature (°C)	31500/200	31500/200	31500/260
Feed concentration (%TS)	48	48	48
Final moisture content (%H$_2$O)	3.5	3.5	3.5
Product rate (kg h^{-1} of powder)	1140	1420	2755
Energy requirement per kg powder produced (kJ kg^{-1})	6675	5650	4015

5.3 Spray dryer layouts

5.3.1 Cocurrent drying chamber with rotary atomizer

This layout is selected for non or low fat feed formulations (Figure 5.7) where non-agglomerated powders are required. These dryers are often single-product plants. The most convenient atomizer unit is a high speed wheel. The roof mounted air disperser creates a swirling air flow in the drying chamber. The fine droplets produced at the atomizer quickly dry and move to the chamber walls under the influence of the air flow. Particles are dry by the time they contact the wall and slide down the cone for discharge into a pneumatic conveying system via a rotary valve. The presence of larger droplets in the spray gives rise to the possibility of some particles reaching the wall before drying has been completed. This results in the tendency for wall deposits to form, especially at the wheel level. Knocking hammers are mounted on the wall of the drying chamber to minimize the formation of such deposits and to assist powder movement to

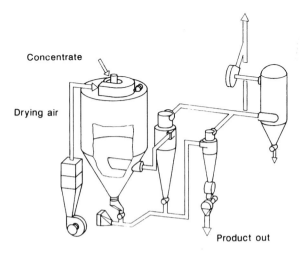

Figure 5.7 Cocurrent spray dryer with pneumatic conveying system for powder transport.

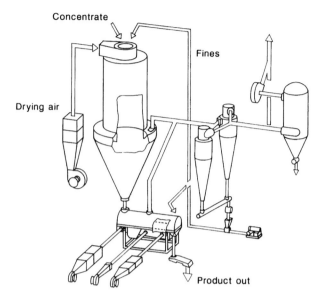

Figure 5.8 Tower type spray dryer in a two-stage layout, with vibrating fluid bed under drying chamber.

the base exit. This layout is used for the more easy to dry products where fairly high outlet air temperatures are used to complete the drying before the powder is discharged at the required moisture content into the pneumatic conveying system. This conveys powder leaving the chamber and cyclone to a central discharge area for packing or bulk storage. Chamber accessories such as air brooms and secondary air flow injection at the chamber walls can be utilized in this type of layout to ensure a greater degree of deposit-free plant operation when a non-agglomerated, fine powder is to be produced from feed formulations that exhibit some sticking and wall deposit tendencies during drying.

5.3.2 Cocurrent drying chamber with nozzle atomization

This layout is selected to produce powders consisting of large, individual particles and virtually no agglomerates. It is recommended for fat-containing formulations since this particle structure assists good powder flowability. The nozzle tower can operate as a single stage layout without pneumatic conveying, or as a two-stage layout with an externally mounted vibrating fluid bed acting as after-dryer or cooler (Figure 5.8).

The air disperser creates a streamlined flow pattern in the drying chamber. Droplets created at the nozzle assembly dry as they gently fall down the drying chamber in an air flow that tends to discourage particle movement towards the wall. Hence this type of design limits wall deposit tendencies and enables lengthy production runs to be achieved between shutdown periods for cleaning. The lower cone of the chamber is enlarged to form an effective settling chamber, enabling powder particles to fall out of the air and to pass out of the chamber base exit, while the air is exhausted at the top of the conical or bustle section. The amount of powder entrained in the exhaust air is very low, thereby reducing the powder loading on the cyclone. Powder handling problems associated with fat-containing materials in highly efficient, high velocity cyclones are thereby significantly reduced.

5.3.3 Cocurrent drying chamber with integrated fluid bed

This is a modification of the standard layout and is shown in Figure 5.9. A fluid bed located in the base of the drying chamber enables both higher moisture contents to be handled and drying to be completed at lower product and exhaust air temperatures. The fluid bed is of an annular design enabling the exhaust air to be ducted out through the centre of the chamber base. This design of spray dryer operates with either rotary wheel or nozzle atomization. The air disperser is designed for operation with a wheel atomizer. However, when using the nozzle mode, the nozzles are positioned so that the initial spray trajectory resembles that of droplets leaving a wheel atomizer.

The presence of powder of higher moisture content in the chamber and the swirling air flow pattern created by the air disperser increase the tendency for wall deposit formation. A secondary air flow is therefore introduced around the chamber wall to minimize deposit formation by introducing an air sweep or air curtain. The fluid bed is fitted with an air distribution plate (Figure 5.6) that gives rise to directional flow of powder within the annular bed. This enables a good degree of powder residence time control in what is basically a totally mixed fluid bed.

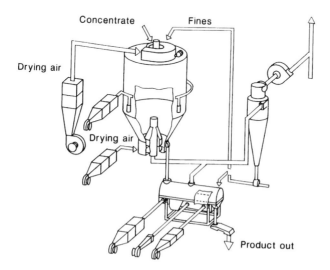

Figure 5.9 Cocurrent spray dryer with integrated fluid bed.

This design of drying chamber operates successfully with a pneumatic conveying system for non-fat, low fat, and non-sticky products forming free-flowing standard powders. The drying chamber can also operate with an externally mounted vibrating fluid bed in a three-stage layout (Figure 5.9) to produce a more agglomerated powder structure.

5.3.4 Cocurrent drying chamber with integrated belt

This layout (Figure 5.10) represents a different drying approach. Moisture removal from particles is completed as the powder lies as a layer on a moving belt. In this way, the particles are not airborne during their most sticky and hygroscopic phase of drying. Deposit build-up usually associated with swirling air is thereby prevented.

This design concept, known as the FILTERMAT® dryer, is well-suited for drying high fat products and formulations containing sugar, where extended residence times are required for completion of both crystallization and drying. It is a two-stage system. Nozzles are used to spray product directly down a tower type chamber equipped with an air disperser that maintains the droplets in vertical motion. The majority of moisture evaporation takes place in the tower section of the dryer but, even so, a moist powder layer is still created on the moving belt. Particles inevitably stick together and form an agglomerated structure. Drying continues as the powder layer moves along the belt, passing sections introducing first warm and then cool air for after-drying and product cooling. All the air is exhausted through the powder layer and the belt itself. The

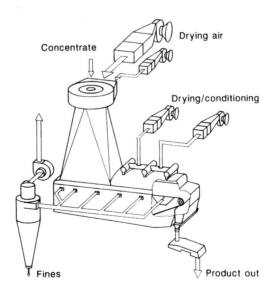

Figure 5.10 Cocurrent spray dryer with integrated belt.

powder layer acts as its own particulate filter medium and is so effective that negligible amounts of fines pass through the belt to the cyclone collector.

5.3.5 Mixed flow drying chamber with integrated fluid bed

This incorporates some of the latest development concepts involving the integration of fluid of beds into drying chambers. As shown in Figure 5.11, the drying chamber has cyclone-like dimensions, i.e. a short cylindrical side in relation to an extended cone section. By adopting the mixed-flow mode with exhaust air off-take at roof level, a different distribution or population density of particles is created in the drying chamber. Elutriation effects from the back-mixed fluid bed plus the upward air flow of the drying air act to fill the entire chamber volume with airborne powder. Sprays of droplets leaving the atomizer dry within a cloud of fine particles. Contact between the evaporating droplets and particles is inevitable. A powder-coating phenomenon occurs, thereby reducing the surface stickiness of the evaporating droplets. With the swirling air flow, contact between the particles and the chamber wall occurs, but deposit formation tendencies are counteracted by the powder-coating described above.

Both fat-containing and hygroscopic products can be handled in both small and large capacity dryers. Drying chambers can have air brooms installed to further assist clean wall operation. This design offers considerable flexibility regarding the type of product that can be handled and the type of powder agglomerate structure produced. All food products can be handled with the exception of very high fat content and sugar based formulations, where the integrated belt dryer described in section 5.3.4 is still preferred.

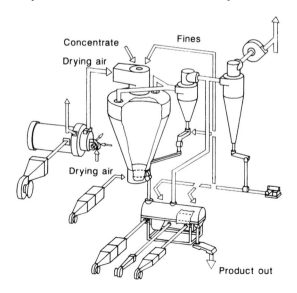

Figure 5.11 Mixed flow spray dryer with integrated fluid bed.

Plant layouts have to be tailor-made to fit the desired product application. In multi-product operations, drying the most difficult product must be given priority in specifying the design. Layouts can be either two-stage or three-stage, the latter featuring an external vibrating fluid bed for agglomerate after-drying and cooling. Fines collected in the cyclone can be returned to either the external fluid bed, the integrated fluid bed, or the atomization zone. As a result of the high powder loading in the drying chamber, a higher proportion of fines is carried out of the drying chamber with the exhaust air as compared to a cocurrent design. This leads to a high fines recirculation and, therefore, the separation and recovery of fines from the exhaust air is essential. Exhaust air cleaning can be carried out in cyclones located in series or in cyclones followed by a bag filter or wet scrubber. Fines recirculation can be avoided by mounting filter cartridges within the drying chamber at the exhaust duct inlet. Powder collected in this manner falls back into the drying chamber. Washable metallic filter cartridges are available for this application.

Both nozzle and wheel atomizers can be used in the drying chamber. However, nozzles are preferred in low to medium capacity plants. For wheel atomization, the air disperser uses high air velocities to restrict the horizontal trajectory of droplets and to direct the spray cloud down the centre of the drying chamber towards the centre of the integrated fluid bed.

The fluid bed is fitted with a special air distributor plate giving the powder a directional flow and thereby control over the residence time needed during the second processing stage to secure sufficient contact between the moist and dry powders to enable agglomeration to take place.

Some typical examples of industrial spray drying installations handling foodstuffs are illustrated in Figure 5.12.

5.4 Meeting powder specifications

One of the most important reasons why spray drying has been adopted for producing powdered foodstuffs, food ingredients, and beverages is the ability of the technology to handle value added products that are sensitive to heat, and to produce an end product continuously to a precise quality specification.

The most commonly quoted specifications involve particulate structure and size, moisture content, bulk density and hygroscopicity.

5.4.1 Particulate structure and size

These are defined by the particle size distribution and particulate structure. Some powdered foodstuffs are required as standard powders consisting of small individual particles. Interparticle binding is so weak that the powder does not exhibit agglomeration characteristics. Such powders are produced in single-stage spray dryers with wheel atomizers (Figure 5.7).

If a powder is required in the form of larger individual particles, a single-stage, nozzle tower spray dryer is suitable, with a second stage fluid bed being installed to act only as a powder cooler (Figure 5.8).

For agglomerated powders produced from liquid feedstocks, spray dryers incorporating either integrated fluid beds or integrated belts are recommended. By involving fluidization techniques, a strong agglomerate structure is produced. In contrast, agglomerates produced from the belt tend to be larger, but more friable. However, the final size and structure of the particulates are totally product specific, and test work in a pilot plant is always recommended to confirm quality and structural predictions. Table 5.4 illustrates the product granulometry expected.

5.4.2 Moisture content

In a single-stage layout with a conveying or cooling system, the specified residual moisture in the powder is controlled by the temperature of the exhaust

(a)

Figure 5.12 Industrial spray dryers operating in the food industry. (a) Tall form spray dryer; (b) spray dryer with integrated fluid bed; and (c) FILTERMAT® spray dryer. (All photographs by courtesy of Niro A/S.)

(b)

(c)

Figure 5.12 *Continued.*

Table 5.4 Granulometry guide

Type of powder	Spray dryer layout	Mean particle size (μm)
Individual particles, non-agglomerated	Figure 5.7	20–200
	Figure 5.8	100–350
Agglomerate with open structure	Figure 5.9	100–200
	Figure 5.11	100–400
Porous agglomerate	Figure 5.10	300–2000
	Figure 5.11	100–400

air leaving the drying chamber. The humidity of the air may also be a factor. Higher ambient air humidity conditions can give rise to a requirement for an increased outlet air temperature in order to maintain the desired residual moisture content.

In two- and three-stage layouts, the temperature of the exhaust drying air is adjusted to achieve the moisture content required for the particles to form the specific agglomerate structure in the second-stage fluid bed and yet not to be too moist so as to cause excessive wall deposits. The moisture content influences many other powder properties such as solubility, bulk density, particle density and shelf life.

5.4.3 Bulk density

To optimize bulk density, attention must be given to particle density, occluded and interstitial air content, which are related to the feed properties, air temperatures, drying times and powder handling procedures.

The effect of the operating variables is highly product dependent, but there are some general trends. Higher atomizer wheel speeds and nozzle pressures decrease droplet size and, for single-stage operations, a small particled powder of higher bulk density is obtained. Spherical shaped particles result in a low degree of interstitial air, especially as the small particles in the size distribution fill the void spaces between the large. Irregularly shaped agglomerates lead to a lower bulk density. Two- and three-stage processes therefore produce powders of lower bulk density than single-stage processes. Lower outlet temperatures, higher feed solids and lower powder temperatures all tend to increase bulk density.

If the use of higher feed temperatures reduces feed viscosity and improves the uniformity of atomization, higher bulk density powders also result.

5.4.4 Hygroscopicity

Many feed formulations that are spray dried exhibit hygroscopic properties in powder form. Spray dried particles can absorb moisture from the surrounding air and, unless necessary precautions are taken, the surface of the powder becomes sticky and powder caking occurs. The ability of a powder to pick up moisture depends upon temperature and humidity conditions. A knowledge of the dependence of equilibrium moisture content on these parameters for the particular food formulation in question is essential for a full understanding of the drying process and optimization of the spray dryer performance.

Hygroscopic food products are best dried and handled in spray dryers with integrated belts as the powder is in a stationary state during the latter phases of drying. Here the powder can be conveniently subjected to local streams of dehumidified, cool secondary air at a time when the particle surface conditions are at their most problematic. The belt also gives strict control over the residence time.

Similarly, dry, cool air can be used to advantage in spray dryers with integrated fluid beds. In very hot and humid climates, for example in the Far East, successful spray drying of hygroscopic food products is only made feasible by dehumidification of the drying air passing to the air disperser.

5.5 Special design features for hygiene and safety

Improvements to the design of spray dryers for food formulations are constantly being made, since powder specifications require the meeting of stricter bacteriological standards and dryers must be safer to operate. However, any design improvement involving hygiene and safety must be regarded as an added value to the spray drying facility. The selection of safer and more efficient plant may not necessarily provide a quick return on investment.

Spray dryers which are 10–20 years old, or which are based on the designs of those days, do not fulfil today's hygiene and safety requirements. Their use therefore poses a real risk of, for example, the production of contaminated powders, loss of market share, and loss of production.

The design features necessary for hygienic production include the filtration of air used for drying, insulation and cladding of drying chambers, and the prevention of fires and explosions.

5.5.1 Drying air filtration

All air supply systems are best placed in a common fan/filter room, to which pre-filtered air is supplied at slight pressure by means of an axial fan. The main drying air, which has to be heated above 110°C, should be coarse filtered (90–95%) whereas all secondary air required on the plant at a temperature below 110°C should be fine filtered. In order to prolong the lifetime of the fine filter (99.995%), a pre-filter (95%) should be installed. Such layouts eliminate the risk of contaminated particles entering the spray dryer.

5.5.2 Insulation and cladding of drying chambers

Drying chambers are fabricated with thin gauge, stainless steel sheet linings, welded together on site during assembly of the spray dryer. During the subsequent operational life of the drying chamber, the lining is constantly subjected to thermal shock during start-up and shutdown, and mechanical shock through the use of knocking hammers. Corrosion can be another source of damage, particularly where chlorides are present in the feed or wash liquids. All this results in the occurrence of cracks and pinholes in the welding and sheet. During subsequent washing operations, product in liquid form penetrates these cracks and pinholes, and wets any solid insulation present. This results in cold areas that further aggravate deposit formation. Moreover, the presence of product in the insulation constitutes a bacteriological risk because of the temperature of that environment during plant operation.

Thin gauge, stainless steel sheeting is still used today, but the last few years have seen the development of removable air insulation panels, which act as the chamber cladding. By eliminating the need for mineral wool and other solid insulation materials, bacteriological risks are avoided. By making the cladding panels removable, it is possible to inspect easily the condition of the drying chamber lining for leaks and cracks. Access for repair is much easier. When the chamber has to be inspected, the appropriate panels are simply lifted out by hand (Figure 5.13).

The design of the panel support acts to prevent cold spots and thermal bridging. These phenomena occur with the mineral wool insulated spray dryer

Figure 5.13 Drying chamber with removable insulation panels. (By courtesy of Niro A/S.)

designs and result in increased local deposit formation on the wall and the need to use higher outlet air temperatures to dry the product. Heat losses from drying chambers with air insulation are no higher than from dryers with conventional material insulation.

5.5.3 Prevention of fires and explosions

Food powders, owing to their organic nature, can burn under appropriate circumstances, and all form mixtures in air that can ignite and explode under specific temperature, ignition source and concentration conditions. As inertizing by drying in nitrogen is not economically feasible in food spray dryers, and self-inertizing air systems require direct heating, which may or may not be acceptable, fire and explosion protection in food spray dryers relies on fire extinguishing systems, venting with or without partial containment to reduce vent area, and suppression.

The safety record of spray dryers for food products is impressively good. Fire occurring in dryers handling fat-containing products is the most commonly reported incident. Extinguishing systems are always installed in these plants and quickly act to prevent damage. However, clean-up and loss of product are always involved.

Explosions are rare events and normally relate to the chamber deposit over-heating through exothermic reaction or overheating in the presence of hot air before the deposit is dislodged and falls through a swirling cloud of dried fine particles in the drying chamber.

Dangers of fire and explosion have increased with the introduction of two- and three-stage dryer layouts because of the added risk of hot particles entering the fluid bed, which provides an ideal situation for flame generation. Other quoted ignition sources include the discharge of static electricity, dust particles passing through the heater as a result of poorly maintained air filters, and heat generation through local frictional effects. However, these are rarely the cause of a fire or explosion.

The safest plant results from the safety evaluation conducted during the design stage by the supplier, and from safety being given the highest priority by the end-user. These may result in extra investment being required and the operating staff being trained more fully to detect the development of an operational situation that could lead to danger. Signs such as change of smell around the plant, deposit formation around the air disperser, a leaking atomizer, a chamber light which has not been switched off, sudden changes to process control, the presence of dark or burned specks appearing in the powder, and inadequate feed liquid in the feed tanks are points of particular importance. Furthermore, there is no substitute to good housekeeping around the plant and, of course, safety interlocks should never be tampered with.

The components of a spray dryer handling powders in air are protected against fire damage by the installation of a fire extinguishing system and against structural damage by explosion or overpressure venting. The required vent area is selected according to the explosion characteristics of the powder being spray dried and the design strength of the components handling the product in powder form, i.e. drying chambers, ducts, cyclones and fluid beds. To minimize the vent area, the components are designed to higher pressure shock resistance ratings. All venting doors and panels require ducting to a safe area. If the safe area is the building in which the dryer is installed, it must be designed to withstand the secondary effects of a possible pressure release from the drying chamber. Furthermore, areas around the vents must always be 'no go' areas for personnel during the spray dryer operation. Venting into the building is therefore not the optimum choice. Ducting the vent to a point outside the building is a better solution. However, duct lengths are best kept short, less than 3 m. This can normally be arranged in planning the layout of a new spray dryer and building so as to protect the drying chamber and cyclones. However, there is always a problem with the vibrating fluid bed installed at the base of the drying chamber. This can de designed to a safe-containment specification, or fitted with explosion suppression systems. These systems detect the onset of an explosion by using pressure or infra-red sensors in the milliseconds available and then act to inject

Figure 5.14 Vibrating fluid bed with suppression protection. (By courtesy of Niro A/S.)

a flame suppressant to extinguish the flame propagation before any significant build-up of pressure occurs. A fluid bed with suppressant protection is shown in Figure 5.14.

There is no universal method for sizing vents on spray drying chambers and components. Directives as stated in the Verein Deutsche Ingenieure (VDI) 3673 and International Dairy Federation (IDF) publications are used as a basis, with the designer adapting the directives to a particular design of spray dryer, i.e. the presence of a fines return system or a fluid bed, and the mode of flow in the spray drying chamber. Attention must also be paid to the selected operating temperatures, the strength codes adopted, the chamber volume occupied by moist and dried particles, and of course any previous operating experience with the product in question.

The mixed flow, integrated fluid bed spray dryer, for instance, having a dried powder cloud throughout the entire chamber volume, has to be designed to a higher pressure resistance code to maintain the vent area in the drying chamber wall, roof area, and cyclones in a practical proportion. These designs are therefore fabricated in thicker gauge stainless steel – an extra cost to provide a safer plant based on the design that gives a better agglomerated powder quality.

6 Contact dryers

D. OAKLEY

6.1 Introduction

Contact dryers are those in which the wet material is dried by direct contact with a heated surface. Heat transfer to the wet material is mainly by conduction from this surface through the bed or layer of wet solids. This is in direct contrast to convection dryers, described elsewhere in this book. Since no hot gas is required as a source of heat in contact dryers, gas flow through the system can be low and ultimately limited to the vapour evaporated from the wet material. The low gas flows possible in a contact dryer lead to substantial benefits over their convective counterparts as follows.

- *High energy efficiency*
  The energy efficiency of contact dryers is higher than that of convection dryers, because the energy lost through the exhaust gas stream is greatly reduced.

- *Ease of exhaust gas clean-up*
  Since the exhaust flow is low, the problems of cleaning the exhaust gas are minimized. Unlike convective dryers, there will be very few entrained particles present and any vapours can be easily condensed.

- *Vacuum drying is possible*
  Unlike convection dryers, contact dryers can be operated under vacuum. This is ideal for heat sensitive materials such as some food products, since drying can take place at much lower temperatures.

- *Suited to handling flammable and explosible materials*
  Drying can take place in a vacuum or an atmosphere of vapour having a low oxygen content. Explosion risks are therefore reduced. If necessary inerting with nitrogen is easy to implement because of the low volumes required.

- *Flexible and controllable*
  Contact dryers are easy to control and give reproducible drying results. They are also highly flexible with the capability of operating satisfactorily at as low as 20% of full capacity.

- *High product quality and integrity*
  The flows into and out of a contact dryer are relatively small and easily controlled. Ultimately, when operated in batch mode, a contact dryer is a 'closed system'. This means that contact drying is well suited to situations

where high product quality and integrity are required, e.g. products that require hygienic processing conditions.

The main drawback with contact dryers is that the drying rate is limited by the available heat transfer area. Since the surface:volume ratio of a dryer decreases with increasing scale, this restricts the maximum size. In addition, many types of contact dryer can only be operated in batch mode. For the above reasons, contact dryers generally have lower production capacities than convection (particularly dispersion) types, although high production rates are possible in some cases (e.g. indirectly heated rotary dryers). Furthermore, they have a higher initial cost than equivalent convective dryers.

Contact dryers are therefore more suited to the drying of relatively expensive, heat sensitive materials at low or medium rates. The use of contact drying may also be dictated by the form of the material or other special requirements (e.g. drum drying for solutions and slurries). Contact dryers are now also finding favour because they are more environmentally friendly than convective dryers, both in terms of energy consumption and emissions to the atmosphere. In this chapter, an account of the types of contact dryer commonly used in the food processing industry is given in section 6.2. This is followed in section 6.3 by a brief outline of theoretical aspects of contact drying. Finally, methods of design are briefly discussed in section 6.4.

6.2 Industrial equipment and applications

6.2.1 Introduction

There is a wide variety of contact dryer designs depending on the form of the material and how it is brought into contact with the heated surface. These can be broadly divided as follows.

- Material held as an unagitated layer on a heated tray, band or plate: vacuum tray, vacuum band and plate dryers.
- Material (liquid or paste) handled as a thin film on a heated surface: drum and thin film dryers.
- Material handled as an agitated bed of solids (or in some cases pastes and sludges): horizontally and vertically agitated dryers, rotating batch dryers and indirectly heated rotary dryers.

The main types of contact dryer are summarized in Table 6.1 and described in more detail later in this section. Freeze dryers, which are a specialized type of contact dryer that use low vacuums and low temperatures to sublime ice directly to water vapour, are described in Chapter 7.

6.2.2 Contact dryer selection

The correct choice of contact dryer will depend on a range of factors, the most important of which are outlined in the following sections.

Table 6.1 Types of contact dryer

Dryer type	Batch or continuous	Vacuum or atmospheric	Suitable feed form	Production rate	Typical applications[a]
Vacuum tray	Batch	Vacuum	Any	Low	Fruit pieces, meat extracts, vegetable extracts
Vacuum band	Continuous	Vacuum	Pastes, solids	Low–medium	Chocolate crumb, meat and vegetable extracts, fruit juices
Plate	Continuous	Vacuum or atmospheric	Free flowing solids	Low–medium	Tea, coffee
Thin-film	Continuous	Vacuum or atmospheric	Liquids	Low–medium	Tomato concentrate, gelatine
Drum	Continuous	Vacuum or atmospheric	Liquids	Low–medium	Instant potato, corn syrup, baby foods
Rotating batch	Batch	Vacuum	Free flowing solids	Low–medium	Gravy mix, pectin, saccharin
Horizontally agitated	Batch or continuous	Vacuum or atmospheric	Liquids, pastes, powders	Low–high	Chocolate crumb, corn meal, confectionery
Indirect rotary	Continuous	Atmospheric	Free flowing solids	Medium–very high	Brewers' grain, starch
Vertical agitated	Batch	Vacuum or atmospheric	Liquids, pastes, powders	Low–medium	Plant extracts, food colours, glucose, starch

[a] In some cases the applications quoted are from successful vendor tests rather than actual installations.

(a) Atmospheric or vacuum drying. For materials which are heat sensitive or easily oxidized, drying under vacuum should be considered. For reasons explained in section 6.3, this allows drying to take place at low temperatures. However, drying under vacuum is more expensive because of the necessity of constructing the dryer to pressure vessel standards and the need for additional equipment such as air locks, vacuum pumps and condensers.

(b) Production rate. The correct choice of dryer will depend partly on the production rate required. In general, batch dryers, particularly the tray dryer, are unsuitable for large production rates, because of inefficiencies associated with loading and unloading batches. However, they are flexible and widely favoured for drying smaller quantities.

(c) Form of the material. In many cases, the choice of dryer will be dictated by the form of the feed. For example, a drum dryer is an obvious choice when the feed is a liquid. Many of the dryers have special requirements relating to the handling properties of the wet material. For example, the plate dryer can only handle free flowing granules and powders. Also, many of the agitated dryers experience problems when handling sticky materials which cake to heating surfaces.

(d) Agitated or non-agitated.   Agitated contact dryers give significantly higher drying rates per unit heating area because the layer of solids in contact with the heated surface is continually renewed. Furthermore, agitation improves mixing and uniformity of the product. However, problems are frequently encountered when trying to dry sticky materials in agitated dryers because of agglomeration, balling-up, caking of surfaces and the difficulty in turning the agitator. Conversely, in a tray dryer, any form of material can be dried with little regard to its handling characteristics. Another important aspect is that agitation may cause undesirable attrition of the product.

6.2.3 Vacuum tray dryers

The basic vacuum tray, shelf or contact oven dryer consists of a cylindrical or rectangular vacuum chamber containing heated shelves. Trays containing the wet material are placed on the shelves and dried on a batch basis. Drying takes place at a low temperature under vacuum and the evaporated moisture is removed from the chamber by a vacuum pump and condenser. In the standard design, all trays are contained in a single chamber, and the entire contents of the dryer must be treated as a single batch. In alternative designs, the dryer is divided into a number of separate vacuum chambers. This allows more flexible operation since individual trays may be loaded or unloaded separately.

Depending on the required operating temperature, the shelves are heated by the internal circulation of low pressure steam, hot water or heating oil. The shell of the chamber may also be heated to avoid condensation. Typically, trays will be loaded up to a depth of about 40 mm giving a loading of approximately 40 kg of wet material per m^2 of tray area. Drying times depend on the material and conditions but typically range from 20 to over 100 h.

Vacuum tray dryers are extensively used for drying small batches of heat sensitive or easily oxidized material. They are also particularly well suited to materials that need very gentle handling, dusty materials, and where material loss must be minimized. Tray dryers will handle virtually any type of feed: moist solids, pastes, slurries, solutions and shaped pieces. The main disadvantage of vacuum tray dryers is the highly labour intensive loading and unloading operations. For this reason the tray dryer is unsuitable for anything but small production rates. For higher production rates, other types of vacuum dryer should be considered (see Table 6.1).

6.2.4 Vacuum band dryers

The vacuum band dryer consists of a vacuum chamber containing a number of moving bands which pass over heated plates. The wet material is continually fed onto the bands at one end of the dryer, carried by the bands over the heated plates, and ultimately discharged as dried product at the other end. Drying takes place at low pressure and evaporated moisture is removed by a vacuum pump

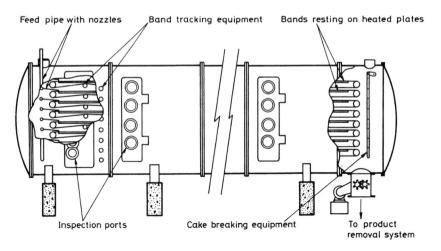

Figure 6.1 Vacuum band dryer.

and condenser. This type of contact dryer is suited to the drying of heat sensitive materials at medium production rates. Feed may be in the form of a thick paste, small shaped pieces or powders. A diverse range of food products have been dried successfully in vacuum band dryers, including chocolate crumb, meat and vegetable extracts, and fruit juices.

A typical vacuum band dryer (Figure 6.1) consists of 8–10 bands located above one another and operating in parallel. A typical band is about 1 m wide with a heated length of 8 m. Each band passes over a sequence of plates (typically three or four) heated internally by steam, hot water or thermal oil. The plates can be controlled independently so that different sections of the dryer may be maintained at different temperatures. For certain materials it is common practice to cool the final plate. This may prevent the material from collapsing under its own weight and ensures that it hardens sufficiently to permit a clean discharge from the band.

If the feed material is a paste, it can be extruded through nozzles on to the bands at the feed end of the dryer. These are located so as to produce a series of closely spaced 'ropes' moving along each band. The nozzles are normally about 10 mm in diameter with a spacing of about 20–40 mm. When extruding, the feed material must be pumpable yet sufficiently viscous to prevent it flowing either through the mesh or over the edges of the bands. Alternatively, non-pumpable materials such as small pieces (e.g. diced foods) and particulates may be fed on to the bands using a suitable airlock arrangement. Dried ropes, formed by extruding on to the band, are broken into short lengths by oscillating breaker bars and then fall down into a rotary breaker, which reduces their size to 15 mm or less. With both feed methods, the product is discharged from the dryer through a double-hopper airlock system.

6.2.5 Plate dryers

The plate dryer is a continuous contact dryer, suitable for drying powders and granular materials, which are both free flowing and non-caking, at medium production rates. Drying may take place under vacuum or at atmospheric pressure. The principle of operation is illustrated in Figure 6.2. The moist solids trickle from a suitable feeder onto the first of a series of heated plates. The rotating raking system turns and conveys the drying product in a spiral path through several revolutions to and over the outer rim of the plate. Here the product trickles to the plate below (which has a slightly greater diameter) where it is conveyed to the centre in the same manner as before. This process repeats itself as many times as is necessary until the dry solids leave the dryer at the bottom. This gentle method of mechanical transport produces very little dust and makes this type of dryer very suitable for friable materials and, together with the low gas velocity in the dryer, permits the treatment of products with fine particle sizes and wide size distributions. However, as noted above, it only works with granular or particulate materials which are free flowing and non-caking.

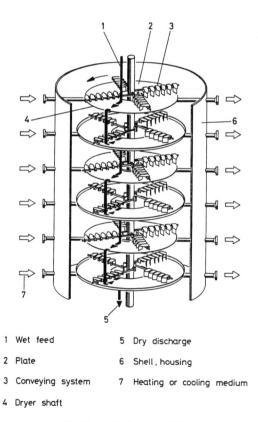

1 Wet feed	5 Dry discharge
2 Plate	6 Shell, housing
3 Conveying system	7 Heating or cooling medium
4 Dryer shaft	

Figure 6.2 Schematic diagram of plate dryer.

The heating plates typically range in diameter from about 1.5–3.5 m. A given dryer may contain between two and 16 such plates. These are of hollow construction and are heated internally by steam, hot water or oil. Each plate may be maintained at a different temperature, allowing the temperature history encountered by the material to be controlled. With some materials it is common practice to pass cooling water through the lower few plates in order to cool the product prior to discharge. Typically, plate dryers range from 2.2–8.8 m in height, 3–4.5 m in diameter, with plate areas of 12.5–175 m^2.

Models are available for atmospheric service and for vacuum operation down to about 200 Pa. In the former case, the vapours are drawn by a fan through a suitable dust collector to atmosphere. In the latter, the dryer is connected to a condenser and vacuum pump. A gas-tight model is also available which may be used in closed cycle systems. If the dryer is operated under vacuum, feeding and discharge of material must take place via airlocks.

6.2.6 Thin-film dryers

Thin-film or wiped-film dryers are generally similar in operation to thin-film evaporators and are suitable for drying solutions or suspensions containing typically no more than 60% solids. The product is a dry powder which usually has a mean size in the range 200–600 μm. The high heat transfer rates and consequent short drying times produced in the thin-film dryer make it suitable for materials such as tomato concentrate and gelatine.

This type of dryer consists of a vertical cylinder, typically up to 1 m in diameter and up to 9 m high. The wet solution or suspension is fed to the top of the dryer where it is spread by the action of a revolving rotor as an even thin film over the internal surface of the heated cylinder. The cylinder is surrounded by an insulated heating jacket. The jacket can be divided into several sections to allow different temperatures to be maintained at different points in the dryer. The liquid film heats rapidly to its boiling point and evaporation occurs. In the course of its short residence time in the dryer, the film is concentrated and eventually becomes a powder which falls off the dryer wall. The evaporated vapour is drawn upwards counter to the product flow by an exhaust fan or a vacuum pump, as appropriate. The operating pressure in the drying chamber can range from about 2.5 kPa to atmospheric.

In contrast to the thin-film evaporator, the wiper blades in the dryer are hinged rather than rigidly attached to the rotor shaft. They are adjusted so as to clear the wall by a minimum of 0.3–0.5 mm. For mechanical reasons, the clearance increases with dryer size. Although the blades are not in direct contact with the wall, they are nevertheless effective in preventing the formation of crusts. The rotor elements also generate turbulence in the product film which results in rapid surface renewal and consequently in high heat and mass transfer rates.

6.2.7 Drum dryers

The drum dryer is a continuous contact dryer widely used in the food industry for drying products initially in liquid form. The variety of feed arrangements available ensures that solutions, suspensions and pastes with a wide range of viscosities can be dried. This type of dryer is suited to many heat sensitive products since exposure to high temperatures is limited to a few seconds. Also, if required, drying may take place under vacuum. Typical examples of food products successfully dried on drum dryers include instant potatoes, corn syrup and baby food. However, drum drying is not suitable for materials which do not adhere to the drum and thermoplastic materials.

In drum drying, a thick film of the feedstock is applied to the external surface of a heated drum which rotates slowly about its horizontal axis. The layer of material remains attached to the drum for about 80% of a revolution during which time moisture evaporates and leaves behind a layer of solids, which is subsequently removed from the drum surface by a scraper or doctor knife. The speed of rotation of the drum will be varied to suit the conditions but speeds in the range 4–12 rpm are typical. The drums are hollow and heated internally by condensing steam or other heating media. Drum sizes range up to about 1.2 m in diameter and 3.5 m in length.

Usually drum dryers will be operated at atmospheric pressure and need only be partially enclosed by a vapour hood to remove evaporated moisture. However, if the material is particularly heat sensitive, vacuum drum dryers may be used. These consist essentially of an atmospheric drum dryer encased in a suitable vacuum chamber. In this case, as well as the usual vacuum raising pumps and condensers, special airlock systems are required to remove product from the chamber.

As illustrated in Figure 6.3, a variety of drum configurations and feed arrangements are possible. The dip feed system (a and b) is suitable for low to medium viscosity liquids. The drum is immersed to a depth of 10–25 mm in the feed and coated with a film. Thick and viscous feeds are usually applied by roller. A single feed roller (c) is commonly employed with medium viscosity solutions, suspensions and emulsions while, with thicker suspensions and pastes, a multiple roller arrangement (d) is more suitable. In the latter case, the thickness of the layer is progressively built up as it passes through the battery of rollers. The repeated application of thin layers of material results in shorter drying times than would have been achieved had the total layer been formed by a single application. Splash and spray feeding arangments (e, f) are especially useful for drying temperature sensitive, low viscosity feeds. This system is claimed to give gentle drying due to the Leidenfrost effect (greatly reduced heat transfer due to a vapour layer between the feed and the heating surface) but the film thickness is difficult to control and this may lead to uneven drying. Spreading rollers are sometimes used to counter this effect.

Double drum dryers employing a nip feeding system (g) are very versatile and

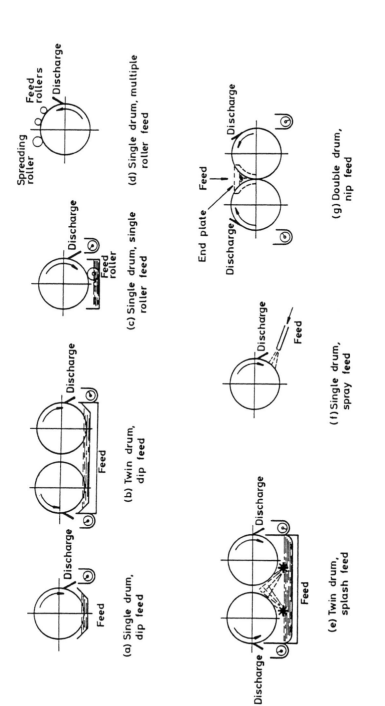

Figure 6.3 Drum dryer configurations.

can handle feeds with a wide range of viscosity. Alternatively, a reverse nip feeding system, where the drums rotate away from the nip, may be used if the feed is lumpy and forces the drums apart, or to avoid breakage of crystals. There is some pre-concentration of the feed in the nip, as a result of boiling. This can increase the capacity of the dryer but may cause feeding problems if the solution contains a solute of limited solubility, which crystallizes as a result. Twin-drum dryers are essentially two single-drum units with a common drive and are used for large scale production.

The product removed from the drum by the doctor knife is frequently a powder but, in certain cases, a flaked or chipped product can be obtained. The outer surface of the drum may be either smooth or grooved. In the former case, the product is frequently a powder. In the latter case, however, it normally consists of rod-shaped strings formed in the grooves.

6.2.8 Rotating batch vacuum dryers

The rotating batch vacuum dryer is essentially a rotating vessel which imparts a tumbling motion to its charge. This class of dryer includes the double-cone dryer (Figure 6.4), common in the UK and USA, and the offset cylinder dryer favoured in continental Europe. Rotating batch vacuum dryers are widely used for drying small to medium sized batches of free flowing powders, granules and crystals. However, the gentle tumbling motion is not suitable for sticky materials which will cake to the wall or ball up.

Rotating batch vacuum dryers consist of a heated vacuum chamber which is able to rotate about a horizontal axis. At the start of the drying cycle, a batch of the wet material is loaded through the charging port (shown at the top of Figure 6.4). The vessel is then closed and evacuated down to the operating pressure. Heat for the drying process is supplied by a heating jacket which surrounds the vessel. During the drying cycle, the vessel rotates and imparts a tumbling motion to the wet solids, aiding heat transfer, mixing and vapour release. The maximum speed of rotation varies from about 5 rpm for a large unit to 30 rpm for the smallest. The total cycle time is typically of the order of 20–24 h. The temperature of the heating jacket may be varied through the drying cycle and it is common practice to replace the heating medium with cooling water in the final

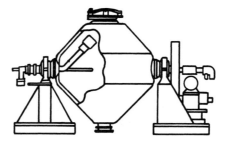

Figure 6.4 Double cone dryer.

stages. The vapour off-take pipe for removing evaporated moisture passes through a union on the side of the vessel. To minimize entrainment of any fines, the end of the off-take is fitted with a filter and remains stationary in the upper part of the chamber while the vessel itself rotates (left-hand side of Figure 6.4).

The rotating batch vacuum dryer is suitable for drying small to medium sized batches. The basic problem with operating at large scales is that, for standard designs, the heat transfer area per unit volume decreases with increasing size. Some designs have attempted to overcome this problem by incorporating internal heating panels. However, these are prone to fouling and are difficult to clean.

With many materials, the gentle tumbling motion can result in the formation of lumps or balls which can greatly increase the drying time required. Also, the material may cake to the walls of the chamber, reducing heat transfer and causing spoiled product. In many cases this can be avoided by taking appropriate measures; for example, changes to upstream processes to modify the initial condition of the solids; or by rotating slowly or intermittently when the material is prone to forming lumps (i.e. at high moisture contents). Also high speed rotating breakers can be fitted to break up any lumps. However, frequently, these measures will be ineffective and hence this type of dryer cannot be used for truly sticky materials.

6.2.9 Horizontally agitated dryers

In this type of dryer, wet material is dried in an enclosed trough or cylinder while being mixed and conveyed by a horizontally mounted agitator. This broad class of dryer may be further sub-classified according to the agitator type: i.e. paddle dryers, screw conveyor dryers. A variety of both batch and continuous models are available, which are employed for processing many different materials including sticky and free flowing solids, pastes and suspensions up to relatively high production rates. Also, since most horizontally agitated dryers can be operated under vacuum, they are suitable for handling heat sensitive materials. Foodstuffs successfully dried in this type of dryer include chocolate crumb, corn meal and confectionery materials.

Figure 6.5 shows a typical horizontally agitated dryer. The drying takes place in a vessel having either a cylindrical or trough-shaped cross-section, which is heated by steam or by a liquid heat transfer medium. Horizontally agitated dryers may be operated either under vacuum or at atmospheric pressure. The solids contained within the vessel are mixed and conveyed through the dryer by means of a rotary agitator(s). Typical designs are shown in Figure 6.5. The agitator consists of a shaft to which specially designed paddles (a), discs (b) or coils are attached. A wide variety of agitator designs is available, the most suitable of which will depend on the nature and handling properties of the material being dried. In some cases, these include scrapers and breaker bars to

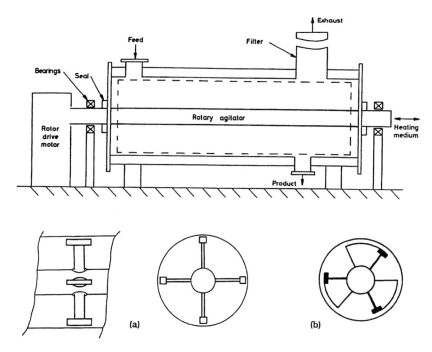

Figure 6.5 Horizontally agitated dryer.

handle sticky materials. In some designs heat is transferred through the wall of the dryer only. However, in many designs, to increase the heat transfer area, both the shaft and its attachments are also heated. Dryers are available with one, two or four agitator shafts.

The wet feedstock enters the top of the dryer at one end of the housing. In the case of continuous units, the discharge port is situated at the end of the dryer opposite to the feed inlet. The hold-up of material in the dryer (and hence its residence time) is normally controlled by means of an adjustable weir at the discharge end. With batch dryers, the discharge port is normally located at the centre of the housing. The wet material is introduced into the dryer by a standard means such as by Mono pump, a screw conveyor or a rotary valve. Discharge typically occurs through a screw conveyor, a rotary valve or a double butterfly valve. These operations are, needless to say, considerably more problematic with vacuum dryers than with atmospheric ones.

The exhaust for removal of evaporated moisture and other gases is located on the top of the chamber. Usually, the upper portion of the vessel remains empty and serves as a vapour take-off channel. This stream will be drawn off to gas cleaning equipment, condenser and vacuum pump or exhaust fan, as appropriate. These dryers are sometimes purged with a slow flow of air or nitrogen, which may be heated, to aid in the removal of water vapour and to aid drying.

The paddle dryer features an agitator, shown in Figure 6.5a, consisting of paddles mounted on the end of radial arms. The purpose of the paddles is firstly to minimize the build-up of solids on the inner surface of the jacket in order to maintain a high heat transfer rate and, secondly, to mix and, frequently, transport the solids. The first objective is achieved by restricting the gap between the vessel wall and the paddles to a few millimetres at the operating temperature. Solids mixing and transport are accomplished by angling the paddles relative to the rotor arm. Some designs feature ploughshare shovels rather than paddles on the end of the radial arms. Basic single-shafted paddle dryers are not particularly suitable for sticky materials, which may ball-up and stick to the internals of the dryer. To overcome this problem, some paddle dryers may be fitted with floating breaker bars which break up agglomerates and help to prevent material adhering to the arms. With some materials, sticking and balling still occur despite their presence.

Some dryers have agitators specifically designed to handle sticky powders, pastes or those materials which tend to bake on. These typically consist of a number of closely spaced agitators on the rotating shaft, interspaced by stationary breaker bars. The clearances between agitators and breaker bars are relatively small, enabling surfaces to be scraped clear of deposits and imparting a 'kneading' action to the material in the dryer.

6.2.10 Indirectly heated rotary dryers

Indirectly heated rotary dryers are continuous, atmospheric contact dryers widely employed in industry for continuously processing large quantities of wet solids. Indeed, the indirect rotary dryer is the one type of contact dryer that can handle very high throughputs. Typically, indirect rotary dryers would be used when direct (convective) rotary dryers are unsuitable. This may be because direct contact with hot combustion gases is not permissible for product quality reasons or because of the presence of fine particles, which can easily become entrained in the higher gas flows found in convective rotary dryers. The indirectly heated rotary dryer can only process relatively freely flowing solids; it cannot handle liquids or slurries or adhesive solids. Materials processed successfully in this type of dryer include brewers' grains and starch feeds.

The most common design consists of a cylindrical shell inclined at a small angle (e.g. 1–5°) to the horizontal. Wet feed is introduced into the upper end and dried product withdrawn at the lower end. Transportation of the solids through the dryer is assisted by slowly rotating the shell. Heat is supplied to the drying solids through as many as three staggered, concentric rows of tubes running the full length of the dryer which contain the heating medium. These are mounted on the shell and rotate with it. The shell is also equipped with flights which shower the solids over the tube bundle from which they pick up heat on impact. The spacing of the tubes must be sufficiently large (e.g. 2 diameters) to ensure that the solids can flow freely around them. In general this type of dryer is suitable for free flowing granules and powders only. Materials which have adhesive properties will quickly foul the tube bundle. To a limited extent, fouling problems may be reduced by increasing the spacing and decreasing the number of tubes. Knockers and hammers on the external surface of the dryer may also have some effect.

6.2.11 Vertically agitated dryers

The class of vertically agitated dryers encompasses all types of contact dryer which consist of a heated vessel fitted with a slow moving vertical agitator. This includes the vertical pan (Figure 6.6) and the Nauta or cone dryer (Figure 6.7). Dryers of this type are used principally for the batch drying of pastes and slurries on a small scale. Both atmospheric and vacuum operation are commonly employed.

The vertical pan dryer consists of a flat bottomed, jacketed, cylindrical vessel with bottom and side surfaces swept by a vertical agitator. The jacket of the vessel can be heated by water, steam or oil. Operation can be under vacuum or at atmospheric pressure. Most small units have a hinged top cover for feeding in the wet material and to allow access for cleaning. In larger units the batch of wet material is charged to the dryer through a manhole in the domed cover. Dry product is removed through a hinged door in the side wall. The contents of the vessel are agitated by a robust, four-bladed paddle which sweeps the sides and bottom of the vessel. The agitator also aids in emptying the dryer by sweeping the dry product towards the discharge door. To keep heating surfaces clean, clearances between the paddle and internal surfaces are kept as small as possible. Stationary breaking bars may be incorporated to help break up agglomerates.

Another common type of vertically agitated dryer is the Nauta or cone dryer shown in Figure 6.7, which is a direct development of the Nauta mixer. The

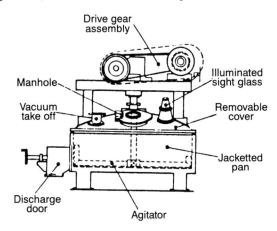

Figure 6.6 Vertical pan dryer.

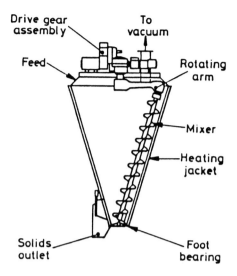

Figure 6.7 Nauta or cone dryer

Nauta dryer is a conically shaped vessel, with capacities ranging from 50 l to 25 m³, surrounded by a suitable heating jacket. These dryers are normally operated under vacuum. Good mixing is provided by a vertical screw which rotates about its own axis and also moves around the chamber on an orbiting arm. The distance between the screw and wall of the vessel in minimized so that the layer of material on the heating surface is continually renewed. The screw may be heated to give a small amount (about 10%) of additional heat transfer area. To avoid contamination of the product, good sealing of bearings is an important consideration. Nauta dryers are also available with combined contact and microwave heating. This is claimed to give more rapid and more uniform drying.

6.3 Theoretical overview of contact drying

In contact drying, the wet material is held as a bed or layer in contact with heated surfaces and the heat required for the evaporation of moisture is transported into the material by conduction. Moisture in the material consists of two types: unbound or free moisture, which is relatively easy to evaporate; and bound moisture, which is physically or chemically bound to the material and is more difficult to remove.

In most practical drying situations, the heating surface temperature will be maintained above the boiling point of the free moisture and drying commences by 'boiling off' unbound moisture close to this surface. If the dryer is unagitated, a drying front will move progressively through the material. If the bed is

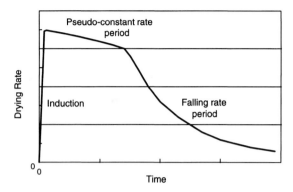

Figure 6.8 Contact drying rate curve.

strongly agitated the drying front will not be evident since the bed will be mixed and the layer in contact with the heated surface continually renewed. While unbound moisture is present, the solids will be maintained at the boiling point of the free moisture by evaporative cooling. However, once the unbound moisture is removed, the solids attain a higher temperature necessary to remove the bound moisture. Ultimately, when drying is complete, the temperature of the solids may rise to the heated-wall temperature.

Figure 6.8 shows a typical contact drying-rate curve. After an initial heating-up period, the maximum drying rate is attained when unbound moisture in contact with the heated surface starts to evaporate. This is followed by the pseudo-constant rate period during which the drying rate slowly decreases. Unbound moisture is still being evaporated at this stage; the decrease in drying rate is caused by a reduction of heat transfer into the bed as the solids dry out. For static beds, this reduction is caused by the drying front receding from the heated surface. Higher drying rates can be achieved with agitation of the bed; however, there will still be a reduction during the pseudo-constant rate period because the thermal conductivity of the bed, and hence heat transfer, will decrease with moisture content. In the falling rate period, the drying rate falls still further as bound moisture is removed. The lowest possible moisture content achievable under a given set of drying conditions is the equilibrium moisture content X^*. Its value will depend on the partial pressure of evaporated vapour in the dryer and the dryer temperature.

An important characteristic of contact dryers is that drying may take place under partial vacuum. The advantage of operating under vacuum is that drying may take place at a much lower temperature. However, this is only justified for certain materials because of the extra equipment involved. As stated above, evaporation of unbound moisture usually takes place at its boiling point. This is the temperature at which the vapour pressure of the moisture is equal to the applied pressure. Hence lowering the applied pressure reduces the drying temperature. For example, while the boiling point of water at 100 kPa (1 bar) is 100°C, at 5 kPa it is approximately 33°C. Operating under vacuum therefore has obvious advantages for heat sensitive materials such as some food products. Note that, while unbound moisture exerts the same vapour pressure as pure liquid, bound moisture, because it is more strongly attached, exerts a lower vapour pressure. Hence bound moisture will boil at higher temperatures. For very tightly bound moisture this may require excessively high temperatures and, in some cases, it is useful to dilute the vapour in the dryer with purge gas (air or nitrogen) so that drying can take place at lower temperatures.

The drying rate in a contact dryer is determined by the heat transfer into the bed (heat transfer controlled). The total heat absorbed by the bed, Q_{bed} (kW), can be divided into two parts: the sensible heat used to raise the temperature of the bed, Q_{sen} (kW); and Q_{ev} (kW), the heat used to evaporate moisture. The drying rate, W_{ev} (kg s^{-1}), is simply determined by dividing Q_{ev} by the latent heat of vaporization of the moisture, λ (kJ kg^{-1}). Except during the initial heating period and when the solids are virtually dry, Q_{ev} is much greater than Q_{sen}. Hence

$$W_{ev} = \frac{Q_{ev}}{\lambda} \approx \frac{Q_{bed}}{\lambda} \qquad (6.1)$$

An overall heat transfer coefficient, U (kW m^{-2} K^{-1}), can be defined which relates heat transfer to the bed temperature, T_{bed} (K).

$$Q_{bed} = UA(T_{hm} - T_{bed}) \qquad (6.2)$$

where A is the heat transfer area (m^2); and T_{hm} (K) is the heating medium temperature which, for all practical purposes, is equal to the wall temperature, T_W (K). In the early stages of drying, when unbound moisture is present, T_{bed} can be taken as the boiling point of the moisture T_{bp} (K). However, in the later stages, when bound moisture is being removed, T_{bed} will rise towards the heated-wall temperature. For design purposes it is frequently useful to rewrite equations 6.1 and 6.2 in the following form:

$$W_{ev} = \frac{Q_{ev}}{\lambda} = \frac{UA}{\lambda}(T_{hm} - T_{bed}) = \frac{U_m A}{\lambda}(T_{hm} - T_{bp}) \qquad (6.3)$$

U_m is a modified heat transfer coefficient (kW m^{-2} K^{-1}) which allows the bed temperature to be taken as T_{bp}. Whereas U is dependent on heat transport properties only, U_m also takes into account the hygroscopic nature of the material. Clearly, U and U_m are important quantities for determining drying rates and the requisite heat transfer area in contact dryers. Unfortunately, their values will vary during the drying cycle and will also be strongly dependent on the nature of the material, its moisture content and the level of agitation. In general, it is not possible to predict U and U_m accurately in advance without experimental tests (see Section 6.4 on design). Nonhebel and Moss (1971) indicated that, for agitated dryers, the mean heat transfer coefficient for the complete drying cycle is usually $10–85 \ \text{kW m}^{-2} \ \text{K}^{-1}$.

6.4 Design of contact dryers

It is not practical to design a contact dryer on the basis of theoretical considerations alone and laboratory or pilot scale tests on the material in question are necessary. These tests are essential for several reasons. The first question that must be answered is whether the selected type of dryer is suitable for the wet material. For example, many dryers can only handle materials that remain free flowing and do not ball up during the drying process. Other dryers require special handling characteristics (e.g. drum dryers – will the material adhere correctly to the drum? vacuum band dryers – is the material a thick paste that can be extruded onto the band?).

At the same time the tests will also be used to determine suitable drying conditions for the process. The heating-surface temperature will generally be selected on the basis of the thermal sensitivity of the material; clearly, having as high as possible a surface temperature maximizes the rate of heat transfer. The pressure within the drying chamber, if it is under vacuum, will then be selected so as to give a reasonable temperature driving force. The chamber pressure may also have some effect on the physical structure of the dried material. For example, drying under low absolute pressure gives certain materials a cellular, porous structure and a low bulk density, which is desirable for some food products.

Once the suitability of the selected dryer and the drying conditions have been established, the pilot scale tests are used to determine a drying curve for the material which can be used to scale up to the desired production rates. In essence, these tests will be used to determine U and U_m throughout the drying cycle so that equation 6.3 can be used to calculate the heated surface area for the required drying rate. It is important in these tests to replicate the intended drying conditions as closely as possible. In unagitated dryers such as tray dryers, the scale-up process is relatively straightforward. Tests are conducted on a layer of thickness equal to that which will be used in the real process. Scale-up is then simply a case of increasing the heat transfer area in proportion to the measured drying rate up to the required production rate. The scale-up of agitated contact dryers is far more problematic and is certainly not an exact science. In general, with this type of dryer, tests will be performed on laboratory and pilot scale equipment of the same design to determine U_m, and then scale-up will be performed on the basis of the speed of agitation.

Consideration should then be given to auxiliary equipment. Devices for transporting wet material and dried solids into and out of the dryer should be sized according to the desired production rate. For atmospheric dryers, the gas exhaust will be handled by exhaust fans and appropriate gas cleaning equipment, while for vacuum dryers the exhaust will be handled by a condenser, a vacuum pump and, if necessary, gas cleaning equipment. Note that exhaust gas clean-up is far easier in contact dryers than in convective dryers. Generally, the total exhaust flow rate is much smaller, with very few entrained particles. Also, any vapours in the exhaust are relatively easy to condense.

In the case of vacuum dryers, the condenser is a very important part of the vacuum-raising equipment. Ideally, all vapours will be condensed, leaving the vacuum pump to remove only non-condensible gases, which may be present in the feed or through leakage into the chamber. The condenser temperature is an important additional consideration in the selection of the chamber operating temperature. This must be below the dew point of the evaporated vapour, which, for low operating pressures, may require refrigeration. However, cooling water is usually sufficient. Sizing of all gas handling equipment should be on the basis of the maximum drying rate. With batch vacuum dryers, the vacuum pump and condenser should be sized to give sufficiently rapid 'pump-down' at the start of the drying cycle. With all vacuum dryers, special attention should be paid to the construction of the chamber which must be manufactured to pressure vessel codes such as BS 5500. Also for continuous operation, special airlock systems will be required to introduce and remove solids from the vacuum chamber.

Acknowledgements

The author acknowledges the cooperation of AEA Technology plc in the writing of this chapter. All illustrations are reproduced with permission from the Separation Processes Service, AEA Technology plc.

References

BS 5500, *Specification for Unfired Fusion Welded Pressure Vessels*, British Standards Institution, London.
Nonhebel, G. and Moss, A.A.H. (1971) *Drying of Solids in the Chemical Industry*, Butterworths, London.

7 Freeze dryers

J.W. SNOWMAN

7.1 Introduction

Drying a wet material is usually achieved by evaporating the liquid water it contains. For foods, this often results in profound changes to appearance and taste. Plums become prunes, and grapes become raisins. Rehydration is difficult when solids agglomerate to form an impervious layer at the surface as the liquid phase retreats. Freeze drying avoids many of the problems of liquid-phase drying. There is good retention of aroma, flavour, and nutrients and little shrinkage. Texture is acceptable, especially with vegetables. Freeze dried food can be stored at ambient temperature for long periods and re-hydration is easy. A simple definition of freeze drying is that it is a means of drying, achieved by freezing the wet substance and causing the ice to sublime directly to vapour by exposing it to a low partial pressure of water vapour. In this context, 'low' means below about 600 Pa (4.6 mm Hg), the triple point of water. It can take place in nature when the weather is very cold but dry. An example is when washing is hung out in freezing weather. If the air is dry so the partial pressure of water vapour is very low, and the sun is shining to provide energy, the washing will dry by sublimation. In industrial applications, the low temperature is created by refrigeration. The low partial pressure of water vapour is most conveniently ensured by carrying the process out under vacuum. The energy for sublimation is provided by electrical, or sometimes steam, heating.

Unfortunately, freeze drying is an expensive process in terms of both capital and operating cost, and it is also time consuming. During the 1960s it was expected that the process would have wide application, and considerable resources were devoted to developing novel equipment and methods. However, in the USA and Europe, highly efficient distribution and storage systems for frozen and chilled food were by then becoming firmly established. This has largely restricted the market for freeze dried food to instant coffee and tea, military, mountaineering, and camping solids, and high cost additives to food dried by other means, such as shrimp in spray dried sauce. In other parts of the world there is much wider application, including chicken, beef, fish, shrimp, eggs and fruit.

7.2 Process overview

Freeze drying takes place in three distinct stages. The first is pre-freezing to a temperature where the material is completely solid. This can be significantly below 0°C. The second stage is sublimation or 'primary' drying in which all the free ice is removed, leaving an apparently dry material which, however, may contain significant bound water. The third and final stage is desorption or 'secondary' drying when bound water is removed.

Selection and preparation of food before freeze drying is important. Only high quality undamaged material should be used. Similarly, proper storage of the dried food before rehydration and use is also important in maintaining quality. It must be kept dry, ideally under an inert gas atmosphere, usually nitrogen. Figure 7.1 shows the stages of the process from initial food preparation to final reconstitution and use. The stages comprising the freeze drying process are labelled and show how the temperature varies during a typical cycle.

In common with almost all materials, water can exist in three different states or phases; a liquid, a solid (ice), or gas (water vapour). By plotting a graph of pressure versus temperature over an appropriate range, the boundaries of the three states can be seen. This is called a phase diagram. Figure 7.2 is the phase diagram for water around the 'triple point' where ice, liquid water, and water vapour can exist together in equilibrium. In evaporative (liquid phase) drying, water at point 'A' is heated to its equilibrium vapour pressure at 'B' where adding more energy supplies the latent heat of evaporation thus causing the water to change phase to a vapour. In freeze drying, water at point A is cooled through its freezing point C to D. When the water is completely frozen, the pressure over the ice is reduced to point E, the equilibrium vapour pressure exerted by the ice. By adding the latent heat of crystallization and evaporation, the ice is sublimed to water vapour.

Pre-freezing, sublimation drying and desorption drying usually overlap to some extent during the cycle. For example, if a low melting point material is present, it may not freeze at the temperatures available in practical freeze dryers. At the commencement of primary drying, it will evaporate before true freeze drying of the product begins. Later, towards the end of the drying cycle, some desorption will begin in the drier parts before sublimation drying is complete.

Unfortunately the apparent simplicity of the process masks the fact that it can cause complex and potentially damaging effects to material being freeze dried (Franks et al., 1991). The nature of the wet starting material exerts a major influence on the process. Pretreating and freezing a product and developing an optimum drying cycle are mutually interacting activities which will often require several iterations before a satisfactory product and process are found.

Being frozen, constituents of the material are kept immobilized during sublimation. The shape of the dried substance is broadly the same as that of the frozen wet substance and migration of solids to the surface to form a skin is reduced or eliminated. Because freeze drying takes place at a low temperature, heat damage is minimized and volatile components are retained. Figure 7.3 shows typical time and temperature ranges for three methods of drying. Spray drying requires exposure to temperatures up to 100°C for periods of seconds, oven drying requires temperatures of typically 60°C for periods of minutes, and

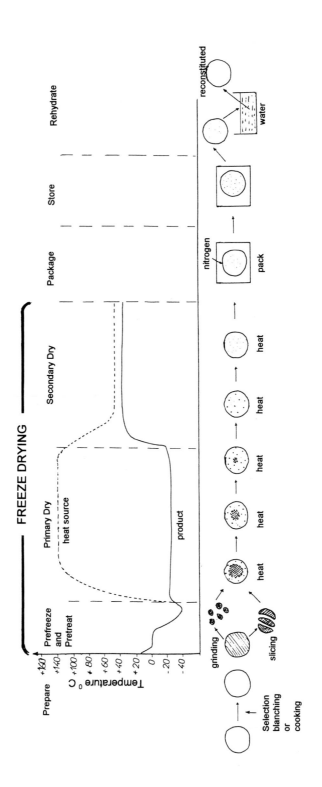

Figure 7.1 Life cycle of freeze dried food.

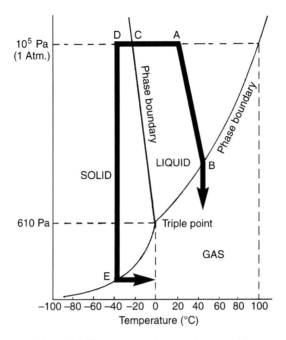

Figure 7.2 Phase diagram for water (not to scale).

freeze drying exposes material to temperatures below 0°C for periods of hours. A comparison of aroma retention in food products was made by Coumans *et al.* (1994).

Because the subliming ice crystals leave cavities, freeze dried material contains a myriad of interstices into which water can penetrate to give rapid and

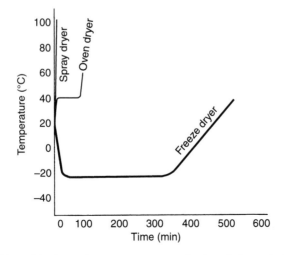

Figure 7.3 Time and temperature exposure for three drying methods.

complete re-hydration when needed. The alternative name for freeze drying favoured by the pharmaceutical industry is 'lyophilization', which comes from the Greek 'to make solvent loving'.

The advantages of freeze drying are:

- minimum damage to, and loss of activity in, heat labile materials;
- creation of a porous, friable structure;
- speed and completeness of rehydration;
- the ability to sterile filter liquids just before dispensing (if needed).

The principle disadvantages are:

- high capital cost of equipment (about three times those of other methods);
- high energy costs (also about three times those of other methods);
- lengthy process time (typically 4 to 10 h per drying cycle);
- possible damage to products due to change in pH and tonicity when solutes concentrate as pure water freezes into ice.

Freeze drying is useful when the product meets one or more of the following criteria:

- it is unstable
- it is heat labile
- quick and complete rehydration is required
- the product is of high value
- weight must be minimized
- frozen or chilled storage is not appropriate.

7.3 Description of the freeze drying process

7.3.1 Pretreatment of food

Since freeze drying essentially removes water, the process is unsuitable for food with a high fat content, which can develop rancidity. With this exception, a wide range of other foods has been freeze dried successfully. The most widely known are coffee and, more recently, tea. Others are lean meats, fish steaks, shrimp, vegetables, fruit, milk and eggs. Solid materials should be cut up to reduce thickness and therefore drying times. A thickness of 20 mm or less is best, with the 'grain' of the material oriented for the shortest vapour escape path. For this reason steaks are preferred to fillets for freeze drying. Cooked meats can be diced or minced, as can root vegetables. Peas, brussel sprouts, and similar vegetables can be dried whole, but leaf vegetables need to be shredded. Fruits with an impermeable skin must have it removed or pricked, or must be sliced. Alternatively they can be puréed. Most part-prepared food is then blanched or scalded to inactivate enzymes.

7.3.2 Freezing

In ideal two-component liquid systems, water and a solute may both crystallize. In this case, freezing proceeds as follows. Cooling from the initial temperature to below 0°C eventually results in the nucleation of ice. The release of heat of crystallization raises the temperature towards 0°C. Crystallization of water then proceeds at a progressively falling temperature related to the equilibrium melting point of the concentrating solution, as shown in Figure 7.4. As the temperature falls the unfrozen solution approaches saturation and crystals of solute are precipitated. Eventually a eutectic point is reached where the material become wholly crystalline. When the material contains more than one crystallizable solute, a similar situation exists but the eutectic point is lower than that of any two components combined.

Unfortunately, in the majority of practical cases, the solute does not actually crystallize. The solution continues to concentrate as water is frozen out, and its viscosity increases rapidly. Eventually it becomes syrupy, then rubbery, as shown in Figure 7.5. Further lowering of the temperature finally causes the formation of an amorphous glass containing some bound water. This occurs at the glass transition temperature T_G. In such cases, there is no eutectic temperature as such, but there is a characteristic maximum product temperature for freeze drying called the collapse temperature, T_c. The collapse temperatures for a number of substances have been determined (MacKenzie, 1976).

Wet materials can be frozen by placing them in a vacuum. The more energetic water molecules escape and the temperature of the material falls. Eventually it freezes. About 15% of the water is lost. The process is called evaporative freezing. Scalded vegetables and minced meats can be frozen in a vacuum chamber by this method provided the vacuum pull-down rate is fast enough.

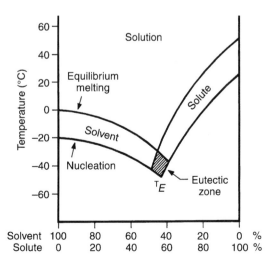

Figure 7.4 State diagram for a solution where both solvent and solute crystallize.

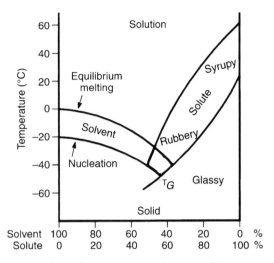

Figure 7.5 State diagram for a solution where the solvent crystallizes but the solute does not.

However, there are disadvantages which have limited the application of evaporative freezing in freeze drying. Some surface shrinkage takes place and particles tend to stick to each other and to trays. Also, a high peak pumping capacity for water vapour is needed. Several ideas to overcome these problems have been tried. Greaves (1966) described a method in which liquids were sprayed on to a rotating disc in a vacuum. The liquids froze on the disc and were continuously scraped off and dropped into a cold hopper. Another approach is to use a continuously scraped cooler producing a mixture of ice crystals, solids and water. A second stage of final freezing in trays is required. A number of systems based on continuously scraped coolers have been investigated. Usually some granulation is necessary after freezing. Meat and fish are not suitable for evaporative freezing as they contain proteins which form a surface skin, inhibiting the escape of vapour. Fruit with a high sugar content is also unsuitable.

The method favoured in most current applications is to freeze trays of product to about −40°C in a conventional blast or flow freezer and then to transfer them quickly to the freeze dryer. The vacuum must be established rapidly before the food melts. In some equipment, blast-freezing facilities are incorporated in the vacuum chamber.

Since the subliming water vapour must pass through the channels left by the ice crystals that formed on freezing, their shape affects the speed of drying. If the crystals are small and discontinuous, the escape path for the vapour is limited; if large crystals are formed, escape is easy and the product can be dried more quickly. The method and rate of freezing are thus critical to the course of sublimation. Typical freezing rates achieved in blast freezers are usually appropriate. Ultra-fast freezing methods, such as the use of liquid nitrogen, must be viewed with caution because of the formation of very small discontinuous ice crystals.

7.3.3 Sublimation (primary drying)

The simplest freeze dryer would consist of a vacuum chamber in which materials could be placed, together with a means of removing water vapour, so as to freeze them by evaporative cooling and then to maintain the water vapour pressure below the triple point pressure. The temperature of the material would continue to fall below the freezing point and sublimation would slow down until the rate of heat gain by conduction, convection and radiation to the material was equal to the rate of heat loss as the more energetic molecules sublimed and were removed. Without a supply of heat the process would eventually stop. A major part of freeze drying technology is concerned with transferring heat energy to the freeze drying front as fast as possible without exceeding a safe temperature.

There have been numerous attempts to apply microwave heating to sublimation but so far none have been successful. A fundamental difficulty is that pressures below 1 mb are needed, but below about 30 mb, the presence of a strong electric field causes gas in the chamber to ionize, resulting in glow discharges. A further practical difficulty arises because it has so far proved impossible to create a sufficiently uniform high power field to avoid charring parts of the non-metallic structures, belts and trays in the chamber. However, there are possible uses for microwave heating in secondary drying, which can be performed at pressures above the triple point (Cohen and Yang, 1995).

Theoretically it is not necessary to provide a vacuum, only a low partial pressure of water vapour. However, it has been found that to provide a continuous supply of sufficiently dry air is uneconomic. At a typical freeze drying cycle pressure, 1 ml of ice produces more than 1 000 000 ml of water vapour. This is too much to be pumped by conventional oil sealed vacuum pumps, and water sealed vacuum pumps cannot reach the necessary pressures. Early food freeze dryers used multi-stage steam ejectors to pump the vacuum system but, because of their low energy efficiency, these have largely been replaced by a combination of a refrigerated trap (called the ice condenser) and a vacuum pump. The ice condenser is fitted between the freeze drying chamber and the vacuum pump to prevent water vapour reaching the pump, which only removes permanent gases (air, nitrogen) from the system. As soon as the required vacuum pressure is established, heating can be started.

During sublimation, vapour is generated from a distinct interface or freeze drying front which moves from the outer surface of the material. As the temperature at this front controls the speed of drying it should be kept as high as possible without inducing melting or collapse. A rough approximation is that reducing the front temperature by 1°C increases drying time by over 10%. A good starting temperature for shelves is about 10°C. Where radiation heating from platens is used, the starting temperature can be as high as 120°C. A number of different drying rate models have been developed for mathematical analysis. Some examples have been given by Lombrana and Villaran (1993), Kumagai *et al.* (1991) and Bruttini *et al.* (1991). A major limitation to drying rate occurs if the material has a skin or contains water in veins. For example grass is very difficult to freeze dry for research purposes if it is in the form of complete blades because water vapour must pass along the whole length of the impervious veins to escape. The grass must be chopped into short lengths to dry it in a reasonable time. For similar reasons, leaf vegetables should also be shredded. Apples and similar fruits are sliced, diced, or pricked to allow vapour to escape. Any restriction of flow would cause an increase in product temperature. If the temperature T_E of the lowest eutectic exhibited by the material is exceeded during drying, then melting can occur, giving rise to gross material faults such as shrinking or puffing. These are sometimes confused with collapse. At temperatures below the collapse temperature T_c, the solute will remain rigid during primary drying. However, if drying is carried out above T_c, the solute will be viscous. While it is supported by the pure ice crystals, it appears to be rigid, but as they sublime away, they no longer support the solute and apparently dry material collapses to form an impermeable mass.

(a) Provision of energy.   Heat to provide energy for sublimation can be transported by conduction, convection, or radiation. Heat from shelves must be transferred to the freeze drying front through the frozen material, which usually has a poor thermal conductivity. The need to avoid melting means that the thermal gradient must be small. Heat from platens transferred by radiation has to pass through layers of dry material, which effectively insulate the drying front. It is possible for product to remain frozen at the centre of a tray with platens at above 100°C. Drying will be fastest when material is disposed in thin layers, which minimize thermal resistance between the heat source and the freeze drying front. Figure 7.6 indicates possible heat transfer routes.

Heat transfer from shelf to product container and subsequently to the freeze drying front during sublimation can be increased by spoiling the vacuum (Pikal, 1990). In one product studied, the vapour flux with a surrounding pressure of 65×10^{-3} mb (6.5×10^{-3} kPa) was around 0.09 g cm^{-2} h^{-1}. When the pressure was increased to 350×10^{-3} mb (35×10^{-3} kPa) the vapour flux increased to 0.16 g cm^{-2} h^{-1}. Spoiling the vacuum is usually achieved by throttling the

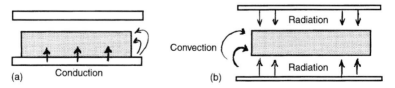

Figure 7.6 Possible heat transfer routes. (a) Product on a shelf at +50°C; and (b) product between platens at +150°C.

vacuum pump. It can also be achieved by bleeding filtered air or dry nitrogen into the chamber. Pressure maintained at around 250×10^{-3} mb (25×10^{-3} kPa) improves convective heat transfer from the heating shelf to the bottom of the product container and reduces the temperature difference required from shelf to product for a given drying rate.

Typical primary drying time for a 10 mm thickness of a 'simple' material is around 6 h with favourable ice crystal structure and optimized temperature and pressure conditions in the drying chamber. Rates of vapour escape from a surface area of 1 m² are often between 3 and 8 kg h⁻¹ depending on the material and its preparation. The approximate variation of drying time with thickness is

$$\text{time} = K \, (\text{thickness})^{1.5}$$

where K is a constant. As primary drying reaches completion, less energy is needed for sublimation and the temperature of the material increases towards that of the heat source.

(b) Sublimation methods. Three different techniques are used to handle products during sublimation drying. The first is to load the product into trays. These may be conventional trays made of aluminium or stainless steel, which work quite well with many granulated, sliced or diced materials provided the depth is limited to about 25 mm. A better alternative is to use trays with internal ribs or fins as shown in Figure 7.7. The improved heat transfer enables depths up to 60 mm to be loaded.

The second technique is to make specialized equipment for particular products to improve heat transfer. Various approaches have included methods of clamping solid product, such as steaks, between perforated metal plates (so called 'accelerated freeze drying') or the 'inner heat' method in which metallic ribs or spikes were dipped in liquid material. The spikes could be refrigerated or heated. They were initially dipped in the product and cooled until a suitable thickness froze on their surface. Excess liquid was then drained and pressure reduced, after which the surfaces were heated to provide the energy for sublimation.

The third approach is suitable for granulated products such as frozen coffee or juices. With such materials, particles are continuously moved in and out of contact with heated surfaces which may be belts, rotating drums, or vibrating surfaces. Figure 7.8 shows a freeze dryer comprising a series of belts mounted in a long thin chamber.

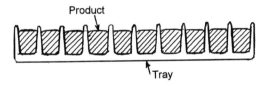

Figure 7.7 Section of aluminium tray with cast-in ribs to improve heat transfer to the product.

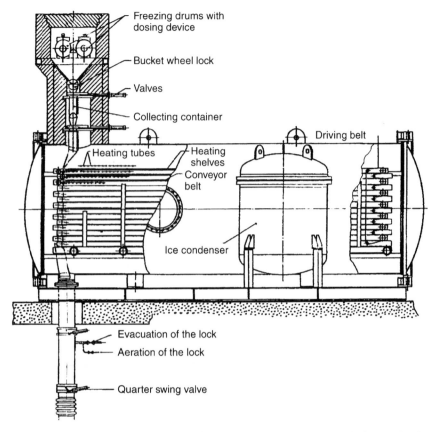

Figure 7.8 A continuous freeze dryer with the product moving on belts. Note the inlet and outlet vacuum locks.

The belts are one above another and move in opposite directions. They are arranged so that material falling from the end of one belt falls onto the beginning of the belt below. Material is fed into the chamber through a vacuum lock at the top and removed through a lock at the bottom. Two ice condensers are fitted to allow continuous operation. The belts run on platens at different temperatures, decreasing towards the bottom. Yet another variant of this approach uses vibrating surfaces as shown in Figure 7.9. Both of these methods are limited to materials which do not produce dust or fines, and do not require temperature adjustments.

7.3.4 Desorption (secondary drying)

This is the removal of bound moisture, which may be water of crystallization, randomly dispersed water in a glassy material, intracellular water, or absorbed water. Primary or sublimation drying accompanied by desorption removes the free water. The remaining bound water usually amounts to below 20% w/w. It can be removed by heating the product under vacuum, usually to between +40°C and +60°C.

7.3.5 Storage after drying

The residual moisture content of dry material is a critical factor affecting storage life. Most freeze dried materials are hygroscopic and must be stored in sealed containers. They are also liable to oxidize. In the early days of freeze drying, products were stored under vacuum. Current practice is to store them under a dry inert gas, usually pure dry nitrogen, in a low permeability sachet. Other gases such as argon and helium have also been used.

7.3.6 Rehydration and use

The method of rehydration affects the final perceived quality of some foods, especially raw meats, which may require the addition of an enzymic tenderizer.

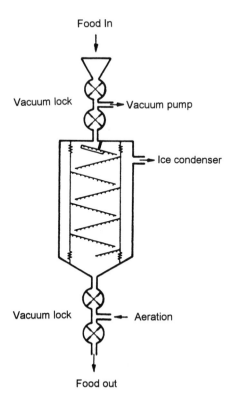

Figure 7.9 Continuous freeze dryer where the product is transported by vibration (after Oatjen, 1966, by permission of Hermann, Paris).

7.4 Equipment

Practical problems have caused the abandonment of most of the novel techniques devised in the 1960s. The 'inner heat' idea foundered on the difficulties of finding products which remained in place during drying without flaking off and then of removing the dried product from the spikes or plates. Equipment based on moving particles can only be used for the few materials which are free of the dust that often originates during freezing or as a result of abrasion in the freeze dryer. Particles smaller than about 1 mm can easily be moved around the system by water vapour from the product and may be deposited anywhere the vapour velocity is low. This usually means in the ice condenser, so the product is lost during defrost. Vapour velocities during fast drying cycles can be very high. Consider a typical situation where the water vapour removal rate is $5 \text{ kg m}^{-2} \text{ h}^{-1}$. If the operating pressure is approximately 0.5 mb (50 kPa) around $6000 \text{ m}^3 \text{ h}^{-1}$ of water vapour have to be removed from 1 m^2 of product area. If the area between particles for vapour escape is 10% of the whole area, the average vapour speed must be nearly 17 m s^{-1}. The water vapour sublimation rate at the beginning of the process is three to four times higher than the average rate and the resulting vapour velocities of around 60 m s^{-1} are more than enough to blow away dust or small particles, especially at the beginning of the drying period.

For these reasons, most general purpose freeze dryers operate with trays of product so that abrasion is avoided. These may be either conventional trays or the ribbed trays which, as well as improving heat transfer, have the advantage of being rigid, a necessary attribute if handling is to be automated.

Small to medium scale food freeze dryers are built for batch processing. Large scale equipment is often built for continuous operation. Considering first batch dryers, these usually comprise a horizontal cylindrical chamber. Full section doors are fitted at one or both ends and a refrigerated surface to condense water vapour from the product is arranged in the chamber. The door is mounted through a wall into a clean area, as shown in Figure 7.10, with machinery located in a separate room, Figure 7.11.

Chambers can be set up with a series of shelves onto which trays of product are placed, the shelves providing the heat energy by condution. However, it is more usual for trays of product to be mounted on a moveable 'conduction', which can be pushed into the chamber, so that they are held between horizontal heating plates, which provide energy by radiation and convection. This is shown in Figures 7.12, 7.13 and 7.14.

For continuous operation trays of product can be arranged to enter a long cylindrical chamber via a vacuum lock. This is an auxiliary small entrance chamber that can be isolated from the main chamber and separately aerated to allow insertion of a product tray. The chamber is subsequently pumped down to allow the isolation valve to be opened and the product tray to be transferred to the main chamber. After drying, trays are extracted by reversing the procedure.

Figure 7.10 A 600 kg per 24 h net weight freeze dryer viewed from the clean room. (Photograph courtesy of Autec Ltd.)

Figure 7.11 A 250 kW refrigeration plant for a freeze dryer with a capacity of 1800 kg per 24 h. (Photograph courtesy of Autec Ltd.)

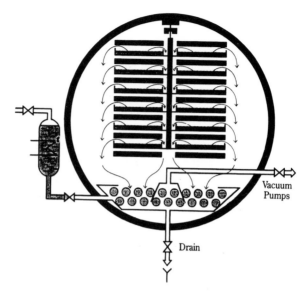

Figure 7.12 Ice condenser for a small batch freeze dryer. (Drawing courtesy of Atlas Industries A/S.)

For small batch plants, defrosting of the ice condenser is carried out after aeration and unloading. The condenser is flushed with pre-heated water. The ice is completely melted after approximately 10 min and the water is drained. Water-flushing is ideal for smaller systems, giving minimum investment cost and simple de-icing. A typical arrangement is shown in Figure 7.12.

Continuous operation requires that the equipment does not have to be stopped for defrosting when the ice condenser is full. Two or more ice condensers in separate compartments are fitted so that each in turn can be isolated and defrosted whilst the other continues to condense water vapour. Figure 7.13 shows a patented double condenser which operates as follows. During de-icing, vapour at 15°C from the de-icing vessel condenses at the cold condenser surface, thus melting the ice. Thereafter, to restore the condenser to operating conditions, the condenser chamber is closed off from the de-icing vessel. The condenser is then cooled to operating temperature, resulting in condensation of any remaining vapour. As the vapour condenses, the pressure decreases until the operating vacuum is achieved. This means that there is no loss of operating vacuum at switch-over between vapour condenser chambers. The system ensures a maximum ice build-up of only 2–3 mm, which results in a low temperature drop over the ice and low energy consumption in the refrigeration plant, constant condenser capacity and high freeze drying capacity per square metre of tray surface. Figure 7.15 shows a modern continuous freeze dryer with a capacity of over 20 000 kg wet matter per 24 h with 28% solids.

7.5 The cost of freeze drying

It is often thought that the cost of freeze drying is dominated by the cost of the energy used. Clearly there are considerable variations between different installations drying different foods. However, the following simplified example shows that amortization of capital is likely to be equally significant. Labour cost may also be significant if extensive pre-preparation is required.

Assume a freeze dryer with a useful shelf or platen area of 50 m², used for drying meat products for 200 days per year. With a product depth of 15 mm, 675 kg can be loaded, allowing for 10% of area being unused. If the equipment depreciates over 10 years, a total of 1 350 000 kg will have been processed during its life. If the capital cost was US$ 250 000, the amortization cost would be 0.18 US$ per kg of wet product. This compares with an energy cost of around

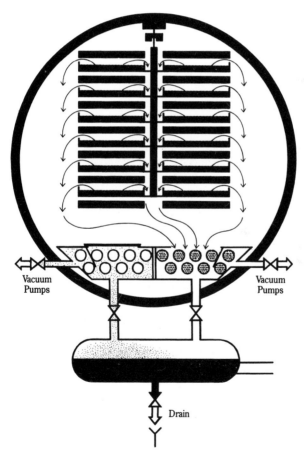

Figure 7.13 Double ice condenser for continuous operation. (Drawing courtesy of Atlas Industries A/S.)

0.16 US$ per kg of water removed. Most meats contain about 80% water before drying, so the energy cost per kg of wet product will be 0.20 US$ per kg, which is of the same order as the cost of capital amorization.

7.6 Freeze drying procedures

The following paragraphs give processing conditions for a few freeze dried foods. The details described are approximate guide figures suitable for initial experiment. Actual process details for any one product depend heavily on product quality, pre-treatment methods, and freezing protocol.

7.6.1 Coffee and tea

Whilst the best known application is for freeze dried coffee, freeze dried tea is also becoming widespread, especially for use in vending machines. Both coffee and tea are brewed, ideally with a solids concentration around 10%. The liquor is dispensed into trays typically to a depth of 30 mm. It is then frozen to below −20°C, before being loaded either onto shelves at 80°C or between radiating platens at about 120°C. Drying to 3% residual moisture takes about 4 h.

Figure 7.14 Heating platens of a small batch freeze dryer, type Ray. (Photograph courtesy of Atlas Industries A/S.)

7.6.2 Raw meat products

There is only limited potential for freeze dried raw meat. It is used by campers and mountaineers along with other freeze dried products. Carefully prepared freeze dried beef steaks or pork chops are difficult to distinguish from frozen samples after cooking. After freeze drying, good quality meat can still be distinguished from an inferior product by taste. However, meat samples prepared and freeze dried in facilities without proper equipment and using incorrect methods can be of very poor quality. Raw meat pieces cut by hand from unfrozen meat and then frozen are slow to dry, have poor appearance, and do not readily rehydrate.

It must be acknowledged that much freeze dried raw meat is not comparable to fresh meat and is tough and dry with a poor flavour. It is difficult to prepare boneless steaks and chops of a uniformly good quality suitable for freeze drying at a reasonable cost on a large scale. However, if the meat is prepared properly and excessively high temperatures are avoided, an excellent freeze dried product can be made. To ensure a reasonably short drying cycle, slices should be cut no thicker than about 15 mm.

The best cut for freeze drying is the loin, which is carefully trimmed, laid out flat, then moulded to fit the final container, and frozen. The steaks and chops are then cut to a thickness of 12 to 15 mm on a band saw. If the slices are cut by

Figure 7.15 Large continuous freeze dryer, type Conrad, before installation. Double moveable racks are shown. At the bottom of the chamber a double ice condenser allows continuous operation. (Photograph courtesy of Atlas Industries A/S.)

hand from unfrozen meat, the fibres may be folded or blocked at the surface by a gelatinous barrier and rehydration is hindered.

For beef to be acceptable, an enzymic tenderizer must be added to the water used for rehydration and a separate packet containing papain and salt in addition to seasonings should be provided. Tenderization occurs during cooking.

7.6.3 Cooked meat products

Freeze dried cooked meats are used in dry soups and dehydrated prepared dishes. They have been produced on an industrial scale since the 1950s. The largest market is for cooked chicken. The quality of freeze dried cooked meats is variable. It is often criticized as being relatively tougher than canned, frozen or freshly cooked meat and is sometimes found to be dry. Lack of flavour is another criticism but this is inevitable when, for instance, meat is cooked and drained free from fat and liquor before dehydration.

7.6.4 Beef

A relatively long high temperature cook is preferred. The drying conditions are not critical. Meat dried to a final temperature exceeding 60°C, sufficient to cause browning, is acceptable, for browning can produce a pleasant 'roasted' flavour in dried beef. Freeze dried cooked beef products are produced for commercial uses and for military rations. They include diced beef, meat balls, oxtail meat and cured smoked beef slices.

7.6.5 Cooked pork and lamb

The quality of dried pork and lamb depends more on freeze drying temperatures than on cooking conditions. Roasted and water-cooked sliced loins should be limited to a maximum temperature of 50°C to avoid browning and to avoid a pronounced reduction in the tenderness of the meat.

7.6.6 Cooked chicken meat

Chicken is processed in larger volume than any other meat. Some turkey is also freeze dried on a commercial scale. Both are used as ingredients of dehydrated soups, stews and casseroles.

Carcasses should be fully cooked, then boned, diced and dried. Great care must be taken to ensure that repeated handling after cooking does not allow meat microbiological growth. However, if good commercial sanitary practices are used, the normal bacteriological requirements may be met.

The carcasses are simmered in an open kettle until the meat can be separated from the bones. The maximum cook time for whole carcasses is sometimes limited to 2.5 h. Prolonged cooking makes it increasingly difficult to cut a clean dice from the meat with a minimum of fines.

After cooking, it is normal to drain the carcasses free from cooking liquor and rendered fat before chilling and boning. Sometimes the liquor from the cooking process is concentrated and combined with the meat prior to freeze drying.

7.6.7 Offal

Only limited quantities of offal are processed, usually for pharmaceutical purposes. Slices of raw liver can be dried but rehydrate slowly and unevenly. Freeze dried kidneys are used as ingredients in combined dishes and military rations. Minced chicken livers are used to prepare pastes.

7.6.8 Shrimp

The pink varieties of shrimp are preferred to the brown for freeze drying. After cooking for 3 min, the shrimp is placed in the drying chamber in trays between platens. Best results are obtained with the platens at 52°C, which gives a drying time of 10 h. Higher drying temperatures give a product which dries in 8 h but rehydrates less well. The product quality is much lower at the higher temperature.

7.6.9 Egg

Liquid egg is mixed with stabilizers to a thick cream. The mix is then filled into trays to a depth of 10 mm and the product temperature reduced to −25°C. When this temperature is reached, cooling is stopped. Evacuation continues to less than 0.7 mb (0.07 kPa); then the platens are heated. The complete drying stage is of the order of 9–10 h, and the egg is left in the trays in the form of a crystalline porous cake. The egg is broken up and then sieved to a powder and packed in sacks (multi-walled paper with separate polythene liners).

7.6.10 Fish

Pre-cooked salmon steaks 13 mm thick dry in about 9 h with a platen temperature of 100°C.

7.6.11 Mushrooms

White varieties are preferred for freeze drying; pigmented varieties tending to shrink more. They should be peeled, sliced and blanched before being loaded into trays. With a platen temperature of 82°C, a drying time of 3 h is possible.

7.6.12 Strawberries

These should be sliced then loaded direct into trays, and dried with a platen temperature of 50°C. A drying time of 2 h is possible.

7.6.13 Milled vegetables

For use in soups and sauces, milled vegetables with a maximum dimension of 0.5 mm can be dried very quickly. With platens at 50°C and a pressure of 0.3 mb, the vegetables can be dried in a few minutes. Large particles require longer drying times. Good results have also been achieved with wheat porridge and rice pudding (Rao *et al.*, 1994).

7.6.14 Difficult materials

While no aqueous material cannot be freeze dried, some are practically impossible because of the extreme conditions or difficult formulations necessary, or because they have poor retention of organoleptic properties. Fruit juices and foods containing high concentrations of sugars will suffer from collapse. It should also be noted that fats cannot be freeze dried and, consequently, any product with a high fat content will go rancid.

7.7 The packaging of freeze dried food

Freeze dried foods have the advantage of being light in weight, so to preserve this advantage, the packaging should also be light in weight. It must also protect the food from moisture and oxygen. Light often affects the colour of the product adversely so opaque materials are preferred.

A freeze dried food will absorb moisture rapidly and must therefore be isolated from the atmosphere if dryness is to be maintained. Also, foods containing fats develop rancidity if they are exposed to oxygen. The package is usually flushed with an inert gas or vacuum packed to prevent exposure to the atmosphere.

The light, porous nature of freeze dried foods makes them rather fragile in many cases, needing protection against crushing and abrasion. Sometimes tissues are hard and sharp, and the package may be punctured. The special characteristics of freeze dried foods must be considered in determining the most appropriate packaging. It is also necessary to consider the shelf life of the product, the conditions under which it will be transported and stored, and the portion size. The following types of containers have been used successfully.

1. Laminated foil and plastic pouches exclude water, oxygen and light and are inexpensive. They are light in weight and easy to open. Their disadvantage is that they give no protection against crushing.
2. Aluminium cans provide the best protection. They combine the qualities of the foil and plastic package with superior protection against crushing, and against the puncturing or tearing of the package by sharp pieces of freeze dried material.
3. Plastic-lined paperboard cartons are useful for products which are used in small quantities and need to be resealed. Bulk quantities can be packed in

plastic sacks, which have been coated with material to reduce moisture and oxygen transmission. For material subject to crushing, these can be protected by fibreboard drums or boxes. Plastic barrels can also be used but are expensive.

7.8 Conclusion

Food freeze drying is an established process for niche markets, especially where transport is difficult.

References

Bruttini, R., Rovero, G. and Baldi, G. (1991) Experimentation and modelling of pharmaceutical lyophilisation using a pilot plant. *Chem. Eng. J. Biochem. Eng. J.*, **45**(3), B67–B77.

Cohen, J.S. and Yang, T.C.S. (1995) Progress in food dehydration. *Trends Food Sci. Tech.*, **6**(1), 20–5.

Coumans, W.J., Kirkhof, P.J.A.M. and Bruins, S. (1994) Theoretical and practical aspects of aroma retention in spray drying and freeze drying. *Drying Tech.*, **12**(1–2), 99–147.

Franks, F., Hatley, R.H.M. and Mathias, S.F. (1991) Materials science and the production of shelf-stable biologicals. *BioPharm.*, **3**(5), 38–55.

Greaves, R.I.N. (1966) *Improved Freezing Method*, British Patent No. 1,026,877.

Kumagi, H., Nakamura, K. and Yano, T. (1991) Analysis of freeze drying of a model system by a uniformly retreating ice front model. *J. Agric. Biol. Chem.*, **55**(3), 731–42.

Lombrana, J.I. and Villaran (1993) Kinetic modelling of sublimation and vaporisation in low-temperature dehydration processes. *J. Chem. Eng. Japan*, **26**(4), 389–94.

MacKenzie, A.P. (1976) *The Physico-chemical Basis for the Freeze Drying Process*. International Symposium on Freeze Drying of Biological Products, Washington, DC. Published in *Developments in Biological Standards*, **36**, 51–7.

Oatjen, G.W. (1966) Freeze drying of food products, in *Freeze Drying Lyophilisation* (ed. D. Rey), Hermann, Paris, p. 172.

Pikal, M.J. (1990) Freeze drying of proteins, Part I. *BioPharm.*, **3**(8), 18–27.

Pikal, M.J. (1990) Freeze drying of proteins, Part II. *BioPharm.*, **3**(9), 26–30.

Rao, D.V., Radhakrishna, K., Jayathilakan, K. and Vasundhara, T.S. (1994) Manufacture of freeze-dried breakfast and dessert foods. *J. Food Sci. Tech.–Mysore*, **31**(1), 40–3.

8 Dielectric dryers

P.L. JONES AND A.T. ROWLEY

8.1 Introduction

Radio-frequency (RF) and microwave heating (both of which are often referred to as dielectric heating) have been used for industrial drying applications for many years, and have a number of advantages over other, more conventional, drying processes. These advantages include: the volumetric dissipation of energy throughout a product, and, in certain cases, the ability automatically to level out moisture variation. However, dielectric heating has its limitations, which are sometimes technical, but more often economic in nature. It therefore can be regarded as an option which can only be justified in certain, special circumstances. These usually are improvements in product quality, increased throughput or the need for compact equipment. It is generally associated with the processing of modest quantities of materials, which have a relatively high value. When combined with other, more conventional, drying techniques, the range of potential applications is broadened considerably.

8.2 The fundamentals of dielectric heating

The term dielectric heating is equally applicable to radio-frequency or micro-wave systems. In both cases, the heating results from the fact that a dielectric insulator (or a material with a small, but finite, electrical conductivity) absorbs energy when it is placed in a high frequency electric field.

Radio- and microwave frequency bands occupy adjacent sections of the electromagnetic spectrum, with microwaves having higher frequencies than radio waves. To the lay person the distinction between the two frequency bands is blurred. For example, some applications at around 900 MHz are referred to as radio-frequency (cellular telephones), and some as microwaves (dielectric heating). However, the technology used to generate and transmit the high frequency electric fields can be used to distinguish them. In RF systems, this is based on high power electrical valves, transmission lines, and applicators in the form of capacitors; microwave systems, on the other hand, use magnetrons, waveguides and cavities.

There are internationally agreed and recognized frequency bands which can be used for RF and microwave heating. These are known as the ISM (industrial, scientific and medical) bands and are defined in Table 8.1. EMC (electro-magnetic compatibility) regulations set limits on any emissions outside these

Table 8.1 The ISM bands

Band	Frequency (MHz)
Radio-frequency	13.56 ± 0.00678 27.12 ± 0.16272 40.68 ± 0.02034
Microwave	~900 (depending on country) 2450 ± 50

bands and, in most countries, compliance is a legal obligation. These limits are substantially lower than those relating to health and safety.

8.2.1 Radio-frequency dryers

There are two distinct methods for producing and transmitting RF power to dryer applicators: 'conventional' equipment, and the more recently introduced '50 ohm' equipment. Although conventional RF equipment has been used successfully for certain applications over many years, the tightening of EMC regulations and the desire for improved process control are leading to the introduction of the 50 Ω technology.

(a) 'Conventional' RF equipment. In a conventional system, the RF applicator (for example the drying chamber) forms part of the secondary circuit of a transformer, which has the output circuit of the RF generator as its primary circuit. Consequently, the RF applicator can be considered as part of the RF generator circuit, and is often used to control the amount of RF power supplied by the generator. In many systems, a component in the applicator circuit (usually the RF applicator plates themselves) is adjusted to keep the power within set limits. Alternatively, the heating system is set up to deliver a certain amount of power into a standard load under known conditions, and then allowed to drift automatically up or down as the condition of the product changes. In virtually all conventional systems, the amount of RF power being delivered is only indicated by the direct current flowing through the high power valve (usually a triode) within the generator.

A typical conventional RF heating system is shown schematically in Figure 8.1.

(b) 50 Ω RF heating equipment

50 Ω equipment is very different, and immediately recognizable by the fact that the RF generator is physically separated from the RF applicator by a high power coaxial cable (Figure 8.2). The operational frequency of a 50 Ω RF generator is controlled by a crystal oscillator and is essentially fixed at exactly 13.56 MHz or 27.12 MHz (40.68 MHz is seldom used). Once the frequency has been fixed, it is relatively straightforward to set the output impedance of the RF generator to

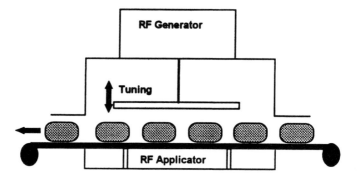

Figure 8.1 Components of a conventional RF heating system.

a convenient value; 50 Ω is chosen so that standard equipment such as high power cable and RF power meters can be used. For such a generator to transfer power efficiently, it must be connected to a load which also has an impedance of 50 Ω. Consequently, an impedance-matching network, which transforms the impedance of the RF applicator to 50 Ω, has to be included in the system. In effect, this matching network is a sophisticated tuning system, and the RF applicator plates themselves are normally fixed to suit the demands of the material being processed.

The advantages of 50 Ω systems are:

- the crystal control of the frequency makes it easier to meet the EMC regulations;
- the coaxial cable allows the RF generator to be sited at a convenient location away from the RF applicator;
- because it is not a part of the tuning system, the applicator can be designed for optimum performance;

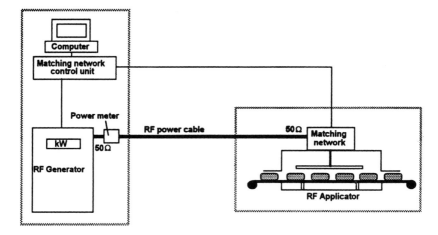

Figure 8.2 Components of a typical 50 Ω dielectric heating system.

- the use of a matching network gives rise to the possibility of an advanced process control system.

The nature of the design and construction of 50 Ω systems means that the dryer can be designed for optimum performance without concern about the inclusion of variable components in the applicator. For example, the applicator plates can be fixed in position in such a manner that the air gaps are kept to a minimum. This, in turn, means that the excessive voltages which lead to electrical breakdown are avoided.

(c) RF dielectric heating applicators

Whether conventional or 50 Ω dielectric heating systems are used, the RF applicator has to be designed for the particular product being dried. Although the size and shape of the applicator can vary enormously, there are three main types: the through-field applicator, the fringe- or stray-field applicator, and the staggered through-field applicator.

'Through-field' applicator. This applicator is the simplest, and the most common, with the electric field originating from a high frequency voltage applied across the two electrodes which form a parallel plate capacitor as shown in Figure 8.3. This type of applicator is used mainly for relatively thick products.

'Fringe- or stray-field' applicator. In this case, the product passes over a series of bars, rods or narrow plates which are alternately connected to either side of the RF voltage supply (Figure 8.4). The major advantage of this configuration is that the product makes complete contact across the bars, and there is no air gap between the RF applicator and the product. This ensures that there will be a virtually constant electric field in the material between the bars (an important requirement to maximize moisture levelling performance). It also reduces the electric field that has to be applied between the electrodes in order to achieve a given rate of heat transfer. Its major disadvantage is that only relatively thin products (typically paper) can be processed.

'Staggered through-field' applicator. For intermediate thickness products, such as biscuits, a bar applicator is again used, but this time the bars are

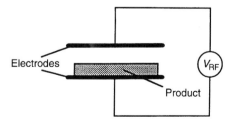

Figure 8.3 Through-field RF applicator.

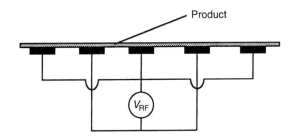

Figure 8.4 Fringe-field applicator.

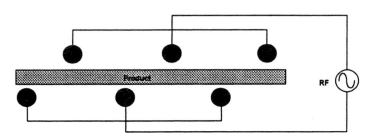

Figure 8.5 Staggered through-field applicator.

arranged above and below the product (Figure 8.5). This arrangement reduces the overall capacitance of the applicator which, in turn, makes the system tuning easier. It also reduces, slightly, the voltage that has to be applied across the electrodes to produce a given RF power density within the product.

8.2.2 Microwave heating systems

Like RF equipment, microwave units consist of a high frequency power source, a power transmission medium, a tuning system, and an applicator (Figure 8.6).

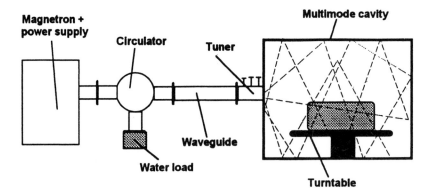

Figure 8.6 A typical microwave heating system.

In this case the power source is usually a magnetron. At 2450 MHz, magnetrons are available with output powers of between 500 W and 15 kW. In the 900 MHz band, magnetrons are usually rated between 25 and 80 kW. Unlike conventional RF generators, the power produced by a magnetron is independent of the state of the load.

The magnetron transfers the power to the rest of the system through an antenna. The electromagnetic waves are transmitted along waveguides (metal pipes) connected to the cavity (applicator). In some applications, the waveguides themselves can form the applicator.

A device known as a circulator is inserted between the magnetron and the load to prevent any power that is reflected back from the applicator reaching the magnetron (which could damage it). The circulator is basically a one-way valve which allows forward power to travel to the applicator, but deflects any reflected power to, for example, a water load. Some form of tuning device (e.g. a stub tuner) is usually inserted between the waveguide and the cavity.

(a) Microwave dryers.   There are many different types of microwave applicators, a few of which are described in the following sections.

Multimode cavity.   The most common form of microwave applicator, which is used in the domestic microwave oven, is the multimode cavity (in essence, a closed metal box). A number of 'standing wave patterns' or 'modes' are established when microwaves enter the cavity. In order to provide some uniformity of heating to the materials in the cavity, there has to be relative movement between the product and the electric field pattern as a result of these standing waves. This can be achieved through the use of a turntable or a device called a mode stirrer or, often, both.

'Leaky waveguide' applicator.   In this applicator, microwaves are allowed to leak in a controlled way from holes or slots cut into the side of the waveguide (Figure 8.7). The product, usually relatively thin, passes over the top of the slots. Such an applicator can be set up to give a relatively uniform power density distribution over product widths of up to 3 m at 2.45 GHz. It is only suitable for treatment of moving products, otherwise local overheating would occur. In order to prevent microwave leakage to the general environment, both the waveguide

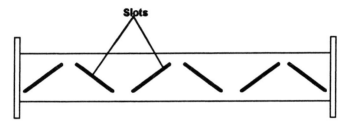

Figure 8.7 A leaky waveguide microwave applicator.

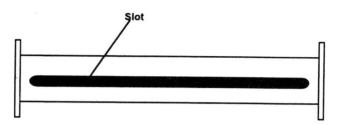

Figure 8.8 A slotted waveguide microwave applicator.

and the section of product being treated have to be enclosed in a metal box (Faraday cage).

'Slotted waveguide' applicator. This applicator is again used mainly for thin products. In this case, the product is drawn through a slot running down the centre of the waveguide (Figure 8.8). In some applications, the waveguide is doubled back on itself several times to form what is known as a serpentine or meander applicator. In this way the applicator gives a much more uniform distribution of heating when the whole length of the product is considered. This type of applicator also has to be enclosed in a Faraday cage.

8.2.3 Safety of dielectric heating equipment

Since the human body is made up of dielectric materials, it will absorb power at both RF and microwave frequencies. RF and microwave radiation have too little energy to cause direct ionization and, therefore, only the thermal effects need to be considered as hazardous. Because of their different dielectric properties, different parts of the body will absorb varying amounts of energy for a given applied electric field strength. The measurement and interpretation of this type of energy are difficult.

Power absorbed is expressed in terms of a specific absorption rate (SAR), which is the power density required (in watts per kilogram) to induce a rise in temperature in different parts of the body. This turns out to be in the range of 4 to $8 \, W \, kg^{-1}$. The safety limits are then set to be a certain fraction of this, typically 0.1, i.e. $0.4 \, W \, kg^{-1}$. In order to arrive at a practical limit which can be measured, this SAR has then to be converted to an equivalent power density (determined from electric and magnetic field strengths): typically $10 \, W \, m^{-2}$ for both RF and microwave radiation. RF and microwave heating equipment has to be constructed so that any leakage is below this value.

8.3 Dielectric drying

8.3.1 The power dissipated within a dielectric

Dielectric drying relies on the principle that energy is absorbed by a wet material when it is placed in a high frequency electric field (RF or microwave).

This energy is principally absorbed by the water present in the material, causing the temperature to rise, some water to be evaporated, and the moisture level to be reduced. Calculation of the actual amount of energy (or power) absorbed by a dielectric body is essential to a full understanding of radio-frequency and microwave drying (Jones and Rowley, 1996).

(a) RF heating. In essence, all applicators used for RF dielectric heating are capacitors. These capacitors can be represented by a complex electrical impedance, Z_c, or the equivalent complex electrical admittance, Y_c ($=1/Z_c$). When empty, an ideal applicator has an impedance which is purely reactive with zero electrical resistance, and no power is dissipated when a RF potential is applied across it. In the absence of a dielectric, the complex impedance of the applicator is given by

$$Z_c = 0 - \frac{\mathbf{j}}{\omega C_0} \qquad (8.1a)$$

and its equivalent admittance is

$$Y_c = 0 + \mathbf{j}\omega C_0 \qquad (8.1b)$$

where $\omega = 2\pi f$ (f = frequency) and C_0 is the capacitance of the empty applicator.

The relative permittivity of a dielectric, ε_r (sometimes called the complex dielectric constant), is given by $\varepsilon_r = \varepsilon_r' - \varepsilon_r''\mathbf{j}$. ε_r' is the dielectric constant and ε_r'' is the dielectric loss factor of the material. If a simple parallel plate capacitor is filled with such a dielectric, then the new admittance is given by

$$Y_c' = \varepsilon_r Y_c = \omega C_0 \varepsilon_r'' + \mathbf{j}\omega C_0 \varepsilon_r' \qquad (8.2)$$

and the corresponding new impedance ($=1/Y_c$) is then

$$Z_c' = \frac{1}{\omega C_0} \frac{\varepsilon_r'' - \mathbf{j}\varepsilon_r'}{\varepsilon_r''^2 + \varepsilon_r'^2} \qquad (8.3)$$

Inspection of equations 8.2 and 8.3 reveals that the impedance of the RF applicator capacitance has been altered in two ways due to the presence of the dielectric (Figure 8.9):

- a finite resistance, R ($=1/\omega C_0 \varepsilon_r''$), has appeared across the capacitor;

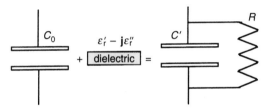

Figure 8.9 The effect of a dielectric on a capacitor.

- the new effective capacitance, C', is greater than the capacitance without the dielectric (C_0) by a factor ε_r' (by definition, ε_r' is always greater than 1).

The increase in capacitance arises from changes in the distribution of electric charge within the RF applicator, and the presence of the resistance gives rise to the possibility of heat generation within the dielectric.

Taking the power, P, dissipated in a resistance to be equal to V^2/R, then for a capacitor containing a dielectric

$$P = \omega \varepsilon_r'' C_0 V^2 \tag{8.4}$$

For a parallel plate capacitor, $C_0 = \varepsilon_0 A_p/d$ where A_p is the plate area, d is the plate separation and ε_0 is the permittivity of free space. The voltage V is equal to the product of the electron field strength $\mathbf{E}$ and d. Equation 8.4 can now be rewritten as

$$P = \omega \varepsilon_0 \varepsilon_r'' \mathbf{E}^2 (A_p d) \tag{8.5}$$

Since $(A_p d)$ is the volume, the power dissipation per unit volume or power density, P_v, is then

$$P_v = \omega \varepsilon_0 \varepsilon_r'' \mathbf{E}^2 \tag{8.6}$$

The power density is proportional to the frequency of the applied electric field and the dielectric loss factor, and is proportional to the square of the local electric field. This equation is crucial in determining how a dielectric will absorb energy when it is placed in a high frequency electric field. For a given system, the frequency is fixed and therefore ω and ε_0 are constants, and the dielectric loss factor ε_r'' can, in principle, be measured. The only unknown left in equation 8.6 is the electric field, $\mathbf{E}$. To evaluate this, the effect of the dielectric on the applied electric field (due to the RF voltage across the RF applicator) must be considered.

(b) Microwave heating. In the case of microwave dielectric heating, the applicator can no longer be considered to be a simple capacitor, and the electric field in the material is now that due to a propagating electromagnetic wave of the form

$$\mathbf{E} = \mathbf{E}_0 \, e^{j(\omega t - kz)} \tag{8.7}$$

where k is the propagation constant in the z direction and t is the time.

The displacement current, $\mathbf{J_D}$, flowing through the dielectric media is defined by

$$\mathbf{J_D} = \varepsilon_0 \varepsilon_r \frac{\partial \mathbf{E}}{\partial t} \tag{8.8}$$

which, in combination with equation 8.7, becomes

$$\mathbf{J_D} = j\omega \varepsilon_0 \varepsilon_r \mathbf{E} \tag{8.9}$$

Substituting $\varepsilon_r = \varepsilon_r' - j\varepsilon_r''$ gives

$$\mathbf{J} = \mathbf{J_D} = \omega\varepsilon_0\varepsilon_r''\mathbf{E} - j\omega\varepsilon_0\varepsilon_r'\mathbf{E} \qquad (8.10)$$

Now, consider a small volume element of dielectric, dv, of cross section dA_x and length dz. The voltage drop across this is then $\mathbf{E}.dz$, and the current through it is $\mathbf{J}.dA_x$. The power dissipated per unit volume is then

$$\frac{dP}{dv} = \langle\mathbf{E}.\mathbf{J}\rangle \qquad (8.11)$$

where $\langle\cdots\rangle$ represents the time average.

If ε_r is real (i.e. $\varepsilon_r'' = 0$) then $\mathbf{E}$ and $\mathbf{J}$ will be always $\pi/2$ out of phase, and dP/dv will be equal to 0 at all times. Otherwise,

$$\frac{dP}{dv} = \frac{1}{2}\omega\varepsilon_0\varepsilon_r''\mathbf{E}^*.\mathbf{E} \qquad (8.12)$$

$\mathbf{E}^*$ is the complex conjugate of $\mathbf{E}$. In the special case when $\mathbf{E}$ can be assumed constant throughout the product, equation 8.12 reduces to

$$P_v = \omega\varepsilon_0\varepsilon_r''\mathbf{E}_{\mathbf{rms}}^2 \qquad (8.13)$$

which is the same as that derived for the RF dielectric heating case (equation 8.6).

8.3.2 The effect of a dielectric on an electric field

A dielectric material consists of an assembly of a large number of microscopic electric dipoles which can be aligned, or polarized, by the action of an electric field. For an evaluation of the interaction of a dielectric with an external field, it is necessary to understand the effect of this polarization.

(a) Electric dipoles and polarization. An electric dipole is a region of positive charge, separated by a small distance from a region of negative charge. Such a dipole has a dipole moment, which is a vector quantity with direction along the line from the positive to the negative charge centres. Electric dipoles can be divided into two types: (1) induced dipoles which only appear in the presence of an applied electric field (e.g. carbon dioxide molecules, and atoms); and (2) permanent dipoles which are present even in the absence of an applied electric field (e.g. the water molecules in drying applications). The polarization of a material is a macroscopic property, and is defined as the dipole moment per unit volume. In the absence of an electric field, the dipole moment of an assembly of induced dipoles is zero. Although permanent electric dipoles always possess a dipole moment, in the absence of an applied field, these moments are randomly oriented in space and the polarization is again equal to zero.

A macroscopic polarization is also possible because of space charge build up at boundaries within the material. Any such separation of negative and positive

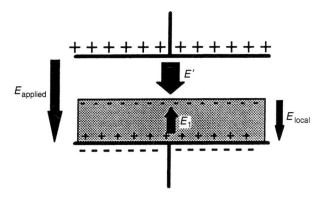

Figure 8.10 The electric fields within an RF applicator.

charges leads to a dipole moment for the whole material (sometimes known as the interfacial polarization).

It is principally the polarization of a dielectric that determines the electric field inside (and outside) the material, and hence the heating rate (the absorbed power density is proportional to the square of the electric field inside the material (equations 8.6 and 8.13)). The fact that the magnitude of the local electric field is smaller than the applied electric field is important. The effect of a dielectric on the electric fields that exist within a RF applicator is shown in Figure 8.10.

Although the local electric field is less than the applied electric field, the electric field in any air gaps surrounding the dielectric, E', is larger than the applied field. This is caused by the development of charge on the surface of the dielectric. In fact, where the surrounding medium is air, E' is approximately equal to $\varepsilon_r' E_{applied}$ and, since ε_r' is always greater than 1, E' is always greater than E_0.

8.3.3 Variation of dielectric properties with moisture content

The material to be dried can be considered to be made up of the dry skeleton solid plus water. The water exists in two forms: free water, which is found in voids, pores and capillaries; and bound water, which is chemically combined with the material. In order to determine how such a material will dry in a dielectric heating application, it is important to have information regarding the dependence of ε_r'' upon:

- moisture content, X;
- temperature, T (the water must first be heated before it can evaporate);
- fibre orientation to electric field (e.g. for textiles, wood, paper).

In both microwave and RF drying applications, the dielectric loss factor of the free and bound water is greater than that of the dry material, and heating is primarily due to energy being absorbed by the water molecules. However, the mechanism which causes energy to be absorbed is fundamentally different for the two frequency bands as follows.

1. At radio-frequencies, energy is absorbed due to the presence of ions in the water, which give it a significant electrical conductivity, σ. Heat is generated as a result of the passage of electrical current through the water in the same way as normal resistance (or ohmic) heating. In this case, ε_r'' is an effective dielectric loss factor, given by $\varepsilon_r'' = \sigma/\omega\varepsilon_0$.
2. At microwave frequencies, energy is principally absorbed by a resonant (frequency dependent) dipolar loss mechanism, and heat is generated due to the rotation of the water molecules being out of phase with the applied electric field.

The total dielectric loss factor of a wet body is derived from that of the skeleton solid, and the bound and the free water (Struchley and Hamid, 1972). Provided the free water is present in sufficient quantity, it largely determines the effective value of the loss factor. As the solid dries, the contributions from the bound water and the solid, although small in absolute terms, start to become important.

Bound water has a loss factor somewhere between that of ice ($\sim$0.05 at RF and $\sim$0.005 at microwave frequencies) and that of free water ($\sim$30 at RF and $\sim$10 at microwave frequencies); the loss factor for many solids can be described in terms of a broad distribution of dipoles. The effective loss factor of an ideal non-hygroscopic solid can be estimated from its constituents by various mixture theories (Tinga and Voss, 1973; Kraszewski, 1977). As a rough guide, the contribution from each component is proportional to the fraction of the total volume which it occupies. Of course, in practice, the solid will interact, to some extent, with the water, and the resulting solid, bound water and free water combination will have a lower overall loss factor than the sum of the individual components.

(a) Variation of ε_r'' with moisture content. The dependence of dielectric loss factor on moisture content (at constant temperature) is similar for most materials. Figure 8.11 shows the variation for a typical material. There are two regions of importance here.

1. In the region OA, ε_r'' is approximately proportional to moisture content. Since $P_v = \omega\varepsilon_0\varepsilon_r''\mathbf{E}^2$, the power density in this region will be directly proportional to the moisture content (given constant $\mathbf{E}$). This is an extremely important result as it means that any moisture variation throughout the product will be automatically levelled in this region, giving a final product with a very small moisture variation through it. This automatic moisture levelling represents a substantial advantage over other drying processes, and often leads to dielectric drying being chosen for a given product. This region corresponds approximately to the removal of free water from the system.

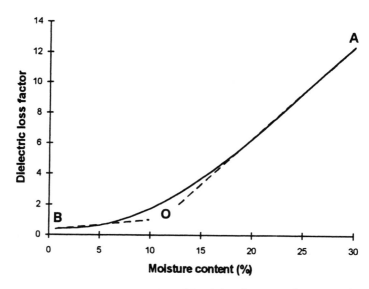

Figure 8.11 The dependence of the dielectric loss factor on moisture content.

2. In the region OB, ε_r'' is only weakly dependent on moisture content, and significant moisture levelling will not take place. This region corresponds approximately to the removal of bound water from the system. The moisture content at O is the optimum level to dry to with dielectric heating and is often quite low (around 4%). For a given material, the gradient of OA is often steeper at radio-frequencies than at microwave frequencies. For this reason, RF dielectric drying tends to be more efficient at moisture levelling than microwave drying.

(b) Variation of ε_r'' with moisture content and temperature.   For water and most dipolar liquids the conductivity, σ, increases with increasing T (relevant to RF heating); and the loss factor due to dipolar loss decreases with increasing T (at microwave frequencies). The consequence of this is shown in Figure 8.12. As the temperature increases, the line OA moves to OA′ if dipolar loss mechanisms are dominating, and moves to line OA″ if conductivity losses are dominating. Since the moisture levelling performance depends on the gradient of OA, then conductivity loss mechanisms are the most significant. This is another reason why RF dielectric heating tends to be better at moisture levelling than micro-wave dielectric heating.

8.3.4 Moisture levelling

One of the main reasons for choosing RF or microwave dielectric drying rather than an alternative, possibly more straightforward, drying technology, is the potential for automatic moisture levelling. The ability of dielectric heating to

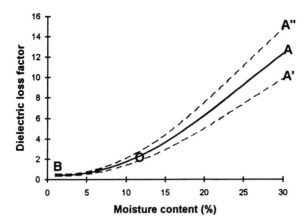

Figure 8.12 The dependence of dielectric loss factor on moisture content with varying temperature.

heat selectively areas with higher dielectric loss factors is a very important advantage of this technology.

Unlike other, more conventional, systems, dielectric drying has an inherent ability to 'moisture level' or 'moisture profile' a given product, with more power being applied to regions with higher moisture content than to regions with lower moisture content. Consequently, the final moisture content variation within a dielectric dried product is often much lower than that within a conventionally dried product, leading to an improvement in final product quality. The degree of moisture levelling possible depends principally on the variation of dielectric loss factor, ε_r'', with moisture content.

8.3.5 Heat and mass transfer within drying solids

When considering the case for dielectric drying, it is perhaps worthwhile looking at some of the constraints which are placed on conventional drying methods. The maximum temperature of the heated surface of the product, which is determined by the requirement to avoid degradation, is normally a limiting factor in the use of such techniques. This, in turn, is related to the length of time the material needs to spend in contact with the drying medium (gas or hot surface). If liquid water cannot move along the capillaries that exist in the material, drying will take place over a smaller effective surface, or along a front retreating from the surface into the interior of the material. This has implications for energy and machine usage efficiencies. This consideration applies equally well to conductive or to convective drying since, once the outer layer of the material becomes dry, its thermal conductivity falls considerably. A further consequence of premature drying of the outer layer is that the local temperature of the product will reach that of the drying medium, i.e. the dry bulb temperature. However, if its surface can be kept wet by liquid movement from the interior of the material, the surface will be maintained at the lower wet bulb temperature, which is often well below boiling point. The actual temperature depends on the moisture content and on the rate at which water is evaporating (Luikov, 1963).

(a) Moisture movement and drying rate.   The ease with which the water can move in the liquid phase depends on the nature of the structure making up the material. In truly capillary porous materials, there is, with time, a natural redistribution of moisture from within the body as the surface water evaporates. However, many materials have structures in which the pores are too big or are too discontinuous for this to take place. In other materials, the water is held in a matrix which makes liquid movement impossible.

The way in which water is held, and the nature of the capillaries or pores, leads to the characteristic drying curves of a material. Typically, when a uniformly wet material is placed in a drying environment, water will evaporate freely from the surface at an even rate per unit of time, machine length or applied energy – the constant drying rate period. Provided the loss of water at the surface can be replaced through internal liquid phase movement, this evaporation at the physical surface will continue. However, at some stage in the drying process, this will cease and drying will then take place from a retreating wet front, or on a reduced effective surface. The former implies that an increasingly thick and dry outer layer becomes established, which is a thermal insulator through which heat has to be conducted to cause any further evaporation. The latter arises from the fact that the larger capillaries cease to bring water towards the surface, and are, themselves, drained by smaller ones.

In this way, only a part of the surface is available for evaporation, and the drying rate falls. This region, the so called falling-rate period, can be very extended; drying becomes progressively more difficult as the dry outer layer increases in thickness. The increase in time, machine length or energy per unit of evaporation is a clear indication of falling efficiencies and ineffective use of capital equipment.

A fundamental aspect of conventional heat transfer is that the temperature gradients and the moisture gradients are of opposite signs, i.e. the temperature is highest at the surface where the moisture is at its lowest.

When conventional heat transfer methods are used to dry a capillary porous body at atmospheric pressure, the internal movement of moisture is due to liquid flow by capillary action and vapour flow by molecular diffusion. Liquid phase movement is related to moisture gradient and temperature, whereas in the vapour phase it is due to a partial pressure or temperature gradient (Perkin, 1983).

The transition from constant rate to falling rate is often gradual and, for some materials, it may be argued that there is no constant rate in the strict sense. Many foodstuffs are examples of such materials.

(b) Dielectric heating in thermal drying. In contrast, when dielectric heating is used for drying, the possibility of surface degradation usually ceases to be a limiting factor on the rate of heat transfer in materials that can withstand the wet-bulb temperature. The limit then becomes the rate at which liquid- and vapour-phase movement can take place through the capillaries, without causing disruption to the structure of the body. Certain materials, such as textiles, are so open structured that there is virtually no limit to the rate at which energy can be applied. However, there are many others, including foods, in which damage of some sort can occur. In such materials, when the rate of heat transferred exceeds a threshold limit, evaporation takes place more rapidly than the structure will permit diffusion to occur. In this way, pressure is built up within the body and it can be ruptured. In the food industry, it is possible to use this to advantage in causing rapid expansion (puffing) without the need to use hot oil.

The major attraction of dielectric heating arises because, as a selective and volumetric technique, the water, in most circumstances, is directly heated wherever and however it is dispersed within a body. Furthermore, the thermal conductivity of the substrate is no longer a limiting factor to the rate at which heat can be introduced to the wet areas within it. Since this form of heating does not depend on transfer through a surface, significant improvements in heating and drying rates can be achieved largely by eliminating temperature and moisture gradients and thereby, in many cases, improving product quality.

As has been shown in section 8.3.3(b), the loss factor of wet material at radio-frequencies rises with increasing temperature and, since drying takes place at elevated temperatures, this may be considered important. However, as soon as the temperature rises, water is lost by evaporation, and, in most cases, there is a self-limiting control making RF dielectric heating very effective for moisture levelling applications.

(c) Combination hot air/dielectric drying. There are many processes in which RF or microwave drying has been shown to be effective but where the economic case cannot be justified, usually because of a high capital cost. The cost of the generators cannot be expected to fall significantly because the basic components are relatively expensive and high standards of engineering are required to meet regulations concerning frequency stabilization and electromagnetic compatibility. Most dielectric drying applications use RF or microwaves as the sole means of heat transfer at some point in the drying cycle, typically after conventional (convective, radiant or conductive) heat transfer has ceased to be efficient. However, if the drying process is considered as a whole, there are a number of opportunities in which combined hot air/dielectric drying is economically viable (Jones, 1989).

Combined heat and mass transfer. Combining volumetric (dielectric) heating and surface (conventional) heating is effective if the dielectric heating causes water to move to the surface. This minimizes the effect of the retreating wet front and maintains the evaporating surface. The actual ratio of dielectric to air heating is important, and will vary greatly from one material structure to another.

The ability of the material to allow internal water movement depends critically on its structure. Some, such as pasta products, have structures too fine for significant liquid phase movement to take place without damage. Others, such as sponge- and bread-like products, have cell structures which are too large for a capillary action to be viable. In the former case combination dryers may be effective, but only at very low power densities. The large cell structures do not have the same mutuality between the surface and volumetric heating, and therefore improvement in drying time is simply a sum of both inputs.

The forcing effect of the RF heating is perhaps best envisaged if the substrate can be thought of as a bundle of very fine tubes lying at right angles to the surface. Initially these will be full of water and, under normal drying conditions, evaporation will take place at the physical surface. As drying progresses, capillary action will cause a certain amount of liquid phase movement, which will keep the water and physical surface coincidental but, eventually, the moisture will begin to retreat into the body, leaving a dry outer layer. If the RF energy is added, the water within the substrate will be heated, and a small quantity evaporated; this will drive the liquid back through the capillaries to the surface.

As described earlier, other mechanisms may be operating. Dielectric drying may be seen as overcoming the problem experienced in conventional drying, where there is a reduction of the proportion of the surface available for evaporation as the larger capillaries cease to function, and only the finer ones continue to bring liquid water to the surface. The volumetric heating, which takes place when a high frequency field is applied, will cause the surface tension and the viscosity of the water to be reduced and thus allow the larger pores to continue to provide a means of liquid transport.

Combined dielectric/convective dryers. If it is accepted that volumetric heating can bring moisture to the surface, then only a proportion of the total drying heat need be provided by the dielectric system; the bulk of the heat is required for evaporation, and can be provided at the surface. The dielectric proportion of the total input energy may be expected to be in the order of 10 to 20% of the total. This arises because it is principally providing 'sensible heat', with the latent heat coming from a conventional heat source. Radiation and conduction have been considered but both present difficulties when combining them with dielectric systems. In contrast, convective heating fulfils the drying requirements, and is compatible with the high voltages associated with dielectric heating. Thus, in most applications, dielectric heating is combined with gas-fired convective heating within the same enclosure, and volumetric and surface heat transfer take place simultaneously.

8.4 Applications of dielectric heat in drying

It is important to emphasize that dielectric heating does not necessarily give rise to a low cost drying solution. The equipment is expensive in terms of initial capital outlay, and can be expensive to run because of the cost of using electricity in the first place, and because of the inefficiency that arises in the conversion from mains frequency to the high frequency required. The established applications, and those likely to succeed in the future, involve the drying of relatively high value products at relatively modest production rates.

8.4.1 Microwave or radio-frequency drying?

It is sometimes argued that, because the absorbed power density is directly proportional to the frequency (equations 8.6 and 8.13), the much higher values of frequency associated with microwaves must be an advantage. Furthermore it is assumed that, because the dielectric loss factor due to resonant dipolar absorption is greater at microwave frequencies than at radio-frequencies, the higher frequency bands are to be preferred for operations involving the heating of water, including drying. However, as shown above, the loss factor at radio-frequencies depends on the ionic conductivity of the water in a substrate and not on dipolar absorption. In most circumstances this is high enough for the choice between the two to be made on engineering, process and economic grounds.

In general terms, RF equipment is easier to engineer into process lines, and can be made to match the physical dimensions of the up-stream and down-stream plant. In the case of microwaves, the electrical engineering of the equipment must be the dominant consideration. In particular, in a continuous process, complex arrangements may be necessary to allow the product to move in and out of the enclosure without giving rise to excessive leakage of energy. This is because the wavelengths at microwave frequencies, (e.g. 13 cm at 2.45 GHz) are very much shorter than those at radio-frequencies (e.g. 11 m at 27.12 MHz). Consequently, the design of the inlet and outlet ports is much simpler at radio-frequencies, allowing it to process a large range of material types; from the thin wide webs of the paper industry to large three-dimensional objects like textile packages.

Any survey of the dielectric dryers installed throughout industry will show that radio-frequency dominates the market. It is only when there are very special circumstances that the case for using microwaves can be justified. For example, for a given power density, the electric field strength necessary for microwave heating is less than that for RF heating (typically by factor of about 10), and the risk of an electrical discharge occurring in a reduced pressure system is much lower. This leads to the use of microwaves in the pharmaceutical and other specialized industries, where drying takes place under vacuum.

8.4.2 Microwave applications

Microwave drying applications are widely described in the literature (Schiffmann, 1987), but relatively few of them can be said to be accepted as processing standards.

(a) Vacuum drying. Some materials are damaged when they are dried at normal wet-bulb temperatures. In such cases, there is an advantage if drying takes place at the reduced boiling points associated with low pressures. This technique, which has been used for many years with conventional heating, is very slow because convection cannot take place in a vacuum and heat transfer is limited by conduction through heated plates. If dielectric heating could be used, the rate of heat transfer would be increased by a factor of approximately 10, with improved product quality. This is not possible using radio-frequency because the high electric field would give rise to glow discharges or electrical arcing within the chamber. In contrast, microwave heating can be used because, at this frequency, the electric fields required to produce the required power density are sufficiently low.

(i) Pharmaceuticals. This is a particularly good example of the use of microwaves in conjunction with vacuum drying. It has all the requirements for success: the products are of high value, and are produced in modest quantities by an industry with a good record of investment. Furthermore, the process requires the precision of control which is achievable by dielectric heating.

During the course of manufacture, ingredients are mixed with water, ethanol, or acetone, and must, subsequently, be dried. Because of the nature of active pharmaceutical materials, any drying must take place at low temperatures. For quality control reasons, legislation requires that manufacture must be carried out in indentifiable batches in sealed stainless steel vessels. These batch-drying vessels used by the industry can, with minimal re-design, become viable microwave multi-mode cavities. Furthermore, the same vessels may combine mixing, granulating, lubricating, and dry sizing in a single step. The principal change required to the design is the insertion of a microwave transparent window compatible with the hygiene and vacuum conditions. Systems as large as 1200 l, requiring 36 kW of microwave power, are in use.

(ii) Food applications. Similar vacuum drying equipment, sometimes with a continuous feed, has been used for a range of temperature sensitive, high value food products such as citrus fruit concentrates. The material, usually in the form of a paste or slurry (often with a higher solids content than can be handled by spray dryers), is dosed onto a conveyor belt which is totally contained within a vacuum vessel operating at pressures of about 2–3 kPa. During drying, the product forms a foam-like structure which has excellent re-hydration properties.

Like all aspects of dielectric heating, commercial success, in the sense of many installations, is only likely to come when the criteria of special properties, modest throughput and relatively high value products are met. For example microwave/vacuum drying of cereals is hard to justify whereas, for fruit juices, there is a case which can be made. The Macdonnell Douglas 'MIVAC' is an example of a continuous processing plant aimed at cereals, soya etc., which, whilst giving excellent technical results, is hard to justify in the commercial sense because of the shear volume of these products.

(b) Atmospheric drying.   Although the literature describes many potential uses for microwave energy in drying at normal pressures, there is little evidence of it in common use for drying in any one particular industry. The exception to this general observation is in the drying of pasta, where a significant number of microwave units have been installed. These systems make use of the volumetric heating effect of microwaves, together with heated high humidity air, to ensure that case hardening and subsequent cracking do not take place. The result has been a reduction in drying time from 8 h to 1 h. A typical dryer handles approximately 1500 kg of product per h using 60 kW of microwave power. The resulting product has reduced microbial activity and insect infestation (Maurer *et al.*, 1972). Other microwave drying applications in the industry have been reported. They include the drying of onions, seaweed, and potato products (Schiffman, 1973). Currently, work is being supported by the European Union on broadening the applications of microwave and radio-frequency drying to a range of fruit and vegetables. The two technologies are being compared, with SIK of Sweden working on microwave techniques and EA Technology in the UK working on RF.

As the evaporative loads are quite large, dielectric techniques can only be justified when used in combination with other heat sources. The work at EA Technology has shown that RF drying in combination with heated air does not give a satisfactory quality because case hardening still occurs. It has been shown that, by controlling the relative humidity as well as the temperature, the quality aspect can be addressed.

8.4.3 Radio-frequency applications

Although the most rapidly increasing industrial market sector for RF heating, in terms of number of installations, is in drying, the greatest use of RF dielectric heating is still the welding of PVC for the manufacture of consumer and industrial products. There are, however, a number of well established RF drying processes (Jones, 1987).

(a) RF post-baking of biscuits.   In this example, RF heating is used to improve the product quality as well as overcome the falling-rate drying. Following the formation of the dough piece, which has a moisture content of about 20%,

baking takes place on a metal conveyor band in a fossil fuel (normally gas) fired tunnel oven, up to 60 m in length. Several processes take place in the oven (such as flavour, shape and crumb development), which involve moisture loss. Drying follows the expected pattern of a retreating wet front, with the consequence that, at the end of the oven, the residual moisture is concentrated in a narrow band in the middle of each piece. The removal and/or redistribution of this is an important part of the production process, since failure to do so leads to a loss of product quality (Holland, 1974).

By adding a radio-frequency dryer (or post-baker) after the conventional oven, increased throughput is achieved at the same time as improvements in product quality. The quality improvement arises from the elimination of unevenness in moisture across the conveyor band. This moisture profile correction is an important feature when total uniformity of product quality (including weight) is required for packing machine operation and for marketing purposes. Equipment manufacturers claim, and users confirm, that increases in throughput of between 30 and 50% can be achieved. Products which can benefit from this technique include crispbreads, biscuits, crackers and cookies.

8.4.4 Application of combination dryers

There is a growing number of combination dryers, both RF and microwave based, being installed in the food industry. It is the simultaneous application of the two forms of energy which makes these units viable. In the non-food industry, the behaviour of temperature sensitive products (such as those based on natural fibres) has shown that enhanced drying can take place. This enhancement arises from the increase in air temperature that is possible when a modest percentage of dielectric energy is added. In one example, the product would suffer permanent surface degradation if the drying air temperature exceeded 90°C, but the addition of about 10% RF energy allows the air temperature to be raised to 200°C without damage occurring.

Installations in the food industry include the processing of a meat/cereal strip on a line that has a total of seven zones, each with a radio-frequency generator, together with gas-fired convection drying. As well as producing a more consistent product, the rate of drying is such that the new dryer is about one third of the length of the dryer it replaced. Very large scale units, both RF and microwave, have also been installed for potato processing. The manufacture of frozen 'French fries' benefits from a pre-processing involving RF combination drying in which about 10% moisture is removed before part-frying and freezing takes place. Microwave units with a power of 0.25 MW are being used to cook/part dry potato slices in the snack industry.

Other installations using combinations of dielectric and conventional heating include biscuit baking (i.e. more than standard post-baking). Cereal products, vegetables (and vegetable products) and meats products all are likely to provide examples of the benefits of combination processing.

8.5 Cost of RF and microwave dryers

It is true to say that if the drying operation can achieve its objectives by conventional means, it is usually the case that dielectric techniques will not be economic. Because they are based on relatively sophisticated equipment in the form of high frequency, high power generators, inevitably such equipment is bound to be expensive when compared to conventional dryers strictly in terms of evaporative capacity. The cost is likely to range from £2000 to £5000 per kW installed, depending on the complexity of the equipment and the materials of construction. However, if this is considered in the light of what dielectric drying can achieve, in particular in the falling-rate period when its drying efficiency is very high, it looks less daunting and, given the right combination of circumstances, can be justified.

8.6 Conclusions

When compared with 'conventional' drying, dielectric heating has a number of advantages, which include the following.

1. No limitation is imposed on dielectric heating compared with that imposed by the thermal conduction of heat from the surface to the centre of a product. In conventional drying, it is this conduction which limits the maximum temperature to which the surface can be exposed.
2. Proportionately more power is dissipated in regions of high dielectric loss factor. As this increases with increasing moisture content, dielectric heating is intrinsically self-regulating. This leads to the phenomenon of automatic moisture levelling and, ultimately, to higher quality final products.

The main disadvantage of dielectric heating is an economic one, with the equipment being relatively expensive to install and operate. The production capacity is also limited by the maximum amount of RF or microwave power available for drying. However, combination drying systems have made dielectric heating easier to justify for a wider range of drying and other operations in the food industry.

References

Holland, J.M. (1974) Dielectric post-baking in biscuit making. *Biscuit Industry Journal*, **6**(8).
Jones, P.L. (1987) Radio-frequency processing in Europe. *Journal of Microwave Power*, **22**(3).
Jones, P.L. (1989) Dielectric assisted drying and processing. *Power Engineering Journal*.
Jones, P.L. and Rowley, A.T. (1996) Dielectric drying, in *Advances in Drying* (eds A.S. Mujumdar and Kundra), Marcel Dekker, New York.
Kraszewski, A. (1977) *Journal of Microwave Power*, **12**(3).
Luikov (1963) Heat and mass transfer with transpiration cooling. *Journal of Heat and Mass Transfer*, **6**.

Maurer, R., Tremblay, M. and Chadwick, E. (1972) *Food Processing.*

Perkin, R.M. (1983) Drying of porous bodies with electromagnetic energy, in *Progress in Filtration and Separation*, Elsevier, Amsterdam.

Schiffmann, R.F. (1973) *Journal of Microwave Power*, **8**.

Schiffmann, R.F. (1987) Microwave and dielectric drying, in *Handbook of Industrial Drying* (ed. A.S. Chujumdar), Marcel Dekker, New York.

Struchley, S.S. and Hamid, M.A.K. (1972) *Journal of Microwave Power*, **7**(1).

Tinga, W.R. and Voss, W.A.G. (1973) *Journal of Applied Physics*, **44**, 3887.

9 Specialized drying systems

D.J. BARR AND C.G.J. BAKER

9.1 Introduction

This chapter describes several quite distinct types of dryer that are used to varying degrees in the food industry. These are the pneumatic-conveying dryer, the spin-flash dryer, the rotary dryer, the tray dryer, the tunnel dryer, and the band dryer. They are included within a single chapter only for the sake of convenience. Each has its own specialized uses. For example, pneumatic- and spin-flash dryers are well suited to drying a variety of feeds ranging from particulates to pastes. Rotary dryers, in their various guises, can be used to process medium and high tonnages of relatively free-flowing particlate materials. Band dryers are also widely used in the food industry to dry particulate material, albeit at somewhat lower production rates. At the opposite extreme, in terms of throughput, tray dryers are used for drying small individual batches of material; the drying conditions and drying time can be varied almost at will. Tunnel dryers are commonly employed to dry a range of fresh vegetables, amongst other products.

9.2 Pneumatic-conveying dryers

9.2.1 Introduction

In pneumatic-conveying dryers, a suitable particulate feed is conveyed through the dryer by a fast-moving stream of hot drying air. In the conveying process, heat from the air is transferred to the wet material, causing the moisture to be flashed off. Thus, the drying process is very rapid. The basic type of pneumatic conveying dryer, often termed a flash dryer, consists normally of a straight vertical tube of circular or square cross-section. The air and the wet solids are introduced at the bottom of this tube. The dried solids, which are collected in a suitable particle-separation device, and the exhaust air, both leave at the top.

The Ring dryer is a development of the flash dryer in which the drying tube consists of a loop or ring. As discussed later, this enables larger, slower-drying particles to be recirculated through the drying zone, thereby permitting more versatile operation.

Pneumatic-conveying dryers are popular in certain parts of the food industry for their ability to dry powdered materials rapidly. Examples include food ingredients, such as starches and proteins, and gravy powders. They are particularly suitable for processing heat sensitive materials because of the short residence time within the dryer. They also have low capital and maintenance costs, and fit easily into a small space.

Pneumatic conveying dryers are extremely versatile and have the following key features.

- They are suitable for drying materials in the size range 10–500 μm.
- The short residence time ensures minimum heat damage.
- The low hold-up of material in the dryer ensures maximum safety of operation and less risk of fire or explosion.
- The dryer can accept wet powders, filter cakes or slurries as feed.
- The dryer exhibits a high drying efficiency and product handling is gentle.
- The dryer has a small 'footprint' and fits easily into a restricted space.

9.2.2 Equipment

(a) Flash dryers. Figure 9.1 shows a typical flash-drying installation, giving a clear view of all the major components in such a system. The inlet air is filtered and passes through an air heater into the drying column. At the base of the column, moist product is introduced by the feeder (disperser) into a venturi, and drying takes place in the 2–3 s it takes for the air to carry the product through the column to the primary particle separator, which is normally either a cyclone or a bag filter. In the particular version shown in the figure, the dry material is collected in a cyclone and discharged by a rotary valve. The air is moved through the system by the main fan and last traces of product are collected in the scrubber.

There are many-large scale installations of this type, which are used for a wide variety of more simple drying applications. The flash dryer shown in Figure 9.2 features a bag filter and also shows how a back-mixing system operates. The product from the collector (bag filter or cyclone) is discharged partly to product handling and partly back to the mixer. In the latter, dry product and wet feed are mixed together to form a friable feed suitable for introducing into the drying column. The rest of the system is as described above.

Flash dryers are simple and do not occupy a lot of space. The drying process is rapid and lends itself well to materials in which most of the moisture that needs to be removed is effectively surface moisture, or free moisture that can diffuse rapidly to the exterior. Drying takes place in a matter of seconds and this allows elevated drying temperatures to be used with many products, since the flash process instantly cools the dying gas without appreciably increasing the temperature of the product. In many cases, this ensures exceptionally good product quality. It is one of the reasons why flash dryers are particularly favoured for powdered materials.

(b) Ring dryers. For more difficult products, or where more carefully controlled drying is required, a Ring dryer can be specified. The Ring dryer (Figure 9.3) is a development of the flash dryer, and incorporates a centrifugal classifier (the manifold) that uses deflector blades to select and classify airborne particles on the

basis of their density. As a result of the centrifugal action, heavier and wetter particles follow the ring-duct contours more closely and are recycled back into the drying air stream, while the fine, dried material leaves the dryer with the exhaust

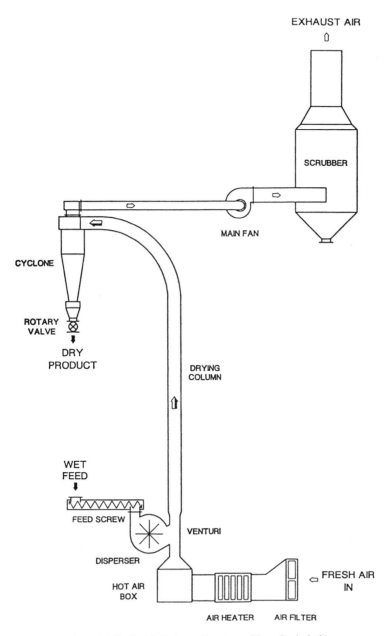

Figure 9.1 Typical flash dryer. (Courtesy of Barr–Rosin Ltd.)

air. This has the effect of lengthening the residence time of larger particles in the dryer, while the finer material, which dries rapidly, leaves in the normal manner. This system promotes even drying in materials in which there is a range of particle sizes, and can ensure that the finished-product moisture content is uniform throughout the dried material.

The air enters through a filter and passes through an air heater into the manifold, where it meets a recirculating stream of partially dried material which it conveys down the first part of the drying duct to the disperser. In the disperser, the wet feed is introduced and meets a stream of partially dried material and hot drying air,

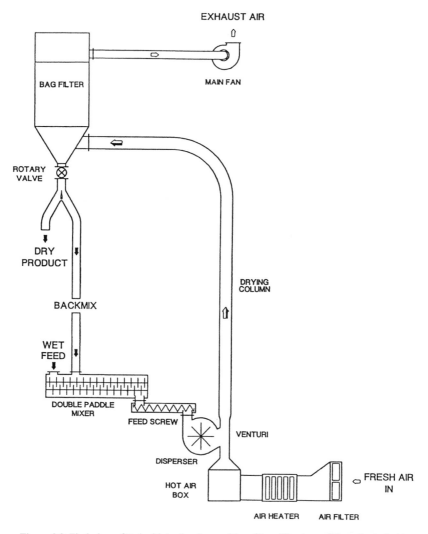

Figure 9.2 Flash dryer fitted with back mixer and bag filter. (Courtesy of Barr–Rosin Ltd.)

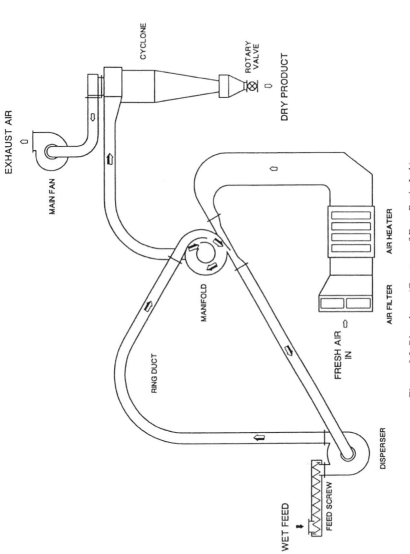

Figure 9.3 Ring dryer. (Courtesy of Barr–Rosin Ltd.)

which helps to disperse it and ensure very rapid drying within the system. Depending upon the application, the disperser can also be used at higher speed as a disintegrator for breaking up material and ensuring the right degree of particle size reduction. From the outlet of the disperser, the material is carried around the upper loop of the ring duct, back to the manifold, where the heavier, partially dried material is recycled back into the hot air stream and to the disperser, while the light, dry material leaves the dryer with the exhaust air and is collected in the cyclone. The main fan draws the air through the entire system and a scrubber or bag filter can be used in exactly the same way as in the flash dryer above.

Where particle-size control is a requirement, Ring dryers are frequently fitted with in-line disintegrator mills, which provide the required particle-size reduction. The disintegrator is located at the base of the drying ring and helps both to increase the surface area of the particles being dried and to produce a degree of turbulence that encourages rapid and efficient flash drying. The hot drying air, together with airborne recycled material, continuously sweeps through the disintegrator at high velocity. This action provides for effective drying of difficult materials at high throughputs. The combined effect of disintegrator and manifold ensures excellent particle-size control, together with efficient and even drying. Several different configurations of Ring dryer are available, making it suitable for a range of different applications.

(c) Feed systems.   In both flash and Ring dryers, feed preparation is paramount. A crumbly or powdered wet material can be fed directly into the dryer by means of a screw conveyor or rotary valve. More frequently, however, the material is sticky or pasty and requires conditioning before drying. This is accomplished by mixing the incoming wet material with recycled dried product in a single- or twin-shaft paddle mixer to produce a suitable feed for the dryer.

It is normally important to seal the feed point of the flash or Ring dryer. This is done by means of a feed screw that runs full, thereby preventing air from passing through it.

The final element in the feed system is often a disperser or kicker mill. This can serve either to break up any agglomerates prior to their entering the dryer, or merely to accelerate the material in finely divided form into the drying air stream.

(d) Product collection.   After drying, the finished product has to be separated from the drying air. The purpose is two-fold: to collect the product and to ensure that a clean exhaust is discharged to the environment.

The simplest type of product collector is a cyclone, in which the solids are separated from the exhaust air by means of centrifugal forces. One or two (in series) high efficiency cyclones, which are normally discharged by rotary valves, are often sufficient. Cyclones are simple, inexpensive, and have no moving parts. Moreover, as the product collects in a high speed air stream at the wall of the cyclone, they are normally self-cleaning. This is a particular advantage in the food industry since there is no product build-up in the collector, which does not therefore require cleaning in normal operation.

Modern high efficiency cyclones have good product collection efficiencies down to as low as 10 μm particle size. Therefore, in many applications, they are capable of meeting environmental emission regulations without secondary collection.

For finer products, where cyclones do not provide sufficiently high collection efficiencies, bag filters are used. These collect product by passing the air through filter bags, leaving the particles as a layer on the surface of the filters, which is regularly cleaned off, normally by a reverse pulse of high pressure air. Bag filters have very high efficiencies and can collect particles down to well under 5 μm while meeting emission standards. However, they are dusty in operation, require regular (usually annual) changing of the filter bags, and are therefore rather expensive to maintain. Furthermore an appreciable amount of material is retained within the bags, so they are not suitable for dryers that handle a number of different products.

In cases where cyclones alone do not achieve the efficiencies required to meet environmental standards, and where bag filters are not suitable, wet scrubbers are frequently used as secondary collectors. Even in cases where there is a lot of fine material, cyclones will usually collect more than 99.5% by weight of the incoming total solids. The scrubber therefore has to deal with less than 0.5% of the total production. This is easily collected but provides the problem of having to handle the scrubber liquor. In most cases, this can be recycled back to the process without material degradation of the product having taken place; the scrubber merely adds a little water to the circuit. Where it is not possible, the scrubber liquor has to be discharged, normally after appropriate treatment in an effluent plant.

9.2.3 Hygiene

One of the other key advantages of flash and Ring dryers in the food industry is their clean operation. The drying system consists of a series of pneumatic conveying ducts handling a mixture of wet and dry material. The velocity of the drying air and material travelling round the dryer tends to ensure that the ducting is kept clean. The product collection system handles only dry material, and rarely requires attention. The only part of the dryer that may need regular attention is the feed system, which handles the wet material. This needs periodic cleaning, certainly every time the system is shut down. It can usually be designed for cleaning-in-place or for easy flush through, with a drain being located at the lowest point of the feed system. Inspection doors are also provided for checking the ducting, particularly around the feed point, to make certain that no deposits have formed. It is certainly most important to ensure that the systems are designed without dirt traps and that access for cleaning and inspection is adequate. However, even in the dairy industry, Ring and flash dryers are considered quite satisfactory and are not cleaned during normal continuous operation.

As well as cleaning, housekeeping is very important. It is necessary to ensure that the areas around any dryer, including pneumatic-conveying dryers, are kept clean and free of dust. Flash dryers can easily be designed to be totally enclosed, so that the only time any part is open to the environment is when the dryer is being inspected. They also normally run under suction, so air leakage is inward and powder leaks are avoided.

Environmental control of the dryer is also important. Firstly, the fan and any other noisy components can be provided with acoustic insulation and exhaust silencers to ensure that ambient noise levels are not excessive. Secondly, the exhaust needs to be controlled so as to minimize particle and odour emissions. This is normally handled by the collection system but, in the case of potential odour problems, scrubbing with additives can be an effective solution. However, water scrubbing in itself is often sufficiently effective to prevent odour nuisance.

9.2.4 Applications

There are many applications for flash and Ring dryers in the food industry, but these are limited by the need to handle and produce a powdered product. Powdered soups, gravies, and mashed potato are normally produced in flash dryers but most of the applications are in the food ingredients area and include materials such as starches, dietary fibres, casein and wheat gluten. There is also a considerable range of food by-products, normally used as animal feed, which are dried in flash and Ring dryers. Examples are distillers' dark grains, yeast residues, potato and maize proteins, olive oil residues and citrus pulp.

One particular industry that uses flash dryers perhaps more than any other is the starch industry, which produces starch from maize, wheat, potatoes, tapioca and a number of other raw materials. It provides a diversity of applications, which illustrate the versatility of flash drying. Moreover, in contrast to much of the food industry, where secrecy surrounds many of the production details, starch-manufacturing processes are well documented worldwide. This makes it easy to record and compare them.

The starches produced, either in natural or chemically modified form, are very important ingredients for the food industry. In wheat-starch manufacture, wheat protein (gluten), is also produced and is normally dried in a Ring dryer. Gluten is an essential additive used in the bread-baking industry to fortify low and medium protein flours. In maize starch production, the maize germ, which is also dried, is the source of corn oil, which is again an important food product.

Other ingredients, which are by-products of the starch industry, include the non-edible proteins, fibrous particles, and concentrated process water. These valuable feeds are also flash dried for optimum product quality.

9.2.5 Design of pneumatic-conveying dryers

In pneumatic-conveying drying, powdered moist product is carried through a duct system by heated air or gas. The product offers a large surface area to the turbulent drying medium, with the result that rapid transfer of heat and moisture takes place and drying occurs within a few seconds. One of the other benefits is high drying efficiencies with low costs.

Pneumatic-conveying dryers can typically handle materials in the size range of 5–500 μm, accept a continuous solid- or slurry feed, with a throughput in the range of 10 kg h^{-1} to 25 t h^{-1} solids. Drying times can range from 2 to 60 s.

Table 9.1 summarizes the operating conditions for a number of pneumatic-conveying drying processes. The operation is particularly easy to illustrate and analyse diagramatically using a psychrometric chart. This may take the form of either a Grosvenor chart (Figure 9.4), which is in general use in the English-speaking world, or a Mollier chart (Figure 9.5), popular in continental Europe. Take for example, a steam-heated flash dryer for corn starch (No. 2 in Table 9.1). Beginning at the dryer inlet, atmospheric air at 10°C (50°F) and 0.5% absolute humidity is heated in a steam-heated air heater to 160°C (320°F). This air is brought into contact with finely divided moist starch from which moisture is evaporated until the air temperature is reduced to 54°C (129°F) and the humidity rises to 4%. This is shown by point 2 on the charts. At this point, the starch is dry at 12% moisture content. The dried starch is separated, and the spent air is exhausted. The exhaust air of the dryer is not saturated. This would have required cooling the air to 39°C (102°F).

The difference between the exit temperature of the air and the temperature at which further absorption of moisture would have achieved saturation, is termed the wet-bulb depression. The wet-bulb depression at the exit of the dryer can be regarded as the 'driving force' of the drying process, in our example 15°C (27°F). This is one of the important characteristics of a drying system. Obviously, the efficiency of the dryer increases as the amount of sensible heat in the drying air that is converted into latent heat for evaporation of moisture increases. This is reflected by a corresponding increase in the temperature difference between the inlet and outlet air streams. The heat that remains in the air at the exit of the dryer is the exhaust loss. Other losses occur in the dryer itself. The change in temperature of the air inside the dryer, as can be deduced from the diagrams, does not follow the lines of constant wet-bulb temperature, because the sensible heat extracted from the air is greater than the heat content of the evaporated water vapour. This is due mainly to two factors. The first is that in evaporating water from a moist material, some heat is required to dissociate moisture from hygroscopic product (the heat of wetting). Starches, and all other products considered here, are, at least to a small extent, hygroscopic. The second reason is the loss of sensible heat within the drying system. For practical purposes, it is usually unnecessary to separate these two factors. This is why the performance of dryers is not usually characterized by an accurate percentage

Table 9.1 Performance of industrial flash-drying systems

Dryer no.	Design	Product	Product moisture content (%)	Output (kg h^{-1})	Evaporative capacity (kg h^{-1})	Air temperature Inlet (°C)	Air temperature Outlet (°C)	Exhaust humidity (%)	Wet bulb depression (°C)	Heating medium	Heat consumption for water evaporation (kJ kg^{-1})	Heat consumption for water evaporation (Btu lb^{-1})
1	Single-pass	Potato starch	20	4000	1350	140	43	4.0	6	Steam	3870	1680
2	Ring	Corn starch	12	2000	900	160	54	4.2	15	Steam	4060	1760
3	Ring	Corn starch	12	10 000	4800	210	60	5.7	16	Steam	3830	1665
4	Single-pass	Corn starch	12	5000	1750	160	60	3.9	21	Steam	4395	1920
5	Ring	Corn starch	6	10 000	4500	230	77	5.5	31	Gas	4475	1950
6	Ring	Corn starch	3	4900	2200	175	100	2.9	58.5	Gas	6695	2900
7	Two-stage	Corn starch	2	4500	450	210	115	4.3	70.5	Steam	5020	2190
8	Ring	Fibre, gluten and corn steep liquor	12	2500	3500	400	100	9.8	45	Oil	4185	1820
9	Ring	Fibre and corn steep liquor	10	5000	7500	445	115	11.0	57	Oil	4395	1910
10	Single-pass	Fibre and corn steep liquor	10	5000	7500	600	133	13.0	72	Oil	5020	2180

efficiency but rather by the heat input required per unit of evaporation or per unit of dried product.

As a first approximation, the heat required for drying may be estimated by comparing the temperature drop across the dryer with the temperature rise in the heater. On this basis, it is clear that, in the interest of high efficiency, the dryer should be operated at the highest possible inlet temperature and the lowest possible outlet temperature. These temperatures are determined by practical considerations.

The first consideration is the maximum temperature that the product can withstand without thermal damage; this is clearly most important as product quality is paramount. The second is the temperature of the heating medium; for drying food products, this is most commonly steam. Frequently, therefore, the pressure (and hence the temperature) of the available steam effectively limits the dryer inlet temperature.

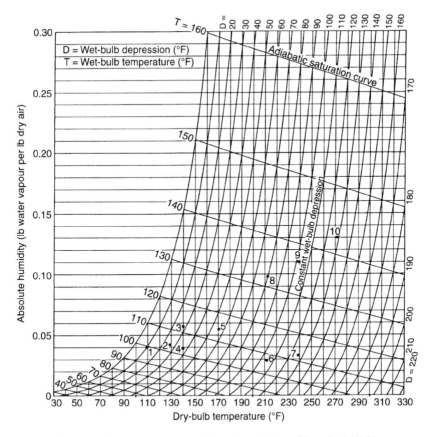

Figure 9.4 Grosvenor psychrometric chart. (Courtesy of Barr–Rosin Ltd.)

Where the temperature of the heating medium (steam) is not limiting and high pressure steam or direct-gas firing is available, dryers can be operated at temperatures up to 250°C for food products. With by-products such as those from starch manufacture, fermentation residues, and other products that are not primarily used in food, but perhaps for animal feed, inlet temperatures up to 400°C (750°F), or sometimes even higher, can be employed, with a marked increase in efficiency. Increasing the dryer inlet temperature proportionately reduces the quantity of air used for drying and hence the size and capital cost. In general, therefore, the highest practicable inlet temperature should be selected for any given application.

The dryer outlet temperature is a function of the required product moisture content and the wet-bulb depression. This, in turn, depends on the system design. Figure 9.6 illustrates the relationship between wet-bulb depression and final moisture content for various starch products and different dryers. In

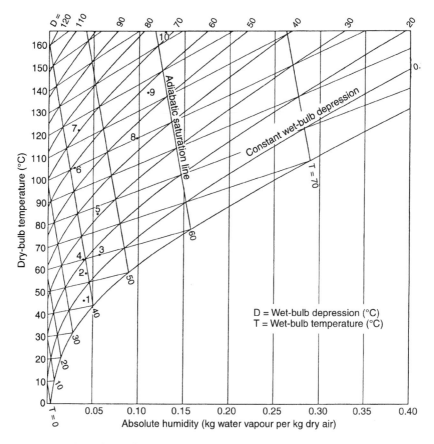

Figure 9.5 Mollier psychrometric chart. (Courtesy of Barr–Rosin Ltd.)

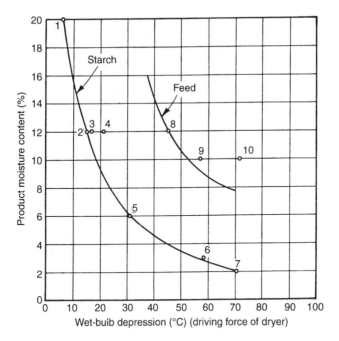

Figure 9.6 Comparison of wet-bulb depression and final moisture content for various products in different dryers. (Courtesy of Barr–Rosin Ltd.)

common with many other food ingredients, corn starch and wheat starch are dried to moisture contents of 12%; potato starch has a final moisture content of 20%, but lower moisture contents are required for many special purposes, and for proteins such as casein and wheat gluten. The temperature at which a required wet-bulb depression is obtained depends upon the inlet temperature and the total moisture content of the exhaust air.

Over recent years, considerable progress has been made in developing a theoretical understanding of the principles underlying the design and scale-up of pneumatic-conveying dryers. Kemp (1994), for instance, described a computerized design and scale-up method based on models of the particle flow, air-to-particle heat transfer, drying kinetics and equilibria, and mass and energy balances. The models were based on extensive experimental data. Their use highlighted a number of interesting effects, which included the possible existence of a significant dead zone near the feed point, the role of agglomeration on the drying process, and the importance of additional drying that may occur within the cyclone.

9.2.6 Comparison of different drying systems

As noted above, the performance of different drying systems is compared in Table 9.1. This table summarizes performance data for ten installed pneumatic-drying systems. The exhaust air conditions of these ten installations are plotted on the psychrometric charts (Figures 9.4 and 9.5) as well as in Figure 9.6. The two curves in Figure 9.6 show immediately that the driving force required for dryers operating at higher temperatures is considerably greater than that required for low-temperature dryers. Among the dryers considered, Nos. 1, 4 and 10 are single-pass flash dryers; the rest are Ring dryers. Case No. 1 is an easy-drying application, with a high outlet moisture content, which lends itself readily to single-pass drying. However, in cases No. 4 and 10, it can be seen that the use of a flash dryer necessitates somewhat higher outlet temperatures as compared with Ring dryers operating under similar conditions. The driving force as such depends upon both the drying conditions and the nature of the product. Ring dryers, with their greater residence time and flexibility, can operate with lower driving forces than simple flash dryers.

9.3 Spin-flash dryer

9.3.1 Introduction

The spin-flash dryer is essentially an agitated fluidized bed that can be used to dry pastes, filter cakes, sludges, and high viscosity liquids on a continuous basis. It provides a useful alternative to the spray dryer (Chapter 5) in cases where the feed cannot be pumped and atomized. There is some overlap between the processing capabilities of the spin-flash dryer and pneumatic-conveying and conventional fluidized bed (Chapter 4) dryers in particular.

9.3.2 Equipment

A typical spin-flash dryer is illustrated in Figure 9.7. The feed material is dropped into the feed hopper in which it is broken up into a cake of uniform consistency by means of a low speed agitator. It is then transferred into the body of the dryer by means of a screw feeder. In cases where the feed is a dilatant liquid, the tank and screw feeder are replaced by a progressive cavity pump. The drying air is heated either in a direct fired natural gas burner or by a steam heater, as appropriate, before entering the hot-air plenum. It then passes through a tangential air inlet, which, together with the axially mounted rotor at the bottom of the drying chamber, causes a turbulent, rotational flow of air within the dryer.

   As lumps of the feed material enter the drying chamber, they become coated with dried powder and fall into the fluidized bed at the bottom of the chamber, where they are kept in motion by means of the rotor. As these lumps dry out, the friable surface material is abraded by a combination of attrition in the bed and the action of the rotor. These light, dry particles are entrained by the rising air stream and travel up the wall of the drying chamber, at the top of which they are

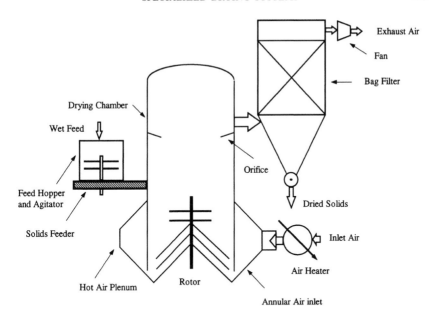

Figure 9.7 Spin-flash dryer.

forced to pass through a classification orifice. This permits the finer material to exit the drying chamber, while the larger particles are retained in the dryer and tend to fall back into the fluidized bed, where they eventually break up. The orifice can be sized to suit the required product particle size.

In the particular embodiment shown in Figure 9.7, the exhaust air and entrained particles are separated in a bag filter. The product leaves via a discharge valve and the air is drawn through an exhaust fan. Alternatively, a cyclone and, if necessary, a wet scrubber may be employed.

In the spin-flash dryer, as in Ring dryers, classification of the material occurs within the dryer and the small, dry particles are carried out of the drying zone as soon as they become sufficiently small and light to follow the air through the orifice.

9.3.3 Applications

Spin-flash dryers are available in a number of standard sizes with quoted throughputs ranging up to 10 t h^{-1}. This figure can be expected to be lower for foodstuffs, however, because of limitations on the inlet air temperature. Food dyes have been dried commercially in a spin-flash dryer. Gum, chitosan gel, crab meat paste, and cocoa cake have also been dried successfully on a small scale.

9.4 Rotary dryers

9.4.1 Introduction

Rotary dryers are widely used to dry relatively large throughputs of granular products and by-products in a number of industries. Like many other types of dryer, they are available in several different forms. Of particular interest to the food industry are the cascading, rotary-louvre, and steam-tube types. The latter is a contact dryer and, as such, is described in Chapter 6.

Rotary dryers are characterized by a slowly rotating cylindrical drum, which is normally inclined at a small angle (0–5°) to the horizontal. Wet feed is introduced into the upper end and dried product withdrawn at the lower end. The internals of the different classes of rotary dryer are, however, quite different, as are the methods used to transfer heat to the drying material.

9.4.2 Equipment

(a) Cascading rotary dryers. Figure 9.8 illustrates a typical cascading rotary dryer, together with possible internal arrangements. In the simplest, and perhaps most common arrangement (Figure 9.8a), the inside of the drum is equipped with a series of lifting flights running along the length of the dryer. As the drum rotates, these pick up solids from the base of the dryer and convey them for a certain distance around the periphery. The solids progressively discharge from the flights and fall as raining curtains through a stream of hot drying air. The cycle then repeats itself. No attempt is made to break the fall of the solids and friable products may therefore shatter on impact.

In order to encourage uniform cascading, the lifting flights are frequently offset at regular intervals along the length of the drum. Straight-, angled- and right-angled flights (Figure 9.9) are most commonly employed in industrial practice. Semi-circular flights are used less frequently. The geometry of the flight affects not only its holding capacity but also the manner in which the solids are shed from it as it traverses the diameter in the upper half of the drum. The design of flights has been discussed in detail by Baker (1988).

Figures 9.8b–d illustrate typical internal arrangements for short-fall or cruciform dryers. These dryers are generally of large diameter, and the function of the internals is to limit the fall of the drying solids to relatively small distances. This makes them well-suited to processing friable materials. These internals also promote better solids–gas contacting and provide an extended residence time where needed. However, cleaning is more difficult and such dryers cannot be used to process sticky feeds.

In most food drying applications, the air and solids flow concurrently through the dryer. This minimizes the chance of thermal degradation of the product. In this case, the axial motion of the solids is aided both by the slope of the dryer and by the drag of the air on the cascading particles. Most of the drying occurs while the particles are in contact with the air stream. The resting period in the flights, however, provides an opportunity for the moisture to equilibrate within the solids.

Drum diameters normally range from about 1 to 4.25 m; length-to-diameter ratios are typically 4 : 1 to 10 : 1. In most cases, the drum is constructed from mild steel. However, where corrosion is a problem, stainless steel or mild steel clad with stainless steel may be employed. Cast iron or steel riding rings are fitted to the drum close to its extremities. These rest on trunnion roller bearings, which support the dryer during operation. Rotation of the drum is effected by an electric motor, normally through a fixed speed gearbox. Peripheral speeds commonly range from 0.25 to 0.7 m s^{-1}.

The two ends of the drum project into breachings where the heated air and the solids enter and leave the dryer. Several types of seal are employed to minimize leakage of solids out of the dryer and ambient air into it. Rotary dryers are normally operated under a slight vacuum in order to prevent dust blowing out

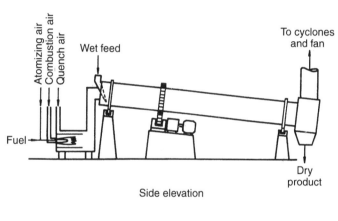

Side elevation

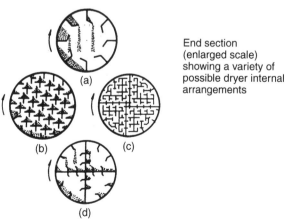

End section
(enlarged scale)
showing a variety of
possible dryer internal
arrangements

Figure 9.8 Cascading rotary dryer. (Courtesy of AEA Technology plc.)

through these seals. This is achieved by employing a dual-fan (blower and suction) arrangement.

The solids feeding system must provide a uniform flow of solids into the dryer. It must also act as an effective seal preventing the leakage of both solids and air. If the solids are free flowing, they may be fed into the dryer via a rotary valve and chute. For pasty or wet materials, a screw conveyor should be used. In such cases, a short length of spiral flights at the feed end of the drum may be required to move the solids rapidly into the body of the dryer, thereby preventing them from sticking to the internals. The dried solids generally flow freely out of the bottom of the discharge breaching.

In common with most other types of convective dryer, the drying air may be heated either by direct combustion, usually with natural gas, or indirectly with a steam heater. Direct firing has the advantage that higher inlet temperatures are possible than with indirect heaters, which are limited to about 175°C, and thermal efficiencies are generally higher. However, it suffers from the fact that

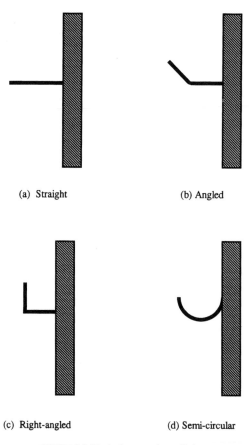

(a) Straight   (b) Angled

(c) Right-angled   (d) Semi-circular

Figure 9.9 Typical rotary-dryer flights.

the products of combustion come into direct contact with the foodstuff, which may be neither permissible nor safe (see Chapter 2). In contrast, indirectly heated air is both clean and food compatible.

The exhaust air temperature is chosen on the basis of practical and economic considerations. In order to eliminate any possibility of condensation in the ductwork, its wet-bulb temperature should be at least 10°C higher than its dry bulb temperature.

Air velocities of 0.5–5 m s^{-1} (under exhaust air conditions) are commonly employed in cascading rotary dryers. Depending on the product being dried and the air velocity employed, the exhaust air may contain some dust, which has to be removed in a cyclone, a cyclone/wet scrubber combination, or a bag filter. Details are given in section 9.2.2(d).

(b) Rotary louvre dryer. The rotary louvre dryer (Figure 9.10) consists of a slowly rotating (2–3 rpm) horizontal cylinder, which contains longitudinal louvres that form a slightly tapered inner drum. Such dryers range in size from about 0.75 to 3.5 m in diameter and from about 2.5 to 10.5 m long. The solids form a partially fluidized rolling bed in the bottom of the dryer. Heated air or

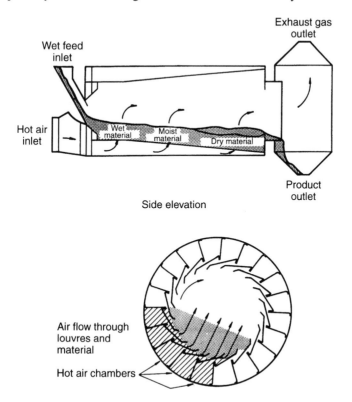

Figure 9.10 Rotary-louvre dryer. (Courtesy of AEA Technology plc.)

combustion gases enter the dryer through those louvres that are covered by the solids.

The major advantage of the rotary louvre dryer over the cascading type is that the solids are handled in a much more gentle manner. As a result, it is well-suited for processing friable material or crystalline solids for which surface scratching would be detrimental to the appearance of the product. An additional benefit is that there is excellent contacting between the gas and the solids, which are exposed to the same drying environment throughout the dryer. Perry and Green (1984) have indicated that the capacity of a rotary-louvre dryer is about 1.5 times that of a cascading dryer of the same size and operating under equivalent drying conditions.

9.4.3 Applications

Rotary dryers are capable of processing a wide variety of materials. They can handle free flowing solids of virtually any size or shape. Fine materials, however, are best dried in indirect rotary dryers, which employ much lower air velocities. Rotary dryers are in a class of their own when it comes to processing very large throughputs. Table 9.2, which is far from exhaustive, summarizes a number of typical food products that have been dried in direct rotary dryers.

9.4.4 Design of rotary dryers

Traditionally, the design of rotary dryers has been more of an art than a science. It relied heavily on prior experience and empirical scale-up rules. Broadly speaking, this is still the case. However, in recent years, significant advances have been made in developing a better scientific understanding of the underlying principles, at least in the case of the cascading type. Few studies of any significance have been undertaken on rotary-louvre dryers.

As is the case with most other dryers, the scientifically based design of rotary dryers requires an equipment model and a material model (Papadakis *et al.*, 1994). The equipment model describes the gas flow, solids transport, and heat transfer to the solids. The material model describes the drying kinetics and equilibrium relations.

Early equipment models were based either on empirical correlations or untested computer simulations. Baker (1983) evaluated a number of models of particle transport in rotary dryers under zero and concurrent airflow conditions. He concluded that, while the correlations were reasonably satisfactory at low air

Table 9.2 Examples of food products dried in rotary dryers

Fish scraps	Apple pomace	Sugar beet pulp
Wheat residues	Sliced potatoes	Sugar
Cocoa beans	Cooked cereals	Starch
Nuts	Flour	Spent grains

flows, they were quite unreliable at air velocities above 1 m s^{-1}, which are commonly employed in industrial practice. Baker also reviewed published heat transfer models, which were derived in most cases by interpreting experimental data in terms of a volumetric heat transfer coefficient based on the effective area of contact between the gas and solids, and the ratio of this area to the dryer volume. Again, these correlations were found to be generally unsatisfactory. When used to estimate the volumetric coefficient at an air velocity of 3 m s^{-1}, for example, the results predicted by the different correlations differed by an order of magnitude.

An improved model of particle transport was formulated by Matchett and Baker (1987, 1988) and extended by Matchett and Shiekh (1990). In this 'two-phase' model, the flow of material through the dryers was considered to consist of two streams.

- The airborne phase, which contains the material falling through the drum, and which is acted upon by gravity and the drag forces exerted by the airflow through the dryer.
- The dense phase, containing the material caught in the flights and resting on the bottom of the drum; this is affected primarily by the rotation of the drum.

In their model, Matchett and Baker analysed the motion in the airborne and dense phases independently. They also distinguished between underloaded, design-loaded and overloaded drums (Figure 9.11). The approach has been tested on a number of industrial scale dryers and found to yield acceptable results. It presupposes a knowledge of the values of a small number of physical parameters, which, in the absence of available data, can be obtained from pilot plant or commercial plant trials. These are used to estimate the particle velocities in the airborne and dense phases, which can then be combined to yield the average axial velocity and hence the solids residence time in the drum.

A more rigorous approach to modelling heat transfer in rotary dryers was first proposed by Schofield and Glikin (1962), who used a theoretically based equation to calculate the surface area for heat transfer and an empirical correlation, applicable to isolated particles, to estimate the convective heat transfer coefficient between the particles and the air. Their model predicted values of the heat load that were much larger than those observed in an industrial unit. Langrish et al. (1988), however, were able to achieve more satisfactory results using modified correlations for the heat transfer coefficient, which were based on the relative velocity between the air and the solids as calculated from their aerodynamic model.

A material model cannot be formulated from first principles because of the complexity of the moisture transport phenomena that occur within the drying solid. Hence the drying kinetics and equilibrium moisture content should be determined from bench scale experiments (Papadakis et al., 1994). These authors also proposed that the characteristic drying curve concept, as described in Chapter 2 (see also Keey, 1992), be used to predict how the drying rate changes with the temperature, humidity and velocity of the air flow. As is the case with other parallel-flow dryers, this is of particular importance in rotary dryers as the temperature and humidity of the drying air change along their length.

Papadakis *et al.* (1994) described how the above particle transport, heat transfer, and material models could be combined with mass and energy balance equations into an overall model, which could be used to undertake design and performance calculations. These calculations were performed on a PC.

9.5 Tray and tunnel dryers

9.5.1 Introduction

Tray dryers, also known as cabinet dryers or stoves, and tunnel dryers, are cross-flow dryers, which account for a significant proportion of dehydration in the food industry (Holdsworth, 1997). In both cases, the material being dried is

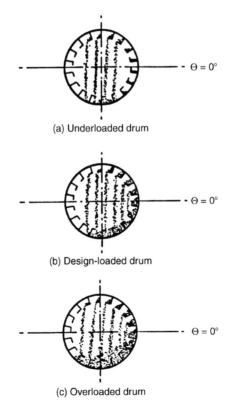

(a) Underloaded drum

(b) Design-loaded drum

(c) Overloaded drum

Figure 9.11 Solids loading in rotary dryers. (After Matchett and Baker, 1987.)

supported on multiple trays, with the hot air being directed at high speed between and across the surfaces. Tray dryers operate in the batch mode. However, two or more of the larger versions of this dryer may be linked to provide semi-continuous operation. Tunnel dryers provide a natural extension to this concept and are continuous.

In both systems, relatively large volumes of air are required to supply heat to the product and to carry away the evaporated moisture. For the sake of economy, it is advantageous to recirculate a proportion of the air. In so doing, the humidity should not be allowed to rise to such an extent as would seriously inhibit drying. These dryers should, therefore, include adequately sized fans and steam heaters, and suitable ducting arrangements capable of providing an even distribution of hot air. Inlet, exhaust and recirculation air flows are controlled by dampers. These may be adjusted manually; alternatively, automatic humidity control may be fitted.

9.5.2 Tray dryers

In the smaller tray dryers of the type shown in Figure 9.12, the foodstuff is spread in thin layers on trays supported by angle slides mounted on the side of the drying cabinet. A typical sized tray is around 800 mm long × 400 mm wide × 30 mm deep; the exact dimensions naturally vary from one manu-facturer to another. The foodstuff is spread in thin layers, usually at 4.75–7.5 kg m^{-2}, though higher tray loadings can be used if the material is re-spread or turned during the drying process.

In large tray dryers (stoves), the trays are usually supported on trolleys which can be pushed into and out of the cabinet, and also turned around so that the air is not always leading on the same edge. Typically, such a dryer would contain one, two, or four trolleys, each holding 40 trays. In some cases, the entire drying

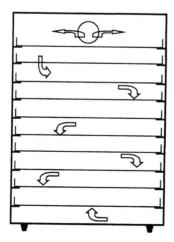

Figure 9.12 Tray dryer.

process is completed within a single cabinet; in others, the cabinets are arranged so that trolley-loads of trays pass from one cabinet to the next, with different flow, temperature and humidity conditions in each successive cabinet. In yet other layouts, only part of the drying process is carried out in cabinets; it is completed in separate finishing dryers, typically aerated bins. Recirculation within the cabinets is practised so far as humidity conditions allow; in multi-stage cabinet systems, air is recycled from the later stage cabinets through heaters to the earlier stage cabinets.

Tray dryers are very flexible in their operation. The velocity, temperature and humidity of the air, and the drying time, can all be varied widely to suit the characteristics of any particular product. Modern dryers are normally fitted with effective instrumentation and control systems. Tray dryers are used primarily for processing relatively small individual batches of foodstuff. Their principal drawback is the high manpower requirement for filling and emptying the trays, loading them into the dryer and, if necessary, turning the product at an inter-mediate moisture content.

9.5.3 Tunnel dryers

Moving trucks individually through successive cabinet doors is labour intensive and wastes energy. As a result, arranging the 'cabinets' in line continuously to form a so-called tunnel dryer was an obvious development. The best designs are able to combine high capacity and high quality of product. The trolleys are normally conveyed mechanically through the tunnel on guide rails. As illustrated in Figure 9.13, various airflow systems are employed.

- Cocurrent flow, in which the air moves in the same direction as the trolleys. This is termed parallel flow in the USA.
- Countercurrent flow, in which the air moves in the opposite direction to the trucks.
- A combination of cocurrent and countercurrent flow. This may be achieved by means of separate tunnels, parallel tunnels, or a centre-exhaust tunnel.
- Transverse flow. This is usually combined with inter-stage reheating of the air and appropriate ducting to provide what is essentially a countercurrent flow system.

(a) Cocurrent flow tunnel dryers.   The hottest and driest air impinges first on the wettest material (Figure 9.13a) and, as a result, the initial rate of evaporation is high. The air is partly cooled as a result, and becomes progressively cooler as it passes over the trays in successive trolleys in the tunnel. At the far end, the dry-bulb temperature of the air may have dropped from an initial value of 90° or 95°C to 60° or 65°C. Its humidity will have risen significantly and much of its drying efficacy will have been lost. As a result, it is not possible to recirculate much of the air, and therefore cool ambient air has to be supplied continuously

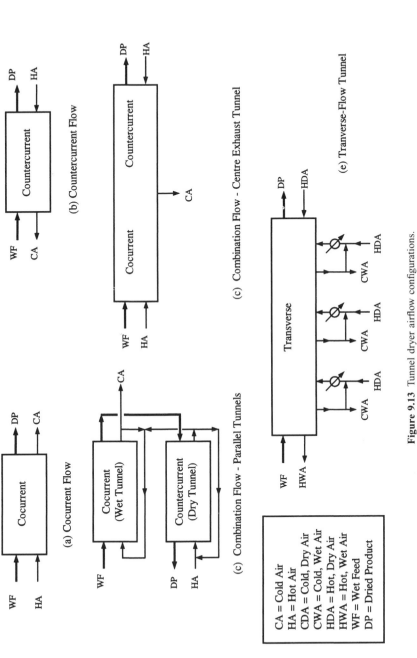

Figure 9.13 Tunnel dryer airflow configurations.

(a) Cocurrent Flow

(b) Countercurrent Flow

(c) Combination Flow - Parallel Tunnels

(c) Combination Flow - Centre Exhaust Tunnel

(e) Tranverse-Flow Tunnel

CA = Cold Air
HA = Hot Air
CDA = Cold, Dry Air
CWA = Cold, Wet Air
HDA = Hot, Dry Air
HWA = Hot, Wet Air
WF = Wet Feed
DP = Dried Product

to the heaters. Consequently, the use of these tunnels is somewhat limited by their relatively high heat (and fuel) consumption.

(b) Countercurrent flow tunnel dryers. In these dryers (Figure 9.13b), the heated air is first blown across the driest material, at a temperature low enough not to cause thermal damage to the product. This is usually about 70°C, which is considerably lower than that used in the cocurrent flow arrangement. As a result of the evaporative cooling along the tunnel, the air temperature will have dropped considerably by the time it reaches the wet end, at which the trolleys containing the wet material enter the dryer. Hence, the initial rate of evaporation is relatively slow. Because of their simple design, countercurrent-flow tunnels have achieved considerable popularity for drying fruit and vegetables, particularly in the USA (Holdsworth, 1997). However, as pointed out by Keey (1992), with cold feeds, moisture from high humidity exhaust gas streams can condense on the incoming feed until the solids temperature has been raised sufficiently to reverse the humidity gradient in the boundary layer. Such difficulties have been reported with countercurrent tea and sugar dryers.

(c) Combination flow tunnel dryers. These are of two types: (a) parallel tunnels, and (b) centre-exhaust tunnels. In parallel tunnels (Figure 9.13c), a cocurrent-flow tunnel and a countercurrent-flow tunnel are built side by side. The cocurrent-flow tunnel acts as the so-called 'wet' tunnel into which the trolleys containing, for example, blanched vegetables are loaded. When the trolleys have passed through the wet tunnel, they are transferred to the counter-current-flow or 'dry' tunnel. In the original 1940s version of this dryer, the trolleys ran alongside each other in pairs, and there was room in each tunnel for seven pairs of trolleys. Commonly, a pair of trolleys was inserted every 30 min, thereby giving a total drying time of 7 h.

Air enters the system by way of an inlet into the recirculation duct of the dry tunnel. It passes through the heater battery and the fan and is forced through the drying section and back into the recirculation duct. Part of this air is directed into the wet tunnel, and part into the dry tunnel. That part which is directed to the wet tunnel mixes with the wet-tunnel recirculating air, passes through a heater and fan, and through the drying section. Part of the exhaust air from this tunnel is discharged and part recirculated. By appropriate manipulation of dampers, quite good control of humidity can be obtained in the tunnel, with substantial fuel savings. Commonly used air temperatures in these tunnels are: wet tunnel, inlet end, 99°C dry-bulb; wet tunnel, exhaust end, 60°C dry-bulb, 43°C wet-bulb; dry tunnel, inlet end, 65°C.

Other systems of parallel tunnels are in use, but the principle is the same; a 'wet' tunnel or tunnels with cocurrent flow and a 'dry' tunnel or tunnels with countercurrent flow.

The same general considerations apply to the centre-exahust version of the tunnel dryer (Figure 9.13d). This consists of a long tunnel with a single large extraction fan located near the centre, forcing air to be drawn in through heaters at each end of the tunnel. The trolleys are inserted at one end and moved along in stages, to be withdrawn, finally, at the other end. Consequently, in the first part of the drying period, the foodstuff experiences cocurrent-flow drying but, after it has passed through the central exhaust region, it is in a countercurrent-flow region. Unless separate finishing dryers are used, the countercurrent flow stage will be longer than the cocurrent flow stage, because of the slow drying rate in this section of the dryer.

(d) Transverse flow tunnel dryers. This type of dryer (Figure 9.13e) can be designed to have a very high evaporation rate, with good thermal efficiency, but the trolleys have to fit very snugly in the tunnel to avoid short-circuiting of the air. It is often followed by a finishing dryer, to which the foodstuff is transferred when its moisture content has fallen to below 20%. The driest air enters at the solids discharge end of the tunnel at a temperature of, typically, 60–65°C. At several stages along the length of the tunnel, it is partially exchanged, mixed with fresh air, and reheated so that its vapour holding capacity is increased as it travels along the tunnel. At the feed end of the dryer, its temperature may have risen to 90–100°C. Such a dryer may contain a load of twelve trolleys, each carrying 180 to 230 kg of wet foodstuff at entry. These trolleys may be moved along one-twelfth of the tunnel at 15 or 20 min intervals, giving a total drying period of 3 or 4 h.

This style of dryer is economical in terms of floor space, and the process is well-controlled. However, it requires fans and air heaters at a number of points, rather than a single fan and heater at one end. As a result of the numerous air direction changes, the power rating can be higher than that for a simpler tunnel.

9.5.4 Design of tunnel dryers

In practice, tunnel dryers are normally designed on the basis of drying curves obtained in simple convective ovens. This approach is sound provided that the data are obtained under similar conditions to those envisaged in the actual dryer. Proper allowance must be made, however, for the changing drying conditions within the tunnel.

Sokhansanj and Wood (1991) designed a tunnel dryer to dry baled forage from 30% to 12% moisture content at 110°C at a nominal throughput of 6 tonnes of dried material per hour. The dryer consisted of three drying zones and one cooling zone with a total length of about 16 m. Each zone was 3.67 m wide and 3.67 m long. Forage bales with a square cross-section (360 mm × 360 mm) and a length of 3.67 m were fed into the dryer crosswise. The bales were packed

Table 9.3 Tunnel-dryer performance (after Sokhansanj and Wood, 1991)

Given drying conditions	Results
Ambient air temperature, 15°C	Forage output rate at 11% moisture content, 5.96 t h^{-1}
Drying air temperature in all three drying zones, 110°C	Total time during which the top layer was at 90°C or higher, 12 min
Airflow rate in each zone, 520 m^3 min^{-1}	Total time during which the bottom layer was at 90°C or higher, 8 min
Belt speed, 0.67 m min^{-1}	Total time during which the middle layer was at 90°C or above, 4 min
Forage input rate at 30% moisture content, 7.6 t h^{-1}	Exit product temperature from the cooler, 17°C
	Heat input, 1 GJ per tonne of dried product
	Overall zone drying efficiency, 59%

against each other as they were transported into the first zone of the dryer over a walking-type floor.

The nominal forward speed of the forage material through the dryer was 0.67 m min^{-1}. Drying air circulated through a series of heating ducts designed to minimize energy consumption and even out the moisture content through the drying bale. The air velocity for each zone was set at about 40 m min^{-1} to minimize the suspension and loss of light fractions such as leaves. The source of heat for each drying zone was a natural-gas burner, located in the recirculating air stream.

Table 9.3 summarizes the calculated performance of the dryer. The product was dried to 11% moisture content and the temperature of the product at exit was 17°C, or 2°C above the ambient temperature. Computer calculations showed that the dryer used almost 1 GJ to dry 1 t of product. The drying efficiency was 59%.

9.6 Band dryers

9.6.1 Introduction

Through-circulation band dryers are widely used throughout many parts of the food industry and within a number of other manufacturing sectors. Their versatile design and construction enable them to process a wide range of different products. In the food industry, these are normally granular particles or extrudates. However, band dryers are also used to dry products as diverse as boards for use in the building trade.

A band dryer consists essentially of a moving, perforated conveyor on which a layer of the drying solids rests. The band passes through a drying enclosure in which heated air is directed through the layer of solids. The air flowrate is often significantly lower than that employed in other types of dryer, for example, fluidized beds. Also, it is frequently directed downwards through the bed of solids. As a result, dust losses can be relatively small and, under these conditions, it may be possible to avoid having to invest in gas-cleaning equipment.

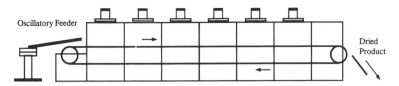

Figure 9.14 Single-pass band dryer.

Band dryers are well-suited to drying medium-to-high throughputs of product on a continuous basis. Integral coolers are frequently fitted. Depending upon the specific requirements, band dryers can be of the single-pass or the multi-pass type.

9.6.2 Equipment

(a) Single-pass dryers. Most band dryers used in the food industry are of the single-pass type, a typical example of which is illustrated schematically in Figure 9.14. The wet feed is deposited onto the band at the entrance to the dryer. It then passes through a single drying enclosure in which it is contacted with hot air. As noted above, the air flow can be either upwards or downwards through the drying layer. The dried product eventually emerges at the opposite end of the dryer.

The band is made up from an appropriate number of stainless steel slats of perforated plate or wire mesh construction. The maximum width of the slats is dictated by their need to pass smoothly over the drive/support rolls at each end of the dryer. The band must be sufficiently robust to support its load and to withstand the tension resulting from its movement through the dryer. In order to prevent maldistribution of the drying air, which would result in uneven drying, it is important to ensure that the material is spread evenly and, as far as possible, to a constant depth over the band. There should be no empty spaces on the band or excessive compaction, which would give rise to channelling or bypassing sections of the bed by the air. A variety of feeders are used, including reciprocating and oscillating feeders, vibratory spreaders, and so on.

Although the band and its load pass through a single drying enclosure, it is common practice to divide this volume into a number of separate stages. This can be accomplished by fitting a series of vertical baffles in the enclosure above and below the band. More commonly, however, the dryer is assembled from a number of standard modules, each of which is fitted with its own integral air heater and fan, inlet and exhaust ducts, and dampers to enable the air flows and hence humidity to be adjusted. It is therefore possible to control the velocity, temperature and humidity of the air in each stage and, if required, to profile the drying conditions along the length of the dryer. For example, many vegetables can be exposed to an air temperature of 100°C at the inlet to the dryer. However, in order to prevent damage to the product, the temperature should be reduced progressively to around 55°C at the exit.

The drying enclosure is constructed from thermally insulated metal faced composite panels. Adequate sealing is provided at either end of the dryer and between stages to minimize air and dust losses and to maximize thermal efficiency. Easy access is provided to the interior of the dryer for cleaning and maintenance purposes. For example, the fans are often mounted on swing-out door panels. Access doors are located at intervals along the dryer, and the panels are normally easily removable. The air heaters (steam or natural gas) and fans are also readily accessible. The modular construction technique provides for ease of assembly on site. It also enables the dryer to be extended readily to meet future production needs.

Band widths vary with the required throughput but range typically from 1 to 2 m. In the larger dryers, the length of a single module is often roughly equal to the width of the band. Frequently, a band dryer may consist of two to six or even more such modules. Typical band speeds are of the order of 1 m min^{-1} but this may naturally be varied to suit any particular application. In band dryers, the product dries either from the top downwards (down-flow of air) or from the bottom upwards (up-flow of air). It may therefore be desirable to 'turn' the product at intermediate points in the drying cycle. This is particularly important if the product shrinks during drying as this can result in channelling of the air flow. In some dryer designs, each stage is fitted with its own individually driven band, the speed of which can be controlled independently. Thus, successive conveyors can be run at slower speeds to compensate for shrinkage and to maintain efficient drying.

(b) Multi-pass dryers. Multi-pass band dryers can be used where floor space is limited or where a long drying time is required. In these dryers, typically three or five conveyors are mounted one above the other. The wet feed is spread on the top conveyor in the normal manner. When it reaches the end of this conveyor, it falls on to the one below. This process repeats itself until the dried product leaves the dryer on the bottom conveyor.

As with the single-pass dryer, it is possible to run the individual bands at different speeds. The drying conditions (air velocity, temperature and humidity) at each level can also be controlled independently. Again, therefore, it is possible to profile the dryer in order to achieve optimum drying conditions.

9.6.3 Applications

Band dryers are well suited to drying relatively high throughputs of a wide variety of granular materials and extrudates. Table 9.4 provides a list of some food products that have been dried on band dryers.

9.6.4 Design of band dryers

In practice, band dryers are designed on the basis of batch drying curves. These are either known from experience of similar products or are measured in the

Table 9.4 Some food products dried on band dryers

Cereals and ceral products	Confectionery
Rice	Pastilles
Corn grit	Nougat
Bran	Chocolate-milk crumb
Flaked breakfast cereals	Jellies
Sugar/honey-coated breakfast cereals	
Puffed breakfast cereals	Vegetables
	Onions
Fruit and fruit products	Garlic
Apples	Potatoes
Grapes	French fries
Coconut	
Peanuts	Other food products
Pectin	Assorted biscuits
Nuts	Sausage rusk
	Breadcrumb
Animal foods	Soya protein
Dog biscuits	Meat
Food concentrates	Gelatine

laboratory. The physical principles underlying the modelling of band dryers are essentially identical to those employed in modelling through-flow dryers for agricultural crops and are described in Chapter 3 (section 3.5). The techniques involved are also discussed at length by Keey (1992).

References

Baker, C.G.J. (1983) Cascading rotary dryers, in *Advances in Drying*, 2 (ed. A.S. Mujumdar), Hemisphere Publishing Co., New York, pp. 1–51.

Baker, C.G.J. (1988) The design of flights in cascading rotary dryers. *Drying Technol.*, 6(4), 631–53.

Holdsworth, S.D. (1997) Food preservation processes, in *Food Industries Manual*, 24th edn (eds M.D. Ranken, R.C. Kill and C.G.J. Baker), Blackie Academic & Professional, London, pp. 499–543.

Keey, R.B. (1992) *Drying of Loose and Particulate Materials*, Hemisphere Publishing Co., New York.

Kemp, I.C. (1994) Scale-up of pneumatic conveying dryers. *Drying Technol.*, 12(1&2), 279–97.

Langrish, T.A.G., Reay, D., Bahu, R.E. and Whalley, P.B. (1988) An investigation into heat transfer in cascading rotary dryers. *J. Separ. Proc. Technol.*, 9, 15–20.

Matchett, A.J. and Baker, C.G.J. (1987) Particle residence times in cascading rotary dryers. Part 1: Derivation of the two-stream model. *J. Separ. Proc. Technol.*, 8, 11–17.

Matchett, A.J. and Baker, C.G.J. (1988) Particle residence times in cascading rotary dryers. Part 2: Application of the two-stream model to experimental and industrial data. *J. Separ. Proc. Technol.*, 9, 5–13.

Matchett, A.J. and Sheikh, M.S. (1990) An improved model of particle motion in cascading rotary dryers. *Trans. Inst. Chem. Eng.*, 68(Part A), 139–48.

Papadakis, S.E., Langrish, T.A.G., Kemp, I.C. and Bahu, R.E. (1994) Scale-up of cascading rotary dryers. *Drying Technol.*, 12(1&2), 259–77.

Perry, R.H. and Green, D.W. (1984) (eds) *Chemical Engineers' Handbook*, 6th edn, section 20, McGraw–Hill, New York.

Schofield, F.R. and Glikin, P.G. (1962) Rotary dryers and coolers for granual fertilizers. *Trans. Inst. Chem. Eng.*, 40, 183.

Sokhansanj, S. and Wood, H.C. (1991) Simulation of thermal and disinfestation characteristics of a bale dryer. *Drying Technol.*, 9(3), 643–56.

10 Solar dryers

L. IMRE

10.1 Introduction

Preservation of human and animal food by open-air drying in the sun was presumably one of the first conscious and purposeful technological activities undertaken by humanity (Imre, 1995). Traditional open-air, solar drying methods are based on long-term experience and continue to be used all over the world to dry plants, fruits, seeds, meat, fish and other agricultural products in order to preserve them.

Over the last few decades, open-air drying has gradually become more and more limited because of the requirement for a large area, the possibilities of quality degradation, pollution from the air, infestation caused by birds and insects, and inherent difficulties in controlling the drying process (Imre, 1993). However, in the future, humanity will need to utilize solar energy for drying to a greater extent because of its advantages. Solar energy is free, renewable, abundant and an environmentally friendly energy source, which cannot be monopolized, and satisfies the global requirements for Sustainable Development (Statement, 1992).

Global requirements can and should be considered from an economic viewpoint, with the main influencing factors being savings and costs (Böer, 1978). Solar drying is not simply a method for substituting fossil fuels by solar energy, but is a technology based process for producing dried materials of the required quality (Imre, 1984, 1992). The quality of the dried product has an effect on the economic performance of the dryer (Beck and Reuss, 1993; Charters and James, 1993). It influences the marketing capability and income generating potential, since a higher price can be obtained for products of improved quality. For this reason, technologically and scientific based efforts have been made over the last three decades to develop the design, construction and operation of solar dryers.

The importance of solar drying is increasing worldwide, especially in areas where the use of abundant, renewable and clean solar energy is essentially advantageous. In the developing countries and in rural areas, traditional open-air drying methods will be substituted by more effective and more economic solar drying technologies (Garg, 1982; Mahapatra and Imre, 1989, 1990; Imre and Palaniappan, 1994).

10.2 Construction of solar dryers

When using the sun's radiation as the energy source for drying, two principal difficulties must be overcome in order to exploit the full potential of solar dryers: the periodic character and the time dependence of the solar radiation. The influence of weather changes should also be considered.

10.2.1 Functional parts of solar dryers

The main functional part of a solar dryer is the drying space, where the drying process takes place. In directly irradiated solar dryers, the energy is absorbed directly by the drying material. In this case the drying space has a transparent (glass or foil) cover.

In solar convective dryers, the drying medium is air preheated by solar energy in a collector, which converts solar radiation into heat. In direct heat transfer systems, the working fluid of the solar collector is the drying air itself. In the case of indirect systems, the working fluid is a liquid (e.g. water) and a liquid–air heat exchanger is used to preheat the drying air. This can be kept in motion by means of a fan driven by an electric motor. The latter is normally connected to the power grid or, in more remote rural areas, to an autonomous photovoltaic system or a wind turbine driven electric generator (Esper and Mühlbauer, 1993; Bansal and George, 1993). A chimney can also be connected to the drying space to induce air flow through a thermosyphon effect (Ekechukwu and Norton, 1993).

The periodic character of the solar radiation can partly be balanced by employing an intermittent drying operation. However, because the number of hours of available sunshine is very much dependent on the prevalent weather conditions, the use of a heat storage unit may be needed. Water can be used as the heat storage medium in indirect systems, and phase-change materials or a rock bed in direct systems.

The dependence of the dryer performance on the weather, especially in large systems, should be minimized by providing an auxiliary energy source, which would normally be based on natural gas, oil or biomass. It is connected to the dryer by means of a heat exchanger and operates in emergency situations only.

Appropriate instrumentation and control devices may be fitted, especially on high performance indirect solar dryers equipped with a heat storage unit and an auxiliary energy source, to facilitate efficient operation.

10.2.2 Main types of solar dryers

Solar dryers can be classified into three main groups, according to the type of energy used for drying and the equipment employed (Imre, 1988a, 1989, 1993, 1995):

- solar natural dryers
- semi-artificial solar dryers
- solar-assisted dryers.

The principal features of these dryers are described briefly in the paragraphs that follow. Fuller details are given in sections 10.3–10.5.

(a) Solar natural dryers.   These devices use only ambient energy and have no active elements. The air flow, if there is any, is maintained by natural convection or, in some cases, by thermosyphon effects induced by a chimney.

Solar natural dryers are mainly used as substitutes for traditional open-air drying methods in areas where no other source of energy is available. In contrast to these traditional methods, however, losses and damage to the product caused by rain, dust, insects, birds and other animals, as well as the pollution from the atmosphere, are avoided by the purpose built construction (i.e. cabinet and tent type arrangements). Their use results in a better quality product, which should be considered as a positive cash contribution (savings) in any economic evaluation.

(b) Semi-artificial solar dryers.   These usually feature a solar collector and a fan for maintaining a specified air flow through the drying space. In the case of directly irradiated solar tunnel dryers, a section of the tunnel may be employed as a transparent plastic covered solar collector (Imre, 1989, 1995; Lutz and Mühlbauer, 1986).

The use of semi-artificial solar dryers is justified by their unsophisticated and fairly cheap construction. They can be recommended for drying materials that are not sensitive to changes in the drying conditions caused by the periodic character of the solar irradiation and by the changing atmospheric conditions.

(c) Solar-assisted dryers.   Solar-assisted dryers are conventional dryers having a solar collector. They are generally fitted with a heat storage system and are always fitted with an auxiliary energy source, such as a thermo-generator fuelled by natural gas or oil, for use in situations where the solar energy collected is insufficient for drying purposes. They are normally fitted with modern control systems.

10.3 Solar natural dryers

10.3.1 Cabinet type dryers

(a) Cabinet dryers.   These are the simplest and cheapest type of solar dryer. They are generally used for drying agricultural products, such as fruits, herbs, vegetables, and spices, in small quantities (Wibulswas and Niyomkarn, 1980; Reuss, 1993).

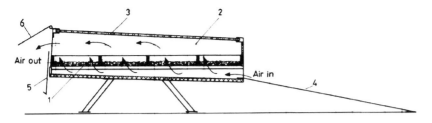

Figure 10.1 Cabinet type dryer for fruit and herbs.

Simple cabinet dryers consist of a south-facing drying space covered by a transparent material (glass or foil), which also serves as a roof to protect against rain and pollution. The material to be dried is spread in a thin layer on a tray having a perforated bottom and is directly exposed to the solar radiation. Holes are opened in the lower part of the front wall and in the upper part of the back wall of the cabinet to induce an air flow by natural convection. This carries away the water evaporated from the material and prevents its overheating.

An improved version of the simple cabinet dryer designed for household use is illustrated in Figure 10.1. The trays with perforated bottoms **1** are arranged under the transparent cover **3**, making it possible to dry simultaneously various kinds of foodstuffs, such as fruits, herbs and vegetables. Each product, in its individual tray, may have its own specific drying time. Air, preheated by a black-painted aluminium sheet **4**, is introduced into the dryer and regulated by the throttling devices **5** and **6**. The open cross-sectional area of the air outlet can gradually be decreased as drying proceeds. At night time, the slide **6** should be closed and the sheet **4** should be fitted to the transparent cover of the cabinet **2**. This dryer is available commercially from P. Fiorentini (Hungary) Ltd.

(b) Tent type dryers. The drying capacity of cabinet type dryers can be increased by enlarging the basic surface area of the dryer. Tent type dryers represent the simplest and cheapest way to achieve this goal. A schematic representation of a tent type solar dryer is shown in Figure 10.2. The drying space **2** is located within a tent, which may have a triangular or semi-circular cross-section. The south-facing surface of the cover **1** should be transparent; the

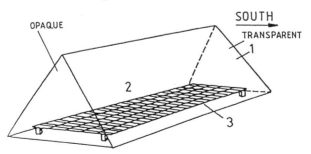

Figure 10.2 Tent type solar dryer.

other sides are opaque. The material to be dried (e.g. coffee beans) is spread on a perforated tray **3** or directly on a concrete floor (Sunworld, 1980).

(c) Shelf type dryers.   Another way to increase the drying capacity, without enlarging the basic surface area, is to employ several trays or shelves, one above the other. Figure 10.3 illustrates a schematic of a shelf type dryer (Wibulswas and Niyomkarn, 1980). Thin layers of fruit, vegetables, etc., to be dried are loaded on the perforated shelves **1**. The south-facing surfaces **2** of the dryer are transparent; the back wall **3** and the floor are painted black and insulated. Since the lower trays or shelves partly shade each other, a separate solar collector **4** should be connected to the drying space to produce the energy required. Air flow is generated by natural convection through the collector and the material layers on the shelves; moist air leaves the dryer through the upper opening **5**.

In the shelf type arrangement, the flow resistances of the separate material layers are coupled in series and the total resistance is their sum. To ensure a satisfactory air flow rate, the height of the dryer should be sufficient to produce an adequate 'chimney effect'.

10.3.2 Cabinet dryers fitted with a chimney

The thermosyphon effect required to induce an adequate flow through a solar dryer can be increased by means of a chimney. A simple version of a chimney type solar dryer is presented in Figure 10.4 (Sunworld, 1980). The lower part of the dryer acts as an integrated solar collector with a transparent cover **2**; the bottom surface **4** functions as an absorber. Material to be dried is arranged in a

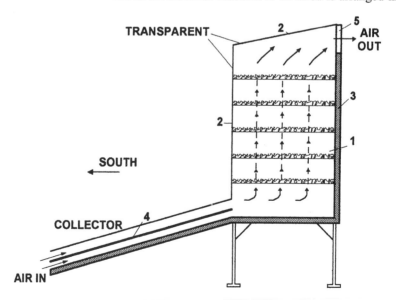

Figure 10.3 Shelf type cabinet dryer.

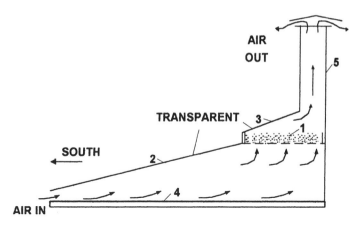

Figure 10.4 Cabinet dryer with chimney.

static bed **1**. Cover **3** over the bed is also transparent. Chimney **5** is connected to the upper part of the drying space. The technical specifications for a solar rice dryer have been given as (Sunworld, 1980):

- mass of rice to be dried, 1000 kg;
- bed thickness, 0.1 m;
- collector surface area, 23 m^2;
- height of the chimney, 5 m;
- drying time (at 15 MJ m^{-2} day^{-1} mean global irradiation), 3–4 days.

10.3.3 Cabinet dryers fitted with a chimney and heat storage

To extend the drying process into the non-sunny hours of the day, heat storage should be employed. Water vessels have been used as a relatively low cost means of heat storage (Puigalli and Lara, 1982).

Sensible-heat storage capacity can be increased by using phase change materials. Figure 10.5 shows a simplified schematic of a chimney type cabinet dryer in which calcium chloride hexahydrate (CaCl$_2$.6H$_2$O) is employed for this purpose (Imre, 1986b, 1989). The phase change material is enclosed in plate shaped containers. The fixed latent heat storage plates **4** are arranged as the absorber of the separate collector **2**, and of the northern wall of the chimney as well. The outer-wall surfaces **6** of the chimney **3** are thermally insulated. Additional heat storage plates **5** are movable and arranged below the cover of the collector. In sunny hours, these plates **5** are pulled out of the collector on both sides and are directly irradiated and charged. During night-time operation, they are pushed back below the collector covering (see figure), forming an air duct with hot upper and lower surfaces. Since the phase change process occurs at constant temperature, a constant thermosyphon effect and a continuous air flow through the dryer will be maintained until the heat storage unit has been fully discharged. The material to be dried is spread in thin layers on trays **7**

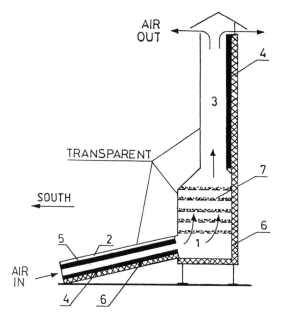

Figure 10.5 Cabinet dryer with chimney and with latent heat storage.

having perforated bottoms located in the drying space **1**. Upper and lower trays can be exchanged during drying to equalize differences in drying rates. The air flow rate through cabinet dryers fitted with a chimney can also be enhanced by using a wind-driven ventilator connected to the upper end of the chimney (Bansal and George, 1993).

10.4 Semi-artificial solar dryers

10.4.1 Solar tunnel dryers

Solar tunnel dryers are frequently used for drying agricultural products (Imre, 1986a, 1995; Lutz and Mühlbauer, 1986; Esper and Mühlbauer, 1993; Bansal and George, 1993). Figure 10.6 shows a simplified arrangement of a serial solar

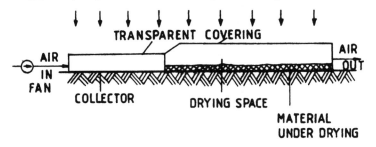

Figure 10.6 Solar tunnel dryer (serial arrangement).

tunnel located on the ground. The required mass flow rate of air is blown into the tunnel by means of a fan. The tunnel has a transparent covering. The first section serves as a solar collector with a black-bottomed surface for preheating the air. The connecting section of the tunnel forms the drying space. Material to be dried is spread on the bottom and is directly irradiated by the sun.

The length of the solar tunnel dryer can be reduced by arranging the collector and the drying section in parallel (Figure 10.7a). The fan is driven by an electric motor connected to the grid or fed by electricity generated by solar photovoltaic or wind-driven equipment. The cross-section of the dryer is shown in Figure 10.7b. A transparent-foil covering sheet is held in place by springs and black foil is used as an absorber and as the bed surface of the material to be dried. Structural elements are supported on concrete foundations.

Solar tunnel dryers are successfully used on farms for drying fruit, crops, vegetables, herbs, coffee beans, cocoa, and coconuts and can be built using materials available locally. The size of the dryer may be adapted to the drying capacity required. For tropical use, a biomass assisted version of the solar tunnel dryer has been developed.

The cross-section of the tunnel may be triangular, semi-circular, or have a flat cover (Mühlbauer *et al.*, 1993; Janjai and Hirunlabh, 1993). A curved or semi-circular shaped cross-section is generally used in the case of the serial collector–dryer arrangement.

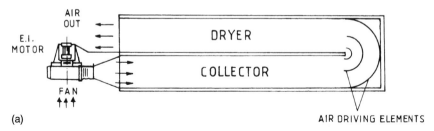

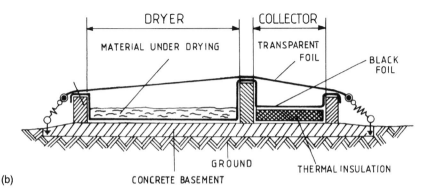

Figure 10.7 Solar tunnel dryer (parallel arrangement).

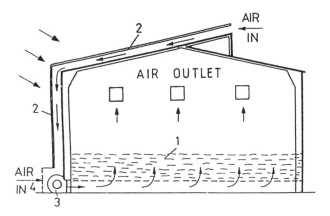

Figure 10.8 Solar room dryer.

10.4.2 Greenhouse-type solar dryers

Foil-covered tunnel type greenhouses can also be used for drying agricultural products (Bansal and George, 1993; Reuss, 1993; Mühlbauer *et al.*, 1993). They always incorporate a fan for maintaining a continuous flow of air, which is sometimes partially recirculated. Where sensitive materials are dried, some kind of shading is needed and continuous feeding can be used. In this case, the material to be dried is moved through the drying space on a belt or a conveyor.

Greenhouse type solar dryers having tilted flat-roof surfaces have been developed for drying herbs and aromatic plants. These feature roof integrated plastic collectors and air recirculation.

10.4.3 Solar room dryers

For drying agricultural products in larger quantities (e.g. grains, fodder), solar room dryers with a static bed are used successfully (Dernedde and Peters, 1978; Wieneke, 1980; Imre, 1995).

Figure 10.8 represents a typical solar room dryer. Air flow through the bed **1** is maintained by a fan **3** selected according to the flow resistance of the material layer. The bottom of the bed is perforated and air is blown into the distribution space beneath it. Preheating of the air occurs in the roof integrated solar collector **2**, using the roof cover as heat absorber. Atmospheric air, entering through the inlet **4**, can be mixed with preheated air to control the inlet air temperature if required. Moist air leaves the room through the air outlets arranged on the side walls. For emergency situations, such as bad weather conditions, a simple auxiliary gas operated air heater can be connected to the suction side **4** of the fan. Solar room dryers with roof integrated collectors have also been used successfully for drying fruits such as peanuts and bananas (Vaughan and Lambert, 1980; Bowrey *et al.*, 1980).

10.5 Solar-assisted dryers

Solar-assisted dryers are conventional dryers to which supplementary equipment is added to enable a significant proportion of the thermal energy required for drying to be replaced by solar energy. In these types of dryer, a planned, and generally optimized drying process can be achieved to obtain superior product quality and good economic performance. Any influence of the weather conditions on product quality and on the performance of the dryer can be eliminated by using an independent energy source, if needed, and proper control facilities.

10.5.1 Solar-assisted dryer for seeds and herbs

In Figure 10.9, the ground plan and the cross-section of the dryer are presented (Imre, 1986a, 1995; Imre *et al.*, 1986a; Catalogue, 1986). For economic reasons, the drying space is divided into two cells. Each cell has its own individual

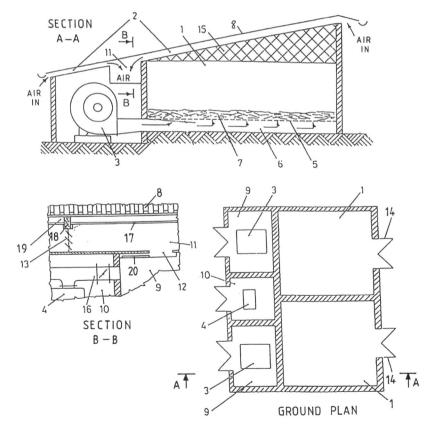

Figure 10.9 Solar-assisted dryer for seeds and herbs.

2-speed fan **3**. Fans are arranged in separate chambers **9**. The conventional energy source of the dryer, a burner **4** fuelled by natural gas, is situated in chamber **10**.

An uncovered air collector **2**, integrated into the roof structure, is used to convert the solar radiation into thermal energy. The conventional roof covering, made of trapezoidal galvanized steel blades **8** with black-painted surfaces, serves as heat absorber (see also Section B-B in Figure 10.9 and Figure 10.17g). A continuous covering of monolith plates **17**, mounted at an appropriate distance beneath the absorber, forms the air ducts **19** between the rafters **18**. These ducts **19** connect into a distribution channel **11** located in the ceiling sections of chambers **9** and **10**. The size of the free openings **12** between the distribution channel **11** and the fan chambers can be changed by means of the sliding-plate dampers **20**. In this way, the total air flow through the collector can be shared between the fan chambers, when valve **13** is open. Fans **3** blow the air into the drying cells. Thermogenerator **4** moves the heated air into the fan ducts in the required ratio through ducts **16**, when required.

In the layer type version of the dryer, the material to be dried is loaded into the drying space **1** through entry doors **14**. It is evenly spread in a bed **7** having a perforated bottom **5**. Air is blown into the space **6** under the holding structure of the bed and passes through the supported solids. Moist air leaves the drying space through the openings **15** in the side walls.

In Figure 10.10, a schematic diagram of the containerized version of the dryer is shown. For fragile materials that may be subject to damage during handling, the use of containers with perforated bottoms **21** is preferred. These are placed on foot stools with proper air stuffing **22** between them.

The main technical specifications for a seeds' dryer are (Catalogue, 1986):

- effective surface area of one cell, 56 m^2;
- dry mass of seed in one cell, 5600 kg;
- mass flow rate of air (one fan),
    at 1090 rpm, 41 000 m^3 h^{-1}
    at 475 rpm, 21 500 m^3 h^{-1};

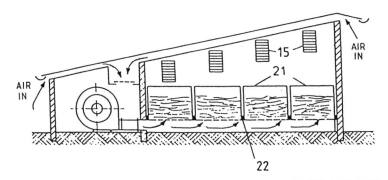

Figure 10.10 Solar-assisted dryer (containerized version).

- surface area of the collector field, 191 m^2;
- output of the thermogenerator, 93 kW.

The use of a two-cell arrangement is motivated by economic reasons. The in-feeding and drying processes in the two cells can be staggered. In this way, the average drying rate and the average energy consumption can be approximately equalized between the cells. The solar energy gained can be distributed between the cells according to actual need and can be used more effectively in the dryer by reducing the exit losses, especially in the falling-rate drying period.

The fans are rated according to the layer thickness employed in the beds and the air inlet temperature required; the fan should be operated at the lower rpm when higher air temperatures are needed.

10.5.2 Solar-assisted dryers with heat storage

Connecting physical heat storage equipment to solar dryers is justified by three main reasons.

- The daily drying time can be extended to those hours of the day when the sun does not shine.
- Overdrying can be prevented by storing the surplus solar energy collected in the peak radiation periods.
- The inlet temperature of the drying air can be controlled.

Natural and artificial materials can be used for heat storage. Natural materials include e.g. water, pebble beds and rock beds, which store only sensible heat. The artificial materials employed are generally salt solutions, which also store latent heat (section 10.3.3).

(a) Solar-assisted dryers with water heat storage.   Figure 10.11 depicts a solar-assisted dryer with water heat storage (Auer, 1980). The working fluid of the solar collector **1** is water, circulated by pump **2** along the pipe **3**. The heated water flows to the heat storage tank **4**. Atmospheric air is drawn by fan **7** through the water–air heat exchanger **5**. The primary working heat exchange medium is the water circulated by the pump **6** from the heat storage tank **4**. An auxiliary air heater **8** can be used if required to maintain the prescribed inlet air temperature. Material to be dried is arranged in a static bed **9**.

(b) Integrated solar-assisted high performance dryers.   Preservation of agricultural products by solar drying is promising, since harvest time, on the whole, coincides with the period of highest-intensity solar irradiation. At the same time, however, seasonal variations limit the annual operating time and the solar energy utilization. In the case of high performance solar dryers with a large collector area and a high heat storage capacity, year-round operation of the collectors is essential in order to achieve an acceptable economic return. This

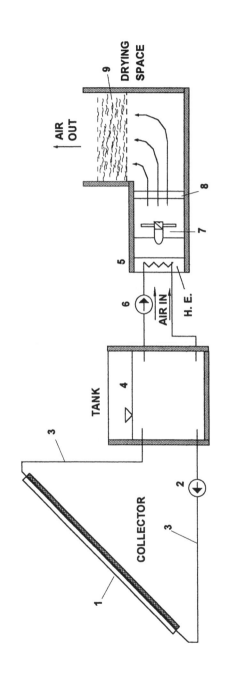

Figure 10.11 Solar-assisted dryer with water type heat storage.

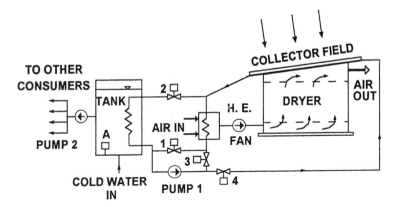

Figure 10.12 Simplified schematic of a complex, integrated solar drying system.

goal can be achieved by the complex use of the solar collectors, which are capable of supplying heat, not only for drying, but also for satisfying other needs such as providing a hot water supply to a farm, and heating or air conditioning of greenhouses or animal houses. For such purposes, indirect solar dryers with a water type collector come to the fore. The water tank can be used as heat storage for the solar dryer as well as for a hot-water system (HWS). At the same time, the conventional energy source of the HWS can be used as the auxiliary heater for the solar dryer.

Figure 10.12 illustrates a complex, integrated high performance solar dryer (Imre, 1986, 1989, 1995; Imre et al., 1986b, 1988, 1991). It was designed to produce dried lucerne hay for dairy farming. During idle periods, the system produces hot water for technological purposes. The drying space is arranged in a barn, which also serves for storing the dried hay. The single-glazed solar flat-plate collector is integral with the roof structure. It can be connected to the water–air heat exchanger of the dryer and to the hot water tank serving as the heat storage vessel by proper manipulation of valves **1** to **4**.

The available drying space in the barn is divided into four cells (Figure 10.13 (Imre, 1989)). Each cell has its own fan and water–air heat exchanger. Hot water produced by the solar collector can be distributed at will between the cells. Auxiliary energy is provided by gas operated thermogenerators. For emergency situations, electrical heating is built into the tank. This can also be used as an additional auxiliary energy source. The principal system specifications are (Imre et al., 1986b):

Solar flat-plate collector
Working fluid, water; surface area, 900 m^2
Maximum volumetric flow rate of the working fluid, 15.6 m^3 h^{-1}

Dryer
System, static bed dryer; area, 800 m^2

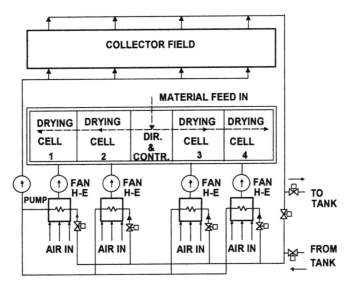

Figure 10.13 Schematic of dryer with four-cells arrangement.

Number of drying spaces (cells), four (200 m^2 each)
Total layer thickness, 6 m
Annual quantity of dried material, 1000 t
Storage capacity, 700 t

Air system
Indirect, preheating the air by water–air heat exchangers
Number of fans, four (one fan per cell). Air flow rate per fan, 106 000 m^3 h^{-1}
Primary auxiliary energy sources, biogas generators

Storage tank
Volume, 100 m^3
System, stratified
Secondary auxiliary energy source in tank, electricity (400 kW) for night-time
operation.

The system has several modes of operation (Imre, 1987). The actual mode is
selected according to the technological requirements and the meteorological
conditions. The four-cell arrangement offers the possibility of various drying
programs. In selecting the mode of operation, a priority is assigned according to
the solar energy collected: 'dryer' has the highest priority, 'heat storage' is
second, followed by the 'hot water consumers' and 'other heat consumers'. The
microprocessor based control system is described below in section 10.8.5 (Imre,
1987; Imre *et al.*, 1983a, b).

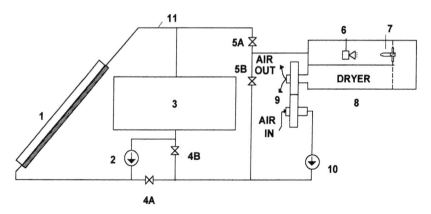

Figure 10.14 Solar-assisted raisin dryer with rock-bed heat storage.

10.5.3 Solar-assisted dryers with rock bed heat storage

In solar-assisted direct type dryers, rock bed, gravel bed and pebble bed heat storage systems are frequently used (Auer, 1980; Read *et al.*, 1974; Contier and Farber, 1982; Niles *et al.*, 1978; Choudhury *et al.*, 1995).

A high performance raisin dryer featuring a gravel bed heat storage system is shown in Figure 10.14. The collector field **1** of 1812 m^2 surface area is located on the ground. Fresh air is blown through the collector loop by the fan **10** connected to a heat recovery unit **9** (dampers **4B** and **5B** are closed). Air, preheated in the collector, flows in the air duct **11** to the location where the gas burner **6** is situated. The burner serves as auxiliary heater and operates when the temperature of the air arriving from the collector is lower than required. Warm air is transported into the drying tunnel **8** by the fan **7**. Moist air leaves the dryer through the heat recovery unit **9**. The rock bed heat store **3** can be charged by the collector when the fan **2** is in operation and air valves **4A** and **5A** are closed.

The dryer can also be operated by stored energy: dampers **4A** and **5B** are closed and fan **2** is switched off. Partial charging of the heat storage unit or partial use of stored energy can be achieved by the proper regulation of dampers **4A** and **4B** and the operation of fan **2**. With this dryer, it proved possible to provide some 69% of the annual drying energy consumption by solar means.

10.5.4 Solar-assisted dryer combined with adsorbent units

Complex and integrated solar-assisted dryers can be combined with adsorbent beds, which may serve for energy storage as well as for auxiliary heating (Imre *et al.*, 1982; Twidell *et al.*, 1993; Imre, 1995). The energy storage capacity of an adsorbent (e.g. zeolite or silica gel) is partly attributable to sensible heat and partly to moisture adsorption. The specific total energy storage capacity is about an order of magnitude greater than that of a physical heat store.

The dynamic adsorption capacity of the adsorbents should be exploited by regeneration at a temperature normally exceeding 150°C. However, a considerable part of the energy used for regeneration can also be used for other purposes, e.g. for hot water production. When solar energy is not available, the adsorption capacity can be used to operate the dryer.

Solar dryers fitted with adsorbent units have proved to be economically competitive with those incorporating rock bed energy storage. However, the economic comparison is greatly influenced by the local conditions, opportunities, and requirements (Imre, 1974; Close and Dunkle, 1977; Pinaga et al., 1981).

10.5.5 Solar-assisted dryers combined with heat pumps

Moist air leaving the drying space has almost the same enthalpy as the dry air had when it entered the dryer. A considerable part of the energy used for drying can be regained as latent heat by cooling the air and condensing its water vapour content. This can be achieved using a heat pump connected to the air outlet of the dryer.

Figure 10.15 illustrates the construction of a complex, integrated solar lucerne dryer combined with a heat pump (Imre et al., 1982; Imre, 1995). Moist air flows from the dryer **7** to the evaporator **9** of the heat pump. Here the air is cooled and moisture condenses. This provides the latent heat required to evaporate the working fluid of the heat pump, which circulates through the compressor **10**, the condenser **11**, the expansion valve **12**, and back to the evaporator. Cold water is heated in the condenser **11** and fed to the water heat

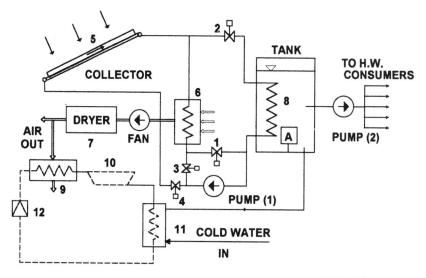

Figure 10.15 Complex, integrated solar-assisted dryer combined with heat pump.

storage tank **8**. Depending upon the ambient conditions, air leaving the evaporator **9** can be recirculated to the dryer heat exchanger **6**. Solar-assisted drying equipment combined with a heat pump and heat storage has been developed for drying peanuts (Auer, 1980).

10.6 Economic evaluation of solar dryers

Evaluation of the economics of solar dryers is more ambiguous than for many other technologies. The main reason for this lies in the special interpretation of the savings (C_s) and costs (C_c) as the main influencing factors (Böer, 1978; Imre, 1988a, 1990).

Various criteria have been proposed and used for characterizing the economics of solar systems (Lunde, 1980; Duffie and Beckman, 1993). A simple criterion frequently used to compare investment alternatives is the payback time. This can be defined as the time (n' years) required for the accumulated savings to equal the sum of the initial capital investment and the maintenance costs. In this case, n' can be determined from the equation (Böer, 1978)

$$\frac{(1+i)^{n'} - (1+i_{eq})^{n'}}{i - i_{en}} C_s = C_c(1+i)^{n'} + \frac{(1+i)^{n'} - (1+i_{eq})^{n'}}{i - i_{eq}} R_m C_c \quad (10.1)$$

where i is the interest rate, i_{eq} and i_{en} are the rates of inflation in the prices of equipment and energy, and R_m is the maintenance–cost factor. The shorter the payback time, the better the return on the investment.

10.6.1 Savings

Savings should be interpreted as the extra income generated by the solar dryer as compared to a conventional dryer. The total annual savings C_s can be calculated as a sum of three components:

$$C_s = \sum_{k=1}^{3} C_{s,k}. \quad (10.2)$$

The principal savings $C_{s,1}$ can be taken as the cost of conventional energy that is displaced by solar energy. In remote areas, this cost should also include any costs associated with transporting the energy to the dryer location.

Savings $C_{s,2}$ are achieved by improving the quality of the dried product and reducing losses as a result of installing a solar dryer. In cases where the product quality, and hence the market price, are lower than those of product dried in a conventional dryer, component $C_{s,2}$ should be taken as negative.

When evaluating solar drying on the level of a national economy, e.g. for decision making purposes, a further component $C_{s,3}$ should be considered. This is the savings achieved as a result of avoiding environmental pollution. In calculating $C_{s,3}$, social, environmental and other external costs arising from energy production, which would be a burden on the public purse, should be added to the utility price (Ottinger, 1991).

Solar natural drying should be compared with open-air drying; in this case, only component $C_{s,2}$ can be regarded as being relevant. Semi-artificial solar dryers should be compared to conventional forced-convection dryers having a similar construction and production capacity. Solar-assisted dryers should be compared to conventional dryers without solar energy converters.

10.6.2 Investment costs

For solar natural dryers, C_c should be taken as the initial investment cost of the dryer. For semi-artificial solar dryers and for solar-assisted dryers, the investment cost of the solar energy converter should be taken as C_c in the calculation, when other parts of the dryer are of the same construction. If not, C_c should be taken as the difference between the initial costs of the two systems.

In Figure 10.16, payback time is shown as a function of C_s/C_c for different values of the parameters i, R_m, i_{en}, and i_{eq}.

10.7 Design of solar dryers

The design of solar dryers is a complex task, with several aspects having to be considered (Imre, 1993). It has three main phases.

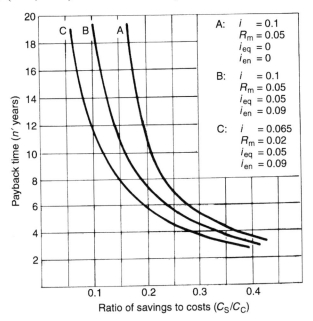

Figure 10.16 Payback time as a function of C_s/C_c for various parameters.

1. Definition of the drying process to be performed by the dryer to achieve the desired end-product quality and dryer performance.
2. Selection of the proper type of solar dryer to be used.
3. Undertaking the detailed design of the solar dryer.

Details of these phases are given in the following sections.

10.7.1 Definition of the drying process

Correct post-harvest treatment of the material to be dried is essential in order to avoid considerable deterioration of the crops. Such damage is irreversible and cannot be corrected by drying. The designer of a solar drying facility should therefore be fully aware of this and make due allowance for it. The initial condition of the material may also have some effect on the drying characteristics.

(a) Material characteristics. The most important data for the designer are the:

- initial moisture content of the material;
- final moisture content of the material required for the safe storage;
- maximum permitted temperature of the material during drying.

Table 10.1 summarizes these data for some common agricultural products (Mahapatra and Imre, 1990). In cases where a temperature range is given, the lower value should be used at the beginning of the drying cycle and the higher value at the end. Further data, such as the drying characteristics of the material, are also needed in order to specify the drying process. In cases where these data are not available in the literature, experimental measurements should be undertaken to determine sorption isotherms and to produce drying curves appropriate to the envisaged range of processing conditions. The main structural characteristics of the material can be recognized from such curves. Numerical methods involving inverse solutions may also be applied to the drying rate data to determine material characteristics such as moisture diffusivity and shrinkage coefficients as functions of moisture content (Imre, 1995). It is also advisable to collect as much information as possible on past practical experience of the product in question from all available sources.

(b) Preliminary considerations. A preliminary drying process can be formulated from consideration of the initial conditions and the drying characteristics of the material. The designer should also consider the time dependence of the solar radiation and the expected effects of intermittent drying. The physical characteristics of the material (e.g. lumpy, granular, thread-like) will also impact on the type of dryer selected. An early initial estimate of the optimum layer thickness should be made.

10.7.2 Selection of solar dryer type

In deciding which type of solar dryer to specify, the designer will have to take the following considerations into account:

- the quantity of the material to be dried;
- the anticipated throughput of the dryer;
- the sensitivity of the material to be dried to temperature and other operating parameters;
- the required quality characteristics of the dried product;
- local conditions (e.g. availability of solar and other sources of energy, investment costs, production and maintenance infrastructure).

The throughputs that can be handled by different types of solar dryers are as follows (Imre, 1993).

- *Solar natural dryers* can handle small quantities (e.g. in household use) of less sensitive materials, under conditions where the local infrastructure is limited.

Table 10.1 Drying data for some agricultural products

Product	Moisture content (%) (wet basis) Initial	Final	Drying temperature (°C)
Bananas	80	15	70
Barley	18–20	11–13	40–82
Cardamom	80	10	45–50
Cassava	62	17	70
Chillies	90	20	35–40
Coffee seeds	65	11	45–50
Copra	75	5	35–40
Corn	28–32	10–13	43–82
French beans	70	5	75
Garlic	80	4	55
Grapes	74–78	18	50–60
Green forages	80–90	10–14	30–50
Hay	30–60	12–14	30–50
Medicinal plants	85	11	35–50
Oats	20–25	12–13	43–82
Onions	80–85	8	55
Peanuts	45–50	13	35
Pepper	80	10	55
Potato	75–85	10–14	70
Rice	25	12	43
Sorghum	30–35	10–13	43–82
Soy-beans	20–25	11	61–67
Sweet potato	75	7	75
Tea	75	5	50
Virginia tobacco	85	12	35–70
Wheat	18–20	11–14	43–82

- *Semi-artificial solar dryers* are recommended for moderate (e.g. farm) quantities of materials that are not sensitive to intermittent drying. A medium level local infrastructure is required.
- *Solar-assisted dryers* are recommended for industrial drying purposes. They can handle large production quantities and yield consistent product quality. A high level local infrastructure is required.

10.7.3 Structural design of solar dryers

In order to undertake the structural design of a solar dryer, it is essential to be aware of the technological capability of the dryer manufacturer. Therefore, the designer should keep close contact with the technical and production staff at the factory. The structural design will greatly influence the investment and maintenance costs as well as the service life of the dryer. These factors have a major influence on the return on investment.

(a) Selection of the materials of construction.   The materials of construction of the structural elements is of great importance. The designer should consider:

- the UV effects of the solar radiation;
- possible counter-effects between the structural elements and the material to be dried;
- the corrosive effects of the atmosphere.

(b) Structural design of the solar collector.   The solar collector serves to provide the primary supply of energy to the dryer. Essentially, the absorber within the collector converts the direct and diffuse solar radiation into heat, which is then transferred to the working fluid of the collector (Lunde, 1980; Wijeysundera *et al.*, 1982; Duffie and Beckman, 1993; Imre, 1995). A great variety of collectors has been developed. These can either be used separately or integrated into the dryer (Wisniewski, 1993).

Figure 10.17 illustrates the constructional features of various kinds of solar collector. Frequently, the working fluid is air, which is used as the drying medium (direct type solar dryers). It flows between the transparent cover **1** and the absorber plate **2** (Figure 10.17a), or between the absorber **2** and the thermal insulation **3** (Figure 10.17b and c). The absorber **2** may have a plane (Figure 10.17a and b), waved, or corrugated surface (Figure 10.17c, f, and g), or a stepwise divided and overlapped surface (Figure 10.17d). A collector having a matrix type absorber is shown in Figure 10.17e, with air flowing through the porous matrix. Figures 10.17f and g illustrate uncovered collectors, which are frequently integrated into the roof of a drying room and used as the roof covering (Imre *et al.*, 1986a; Catalogue, 1986). In the case of solar dryers incorporating an indirect heat transfer system and heat storage, the collector working fluid is usually water or antifreeze solution, flowing in the tubes of the tubesheet type or finned tube type (Figure 10.17h) absorber.

In hybrid (air–liquid) collectors, air is directed between the transparent covering and the liquid absorber. In combined collectors, latent heat storage elements serve as absorber (Imre and Kiss, 1983; Imre, 1995).

The materials of construction commonly used in solar collectors are as follows.

- Transparent covering: glass, UV stabilized plastic or foil.
- Absorber, with air as working fluid: aluminium, copper, steel, UV stabilized plastic; with water as working fluid: copper, UV stabilized plastic.
- Coating of the absorber: selective black coating or paint.

The designer should determine the energy flux required by the collector, ϕ_u (W), from the equation (Imre, 1995):

$$\phi_u \cong \xi A_c I, \tag{10.3}$$

where A_c (in m^2) is the required surface area of the collector, I (in W m^{-2}) is the expected solar energy flux, and ξ is the expected effectiveness of the collector.

Various methods have been recommended for the simplified calculation of the collector efficiency (Lunde, 1980; Duffie and Beckman, 1993; Imre, 1995).

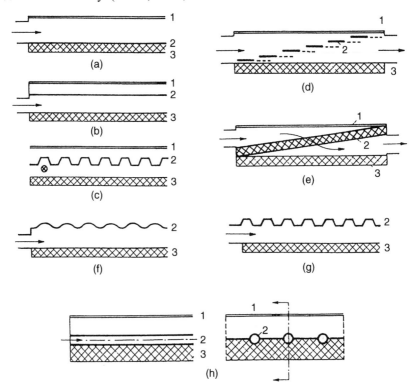

Figure 10.17 Solar collectors: (1) transparent cover; (2) absorber; and (3) thermal insulation.

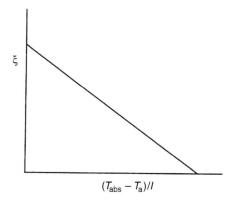

$(T_{abs} - T_a)/I$

Figure 10.18 Instantaneous efficiency of a solar flat-plate collector.

These methods are based on calculation of the heat loss from the collector, which is a function of the temperature difference between the absorber (T_{abs}) and the atmosphere (T_a).

The instantaneous efficiency diagram for a flat-plate collector is shown in Figure 10.18 (Imre, 1995). It can be seen that the efficiency decreases as the absorber temperature T_{abs} increases. The maximum effectiveness ξ_{max} is obtained when $T_{abs} = T_a$. Under these conditions, $\xi_{max} = \tau\alpha_c$ where τ is the transparency of the cover and α_c is the absorptance of the absorber coating.

In the simplified calculation methods, the 'overall heat transfer coefficient' of the collector (U) is used. The value of U is a function of several other parameters and describes the resultant effect of different and complex non-linear heat transfer processes.

The simplified calculation methods do have several weak points and their use can only be recommended for determining a first approximation of the required collector surface area. In the detailed design phase, the performance of the collector should be checked using computer simulation methods (Imre, 1974, 1995; Imre and Kiss, 1983; Mahapatra and Imre, 1995). These methods can also be used to undertake a sensitivity analysis and to optimize the design. Details of the solar radiation and meteorological data relating to the given geographic location are required as inputs to the simulation. For commercially available solar collectors, efficiency diagrams are given in the catalogues.

10.7.4 Modelling and simulation of solar dryers

Simulation of solar dryers offers a valuable tool for the designer to:

- check the performance and the effectiveness of the dryer at the different stages of the design, and to optimize the design;
- analyse the sensitivity of the system to changes in the various input parameters;

- specify a control strategy for the dryer by considering the effects of anticipated disturbances.

An in-depth discussion of the physical and mathematical details underlying a typical simulation of a solar dryer is beyond the scope of this chapter. The interested reader is referred to articles in the literature published on this topic (Imre, 1974, 1981, 1986b, 1987, 1989, 1993, 1995; Imre et al., 1980, 1983a, b, c, 1984, 1986b, 1988; Imre and Szabó, 1974, 1978; Imre and Kiss, 1983; Mahapatra et al., 1994; Mahapatra and Imre, 1995).

The design of solar dryers is principally a synthesis problem, which may have a number of solutions. Demands and requirements related to the drying process, as well as to the economic performance of the system, should be considered and, in this way, the number of the acceptable solutions can and should be reduced.

Practically, the designer approaches the solution to the problem through a series of successive analyses. After evaluating the results obtained in one analysis, the parameters are modified in the subsequent iteration. This process is continued until a satisfactory result is achieved. The 'guess and try' method can be made more efficient if the expected effect of the actual changes on the operation of the dryer can be forecast. This is made possible by means of a sensitivity analysis in which the effect of each variable on the output parameters is assessed.

10.8 Dryer operation and process control strategies

10.8.1 Operational aspects

The performance of a solar dryer is determined largely by its design and construction, but is also affected by material characteristics and operating procedures. The effectiveness of any process control strategy adopted will also impact on performance. Such a strategy can be formulated by due consideration of the possible operating parameters amenable to regulation. Sensitivity analysis may be employed to evaluate the anticipated effects of different interactions (Imre and Szabó, 1974, 1978; Imre, 1981; Mahapatra and Imre, 1995).

Even in the case of the most simple solar dryers, the drying process should be continuously checked and appropriate regulation undertaken when needed. The principal actions involved in formulating an appropriate strategy for the direction of the operation are (Imre, 1988b, 1992, 1995):

- feeding a new layer of fresh material into the dryer;
- turning the layer of drying material, in the case of simple solar dryers;
- regulating the air flow rate;
- regulating the recirculation of air;
- separating the drying space from the atmosphere (e.g. at night);
- regulating the operation of the auxiliary energy source;

- regulating the break interval in the case of an intermittent drying process;
- regulating the operation of the heat storage system;
- distributing the solar energy collected between the drying spaces of multi-cell dryers;
- regulating the mode of operation of complex solar dryers.

Process control actions aim at maintaining the values of targeted operating parameters at their required set points. The principal parameters that require to be controlled are:

- temperature of the working fluids;
- mass flow rate of the flowing fluids;
- relative humidity of the drying air;
- limiting (maximum or minimum) values of certain parameters.

Proper control of the above parameters will result in control of the:

- rate of drying
- intermittent drying
- recirculation
- charging of the thermal storage.

10.8.2 Less sophisticated solar dryers

In the less sophisticated tent type, cabinet type, greenhouse type and chimney type solar dryers, air flow is induced by natural convection or maintained by means of a fan. Material is generally arranged in a thin layer and it is partly or totally irradiated. On the surface of the layer, material will dry quickly, while at the bottom of the layer, drying will start only after some delay. In the case of sensitive materials, deterioration may occur in the lower layers. A possible means of avoiding this is to turn the layer over from time to time. In the case of solar dryers without heat storage, cool-down effects causing possible moisture condensation should be prevented by closing up and covering the drying space during the night. In chimney type solar dryers with heat storage, the air flow can be maintained during the night (sections 10.3.2 and 10.3.3).

10.8.3 Solar-assisted dryers

The operation and control of solar-assisted (industrial) dryers aim at achieving, as far as possible, an optimized drying process, to ensure high product quality, efficient energy utilization, and economic performance.

(a) Energy effectiveness of solar dryers. The solar energy actually utilized in the solar dryer is not equal to the heat flux collected from the incident solar irradiation (ϕ_o) and transferred into the drying air (ϕ_a) but, rather, the energy flow effectively used in the drying process, ϕ_e. The total energy effectiveness e_t

of the solar dryer can be interpreted as the product of the collection efficiency
($e_c = \phi_a/\phi_o$) and the drying efficiency ($e_d = \phi_e/\phi_a$). Thus:

$$e_t = e_c \cdot e_d = \frac{\phi_e}{\phi_o}. \tag{10.4}$$

While the collection efficiency e_c is determined practically by the design of the
dryer and its collector, the efficiency of the drying process e_d can highly be
influenced by the control strategy adopted.

The energy consumed in the drying process depends on the drying rate of the
material, which is a function of the moisture content distribution in the layer or
bed. In the case of materials exhibiting a constant-rate drying period being dried
in a static bed with through-flow of air, the drying front will move upwards
through the bed. When it reaches the upper surface, the drying rate will start to
fall. As a consequence, the energy consumption is a function of time:

$$\phi_e(\Delta t) = \lambda \Delta Y(t) G', \tag{10.5}$$

because ΔY (the increase in the absolute humidity of the drying air) is changing
with time, depending on the drying rate. In this equation, λ denotes the latent
heat and G' the air flowrate per unit area. In the falling-rate period of drying, ΔY
and e_d decrease. The average value of the drying effectiveness

$$e_{d,avr} = \frac{1}{t_f \phi_a} \int_0^{t_f} \phi_e(t)\, dt, \tag{10.6}$$

where t_f is the drying time, can be improved by increasing the thickness of the
bed since, in this way, the time spent in the falling-rate period will decrease.
However, it should also be considered that the increase in bed thickness will
cause a delay in drying the surface layer, with a consequent added risk of
deterioration in the case of sensitive materials.

A really effective way to increase the drying efficiency is to employ
'multiple-feeding'. Its basic principle is to feed a new fresh layer in when the
drying front has reached the surface of the bed. This method is particularly
useful in the case of materials not exhibiting a constant-rate drying period.

(b) Intermittent drying operation. In the falling-rate drying period, the drying
rate, and thus e_d, decrease. This is caused by the increasing internal mass
transfer resistance of the material as it dries out. Intermittent drying is a strategy
designed to counteract this effect. Its basic principle is that the first drying
period $\Delta t_{l,d}$ is followed by a break, $\Delta t_{l,b}$, and so on. During the break, the fan is
switched off, and the uneven moisture distribution inside the material tends to
equalize. That is why, in the next drying period, $\Delta t_{2,d}$, the drying rate, and also
e_d, will be higher than the corresponding values at the end of the previous
drying period. Appropriate values of Δt_d and Δt_b can be determined by optimization (Imre, 1988b, 1992; Imre *et al.*, 1983a–c).

In the case of solar dryers not having any heat storage, the drying process is necessarily intermittent, with breaks during the night. However, these do not generally satisfy the optimal strategy.

In the case of multi-cell solar dryers, the intermittent drying operation can be effected by staggering the drying processes in the individual cells. In this manner, it is possible to utilize more solar energy in the cells in which drying is occurring, while the other cells are in the break phase (Imre, 1989).

10.8.4 Solar dryers with recirculation

When drying materials that do not exhibit constant-rate drying, the outlet air may still have some drying potential. By recirculating a part of the outlet air and mixing it with the inlet air (before or after the solar collector), the effectiveness of the solar dryer can be improved, as it is possible by this means to increase the drying temperature.

In the case of indirect solar dryers having a liquid type solar collector (Figure 10.19), it is possible to supply other heat consumers with hot water. Air is preheated by a liquid–air heat exchanger. The proportion of the air that is recirculated can be determined by consideration of the wet-bulb temperature T_{wb} of the inlet air. A pair of linked dampers, V_2, is used as the actuator element. For the inlet air temperature T_1, the controlling signal is the dry-bulb temperature T_{db}, and the actuator element is the valve V_1.

10.8.5 Solar dryers fitted with a water storage tank and auxiliary heater

The system shown in Figure 10.20 consists of three flow-loops:

- collector–storage tank loop with pump P_1;
- storage tank–heat exchanger (H.E.) loop with pump P_2;
- open air loop with the fan transporting the heated air into the dryer.

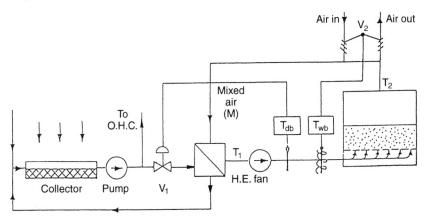

Figure 10.19 Control scheme of an indirect solar dryer with recirculation.

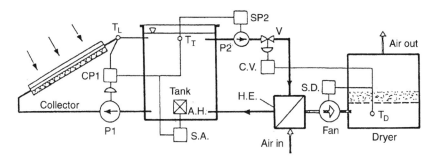

Figure 10.20 Control scheme of a solar-assisted dryer with water storage and auxiliary heater.

The pump P_1 and the water type collector are switched into operation by the control element CP1 when the collector outlet temperature $T_L \geq T_T$, where T_T is required water temperature in the tank. The control signal for CP1 is $(T_L - T_T)$. The auxiliary energy source in the tank (A.H.) is operated by the on–off control device S.A. when T_T is lower than its set-point value T_{TS}.

In the case of a combined solar drying–hot water supply system, where the collector outlet can be connected directly to the liquid side of the heat exchanger, various operating modes, including intermittent drying, can be employed with the aid of microprocessor based control (section 10.5.2(b)). The temperature of the inlet air required for drying (T_D) is controlled by the device C.V. in the loop connected to the liquid side of the heat exchanger. The actuator element is the valve V. In the air loop, the fan is turned on by means of the thermostat S.D. Operation of the pump P_2 is induced by the thermostat SP2. For emergency situations in large industrial drying systems, a second auxiliary heater, such as a gas burner, can also be connected to the air inlet duct.

References

Auer, W.W. (1980) Solar energy systems for agricultural and industrial process drying, in *Drying '80* (ed. A.S. Mujumdar), Hemisphere, New York, p. 280.

Bansal, N.K. and George, O.P. (1993) Solar drying in India and other Asian countries, in *Drying '80* (ed. A.S. Mujumdar), Hemisphere, New York, p. 15.

Beck, M. and Reuss, M. (1993) Performance measurements on a solar drying plant for grass seeds, in *Drying '80* (ed. A.S. Mujumdar), Hemisphere, New York, p. 117.

Bowrey, R.G. *et al.* (1980) Use of solar energy for banana drying. *Food Technology in Australia*, **32**, 290.

Böer, K.W. (1978) Payback of solar systems. *Solar Energy*, **20**, 225.

Catalogue (1986) Solar assisted dryer for seeds and herbs. S.T.S.D. and T.U.B., Budapest.

Charters, W.W.S. and James, K.R. (1993) *Solar Crop Dryers*, in Proceedings of ISES Solar World Congress, Budapest, Vol. 8, p. 31.

Choudhury, C., Chauhan, P.M. and Garg, H.P. (1995) Economic design of a rock bed storage device for storing solar thermal energy. *Solar Energy*, **55**, 29.

Close, D.J. and Dunkle, R.V. (1977) Use of adsorbent beds for energy storage in drying and heating systems. *Solar Energy*, **19**, 233.

Contier, J.P. and Farber, E.A. (1982) Two applications of a numerical approach of heat transfer processes within rock-beds. *Solar Energy*, **29**, 451.

Dernedde, W. and Peters, H. (1978) Effectiveness of solar air-collectors for dryers (in German), *Landtechnik*, **1**, 29.

Duffie, J.A. and Beckman, W.A. (1993) *Solar Engineering of Thermal Processes*, Wiley, New York.

Ekechukwu, O.V. and Norton, B. (1993) *Design and Measured Performance of a Solar Chimney for Natural-circulation Solar Dryers*, in Proceedings of ISES Solar World Congress, Budapest, Hungary, Vol. 8, p. 78.

Esper, A. and Mühlbauer, W. (1993) Development and dissemination of solar tunnel dryers, *Proceedings of the Expert Workshop on Drying and Conservation with Solar Energy*, Budapest, Hungary, ISES, Budapest, p. 63.

Garg, H.P. (1982) *Solar Drying – Prospects and Retrospects*, in Proceedings of the 3rd International Drying Symposium (ed. J.C. Ashworth), Wolverhampton, UK, p. 353.

Imre, L. (1974) (ed.) *Handbook of Drying* (in Hungarian), Müszaki Ki, Budapest.

Imre, L. (1981) Heat transfer simulation of composite devices, in *Numerical Methods in Heat Transfer*, Vol. I (ed. R.W> Lewis, K. Moragn and O. Zienkiewicz), Wiley, Chichester, Chap. 3.

Imre, L. (1984) *Aspects of Solar Drying*, in Proceedings of Fourth International Drying Symposium, Kyoto, Japan.

Imre, L. (1986a) Technical and economical evaluation of solar drying. *Drying Technology*, **4**, 503.

Imre, L. (1986b) (ed.) *Development, Construction and Demonstration of Solar Dryers for Herbs* (Report), Technical University, Budapest, Hungary.

Imre, L. (1986c) Construction of a large roof-integrated collector, FAO (CNRE Workshop on Heating of Animal Houses, Section 1, Lund).

Imre, L. (1987) Solar lucerne dryer of high performance, FAO Solar Drying Working Group Meeting, Hochenheim.

Imre, L. (1988a) *Interpretation of the Economy Evaluation of Solar Dryers*, in Proceedings of the Tenth Congress on Energy, Opatija, pp. 23–28.

Imre, L. (1988b) Direction and Control of Solar Dryers, in *Manual of Industrial Solar Dryers*, UNESCO Report (ed. M. Daguenet), Chap. 2.

Imre, L. (1989) *Solar Drying in Hungary*, in Proceedings of ASRE '89 Symposium, Cairo, Egypt, Vol. 2, p. 1029.

Imre, L. (1990) *Economic Evaluation of Solar Drying*, in Proceedings of Utilization of Renewable and Solar Energy for Agriculture (ed. I. Segal), Jerusalem, p. 55.

Imre, L. (1992) *Analysis and Direction of the Operation of Solar Dryers*, ASRE '92, No. 22, Cairo, Egypt.

Imre, L. (1993) *General Aspects for Designing Solar Dryers*, in Proceedings of an Expert Workshop on Drying and Conservation with Solar Energy, Budapest, Hungary, p. 7.

Imre, L. (1995) Solar drying, in *Handbook of Industrial Drying* (ed. A.S. Mujumdar), Marcel Decker Inc., New York, 2nd edn, Chap. 12, p. 373).

Imre, L. and Szabó, I. (1974) Static and dynamic analysis of dryers by the mass and energy flow network method. *Periodica Polytechnica, Elec. Eng.*, **18**, 145.

Imre, L. and Szabó, I. (1978) *Role of Relative Transfer Factors in Controlling the Operation of Dryers*, in Proceedings 1st International Symposium on Drying, Montreal (ed. A.S. Mujumdar), Science Press, Princeton, p. 76.

Imre, L. and Kiss, L.I. (1983) Transient thermal performances of composite devices, in *Numerical Methods in Heat Transfer*, Vol. II (eds R.W. Lewis, K. Morgan and B.A. Schrefler), J. Wiley, Chichester, Chap. 15.

Imre, L. and Palaniappan, C. (1994) *Development of Solar Drying*, Planters Energy Network, India.

Imre, L., Kiss, L.I., Környey, T. and Molnár, K. (1980) Digital simulation of solar hay drying system, in *Drying '80* (ed. A.S. Mujumdar), Hemisphere, New York, p. 446.

Imre, L., Kiss, L.I. and Molnár, K. (1982) *Complex Energy Aspects of Solar Agricultural Drying Systems*, in Proceedings of the 3rd International Drying Symposium (ed. J.C. Ashworth), Drying Research Ltd, Wolverhampton, UK, p. 370.

Imre, L., Farkas, I., Kiss, L.I. and Molnár, K. (1983a) *Numerical Analysis of Solar Convective Dryers*, in 3rd International Conference on Numerical Methods in Thermal Problems, Seattle, WA.

Imre, L., Molnár, K. and Farkas, I. (1983b) *Some Aspects of the Intermittent Solar Drying of Lucerne*, in Proceedings of the International Conference on Solar Drying and Rural Development, Bordeaux, p. 93.

Imre, L., Farkas, I., Kiss, L. and Molnár, K. (1983c) Numerical analysis of solar convective dryers, in *Numerical Methods in Thermal Problems*, Vol. 3 (eds R.W. Lewis, K.A. Johnson and W.R. Smith), Pineridge, Swansea, p. 957.

Imre, L., Molnár, K. and Szentgyörgyi, S. (1984) Drying characteristics of lucerne, in *Drying '84* (ed. A.S. Mujumdar), Hemisphere, Washington, p. 428.

Imre, L., Fábri, L., Gémes, L. and Hecker, G. (1986a) Solar assisted dryer for seeds, *Drying Technology*, **4**, 503.

Imre, L., Farkas, I. and Gémes, L. (1986b) Construction, simulation and control of a complex and integrated agricultural solar drying system, in *Drying '86* (ed. A.S. Mujumdar), p. 676.

Imre, L., Fábri, L., Farkas, I. and Hecker, G. (1988) *Direct and Indirect High Performance Solar-Assisted Dryers*, in Proceedings on the Development and Use of Efficient Solar Systems, H-ISES, Budapest, p. 273.

Imre, L., Hecker, G. and Fábri, L. (1991) *Solar Assisted High Performance Solar Dryer*, in Proceedings of ISES Solar World Congress 1991, Denver, CO, Vol. 2, Part II, Pergamon, Oxford, p. 2184.

Janjai, S. and Hirunlabh, J. (1993) *Experimental Study of Solar Fruit Dryer*, in Proceedings of ISES Solar World Congress, Budapest, Hungary, Vol. 8, p. 123.

Lunde, P.J. (1980) *Solar Thermal Engineering*, Wiley, New York.

Lutz, K. and Mühlbauer, W. (1986) Solar tunnel dryer with integrated collector. *Drying Technology*, **4**, 583–604.

Mahapatra, A.K. and Imre, L. (1989) Energy alternatives: the Indian perspective. *International Journal of Ambient Energy*, **10**, 163.

Mahapatra, A.K. and Imre, L. (1990) Role of solar-agricultural drying in developing countries. *International Journal of Ambient Energy*, **11**, 205.

Mahapatra, A.K., Imre, L., Barcza, J., Bitai, A. and Farkas, I. (1994) Simulation of a directly irradiated solar dryer with integrated collector. *International Journal of Ambient Energy*, **15**, 195.

Mahapatra, A.K. and Imre, L. (1995) *Parameter Sensitivity Analysis of a Directly Irradiated Solar Dryer with Integrated Collector*, in Proceedings of ISES Solar World Congress, Harare, Zimbabwe, Maha 0044.

Mühlbauer, W., Esper, A. and Müller, J. (1993) *Solar Energy in Agriculture*, in Proceedings of ISES Solar World Congress, Budapest, Hungary, Vol. 8, p. 13.

Niles, P.W.E., Carnegie, E.J., Pohl, J.G. and Cherne, J.M. (1978) Design and performance of an air collector for industrial crop dehydration. *Solar Energy*, **20**, 19.

Ottinger, R.L. (1991) *Environmental Costs of Electricity*, Pace University Center for Environmental Legal Studies, New York.

Pinaga, F., Carbonell, J.V. and Pena, J.L. (1981) *Experimental Simulation of Solar Drying with Energy Storage in Adsorbent Bed*, in 7th International Congress of Chemical Engineering, Prague, Czech Republic, pp. 5–8.

Puiggali, J.R. and Lara, M.A. (1982) *Some Experiments about Small Country Solar Dryers*, in 7th International Congress on Chemical Engineering, Prague, Czech Republic, p. 390.

Read, W.R., Choda, A. and Cooper, P.I. (1974) A solar timber kiln. *Solar Energy*, **15**, 309.

Reuss, M. (1993) *Solar Drying in Europe*, in Proceedings of an Expert Workshop on Drying and Conservation with Solar Energy, Budapest, Hungary, p. 89.

Statement (1992) Environmental Policy Statement, '92 Global Forum, Rio de Janeiro, Brazil.

Sunworld (1980) *Sunworld*, **4**, Special Issue on Solar Drying and Evaporation.

Twidell, J.W., Muniba, J. and Thornwa, T. (1993) *The Strathclyde Solar Crop Dryer: Air Heater, Photovoltaic Fan and Desiccant*, in Proceedings of ISES Solar World Congress, Budapest, Hungary, Vol. 8, p. 55.

Vaughan, D.H. and Lambert, A.J. (1980) An Integrated Shed Solar Collector for Peanut Drying, *Transactions of ASAE*, 218.

Wibulswas, P. and Niyomkarn, C. (1980) *Development of Solar Air-heaters for a Cabinet-type Solar Dryer*, in Regional Workshop on Solar Drying, CNED–UNESCO, Manila, Phillipines, p. 1.

Wieneke, W. (1980) Bin drying of grain and grass with solar heater air. *Agricultural Mechanisation in Asia*, p. 14.

Wijeysundera, N.E., Ah, L.L. and Tijoe, L.E. (1982) Thermal performance study of two-pass solar air heaters. *Solar Energy*, **28**, 363.

Wisniewski, G.D. (1993) *Methods of Rational Solar Collectors Selection for Crop Drying*, in Proceedings of ISES Solar World Congress, Budapest, Hungary, Vol. 8, p. 97.

11 Dryer selection

C.G.J. BAKER

11.1 Introduction

In many cases, it is still true to say that the selection of equipment for the drying of food products and ingredients remains predominately an art, in which knowledge, experience and science all play important roles. Often, there is no 'right' answer in the absolute sense, as more than one solution is both technically and economically viable. However, a careful evaluation at the outset of as many as possible of the factors influencing the choice will help to narrow down the options.

This chapter presents a systematic approach to the selection of dryers, which, it is hoped, will provide the reader with the necessary guidance and assistance, particularly when faced with the challenge of choosing a dryer for a new product or application. It does not seek to replace, but rather complement, the advice, assistance, and use of test facilities that can be offered to potential clients by reputable dryer manufacturers, who have amassed considerable experience in the drying of foods. It will, however, be particularly valuable in circumstances where outside advice cannot be sought at the time for reasons of confidentiality. A prior knowledge of the probable type of dryer to be employed will also help the user select an appropriate vendor to work with.

It should be appreciated that the choice of a dryer and the manner in which it is operated can affect not only the process of moisture removal but also other characteristics of the dried product. This is of particular importance in the drying of food materials, where factors such as appearance, flavour, colour, texture, dispersion and dissolution, and rehydration properties can all influence the consumer's perceived view of product quality. Food functionality can also be affected by the drying process. For example, Bruce and Nellist (1992) reported that seed viability and loaf volume were both diminished when the original wheat was damaged by subjecting it to drying conditions that lay outside a given moisture content–temperature envelope. Vitamin loss is also associated with drying under conditions that are too severe for the product in question.

In this chapter, the reader is presented with a structured method of dryer selection. This involves an iterative approach (Figure 11.1), which includes the following steps: drawing up the process specifications, making a preliminary selection, planning and conducting directed bench-scale tests, making an economic comparison of alternatives, conducting pilot-scale trials, and, finally, selecting the most appropriate dryer type. Each of these steps is described in some detail in sections 11.2 to 11.6. Worked examples of the selection procedure are presented in section 11.7.

The selection procedure does not necessarily apply in cases where particular circumstances point to the use of a specific type of dryer from the outset. Examples include on-farm drying of wheat, for which a number of specific dryers have been developed (see Chapter 3), and the drying of fruit in hot climates, where solar dryers (Chapter 10) could well be the most appropriate.

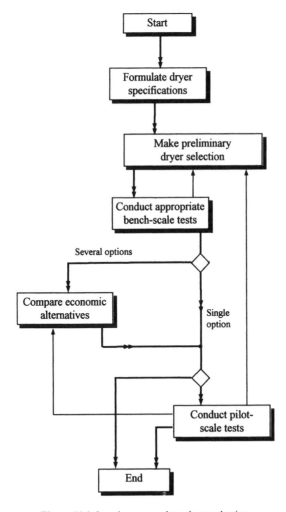

Figure 11.1 Iterative approach to dryer selection.

11.2 Specification of the drying process

The first step in selecting a dryer is to draw up the detailed requirements for the process. Accurate qualitative and, where possible, quantitative information on the following aspects should therefore be specified:

- the form of feed and product (section 11.2.1);
- the nature of the upstream and downstream processing operations (section 11.2.2);
- moisture contents of feed and product (section 11.2.3);
- dryer throughput (section 11.2.4);
- physical properties of the feed and product (sections 11.2.5 and 11.2.6);
- quality changes during drying (section 11.2.7);
- drying kinetics (section 11.2.8).

Any one or more of the above factors may, depending on circumstances, influence the choice of dryer. In the case of foods and other natural biological products, it is normal for quality considerations to override economic factors. In many cases, drying is the final processing step prior to packaging and can therefore exert a major influence on the ultimate quality of the product. In others, it can be an important intermediate operation, which is vital to subsequent processing operations.

A checklist is very useful for recording the information summarized above. Figure 11.2 illustrates a typical example of such a checklist. As well as helping to quantify the principal parameters, it will highlight areas where there is a deficiency of knowledge and where subsequent physical tests should be performed.

11.2.1 Form of feed

An obvious first question to ask is 'Is the feed a liquid or a solid?' If the answer to this question is 'a liquid', then the number of possible choices of dryer will be restricted. For present purposes, the term 'liquid' can include solutions, suspensions, slurries, gels and thin pastes.

In certain instances, it may be advantageous to modify the character of the dryer feed, as follows.

- Liquid feeds, such as milk and fruit juices, are normally preconcentrated by evaporation prior to spray drying, as the energy and capital requirements are significantly lower.
- Solid feedstocks may be subjected to size reduction, flaking, pelletization or extrusion prior to drying. The principal reason for such treatment is normally to transform the feed into the ultimate product structure required while it is still relatively pliable. In other cases, its purpose is to facilitate an increase in the drying rate or to manipulate the feed into a form that can be dried by a specific type of dryer.

- Where the feed has a high moisture content, as is commonly found with gels and colloidal suspensions, it may be desirable to blend it with dried product in order to make it more amenable to processing in a wider range of dryers. Wet particulate solids can be very sticky. This often gives rise to blockage of the dryer at the feed end. Again, this problem can be alleviated by blending feed and dried product.

1.	Name of food material to be dried	....
2.	Form of feed	Liquid / Paste or Gel / Sticky Particles / Free-flowing Particles / Granules, Pellets or Extrudates / Other (...............................)
3.	Mean throughput	....(Wet/Dry Basis)
4.	Likely variation of throughput	....
5.	Upstream equipment	Batch / Continuous
6.	Downstream equipment	Batch / Continuous
7.	Mean particle size of feed	....
8.	Size distribution of feed	....
9.	Density of feed	....
10.	Bulk density of feed	....
11.	Inlet moisture content of feed	....(Wet/Dry Basis)
12.	Likely variation of inlet moisture content	....(Wet/Dry Basis)
13.	Maximum temperature to which feed can be exposed	...................................°C / K / °F
14.	Form of product	Free-flowing Particles / Sticky Particles / Granules, Pellets or Extrudates / Other (...............................)
15.	Mean particle size of product	....
16.	Size distribution of product	....
17.	Product shape	Spherical / Irregular / Plates / Needles / Other
18.	Bulk density of product	....
19.	Outlet moisture content of product	....(Wet/Dry Basis)
20.	Permissible variation of outlet moisture content	....
21.	Maximum temperature to which product can be exposed	...................................°C / K / °F
22.	Estimated drying time	...................................(s/min/h)
	Basis of estimate	Prior experience / Oven tests / Small-scale equipment tests / Other (...........................)
23.	Drying kinetics data available	Yes / No
24.	Equilibrium moisture content	....(Wet/Dry Basis)
25.	Other pertinent information relating to product, e.g. quality parameters, critical processing conditions, fire / explosion hazards.	....

Figure 11.2 Drying process checklist.

11.2.2 Upstream and downstream processing operations

With some notable exceptions, such as the drying of grain on the farm, the dryer will form an integral part of a production line. In general, therefore, if production is carried out on a batch basis, a batch dryer would normally be selected. Conversely, if the production is continuous, the dryer chosen would, in all probability, also operate continuously. Where a production line operates in a semi-continuous manner in that it consists of a mixture of batch and continuous operations, or where the dryer constitutes the last processing operation, there is obviously some room for discretion in making the choice.

Batch dryers are more suitable than continuous dryers in situations where there are frequent product changes. This is particularly the case where hygiene or contamination considerations dictate that it is necessary to clean the dryer between batches. Operating labour requirements are generally significantly higher than those of continuous dryers.

A number of supplementary considerations could influence the flow sheet configuration and/or the choice of batch versus continuous dryer operation. Questions that must be answered include the following.

- Will the feed deteriorate if there is a delay between upstream processing and drying? (This would tend to favour the use of a continuous dryer.)
- Should the product be allowed to cool between drying and subsequent processing, such as milling or flaking, or packaging? (This would necessitate the provision of a continuous cooler or an allowance for batches of dried product to cool after drying.)
- Does the dryer throughput match that of the other unit operations in the production line? (Batch dryers tend to have a lower capacity than continuous dryers.)
- Is the production rate reasonably constant or is it variable or intermittent? (Continuous dryers are best suited to constant-processing conditions.)

11.2.3 Moisture content of feed and product

(a) Definitions.   Moisture content is commonly defined on a wet or a dry basis, and care must be taken to specify which system is being used as failure to do so can result in much confusion. The wet-basis moisture content (X_w, kg moisture per kg wet solids) is defined as

$$X_w = M_w/(M_w + M_s) \qquad (11.1)$$

where M_w is the mass of moisture and M_s the mass of dry solids. It is frequently expressed as a percentage.

The dry-basis moisture content X is normally expressed as a ratio (kg moisture per kg dry solids):

$$X = M_w/M_s. \qquad (11.2)$$

The relationship between X_w and X is:

$$X = X_w/(1 - X_w). \tag{11.3}$$

At low moisture levels, X_w and X are numerically very similar. For example, a wet-basis moisture content of 5% ($X_w = 0.0500$ kg moisture per kg wet solids) is equivalent to a dry-basis value of 0.0526 kg moisture per kg dry solids. In contrast, there are large differences at higher moisture contents; $X_w = 0.50$, for instance, is equivalent to $X = 1.00$.

(b) Influence of moisture content on dryer selection

Food products and ingredients range from liquids (e.g. milk), through gels, to coarse and fine moist solids (e.g. crystalline sugar and starch, respectively). Both the inlet and outlet moisture content will influence the choice of dryer. For example, only a limited number of dryers are capable of handling liquid feeds, and certain continuous dryers are prone to blockage, particularly at the inlet, if the feed is too wet and cohesive. More subtly, some dryers are capable of removing only relatively small amounts of surface moisture, whereas others are able to remove much large quantities of bound bulk moisture.

Note that relatively small variations in the inlet moisture content can have a significant effect on the evaporative load. For example, consider a dryer designed to reduce moisture content from 30% to 5%. If the moisture content of the feed rises to 35%, the evaporative load will increase by nearly 30%.

The outlet moisture content specification is dictated by a number of factors including handling and storage characteristics, quality considerations, equilibrium moisture content, and hygroscopicity. Ideally, the product should be dried to a level beyond which there is no benefit to be derived from removing additional moisture. The desired product moisture content can also dictate the choice of dryer, since certain types are better suited to yielding low moisture products than others. It is important to remember that excessive drying is very costly in terms of both capital requirements and fuel consumption. Therefore, careful specification of the maximum acceptable product moisture content is a very important factor in minimizing costs.

It should be borne in mind that there is absolutely no advantage in drying the product to a moisture content below the equilibrium moisture content pertaining under the envisaged storage conditions, as it would ultimately regain moisture from its surroundings.

The moisture content distribution should also be considered; maximum and minimum values may need to be specified as well as the mean value. In certain cases, the moisture will distribute itself evenly throughout the product in an acceptable length of time after it has emerged from the dryer. However, if moisture transport is impeded as a result of, for example, case hardening, microbial growth may occur in the wetter regions, rendering it unusable. If, on the other hand, part of the product is overdried, its quality may deteriorate. For example, agglomerates may break up and generate unwanted fines, which will detract from the appearance and functionality of the product.

In the case of foods, water activity (a_w), which is discussed in some depth in Chapter 2, is as important as, or even more important than, moisture content *per se*. It is defined as:

$$a_w = p/p_w,\qquad(11.4)$$

where p denotes the partial pressure of water exerted by the wet food and p_w is the vapour pressure exerted by pure water at the same temperature. Water activity is generally considered to provide a measure of the availability of water for microbial growth, and for enzymic and chemical reactions. It can therefore be used as a guide for defining an appropriate product moisture content, which is likely to prove effective in retarding quality deterioration. Note, however, that there is no unique relationship between these two parameters as a_w is strongly influenced by the composition of a particular food. Thus, at a given moisture content, a_w can be lowered by the addition to the food of humectants such as glycerol, sorbitol or other polyhydric alcohols, sodium or potassium chlorides, or certain organic acids (e.g. formic) or bases (e.g. urea) (Ranken *et al.*, 1997a). Plots of moisture content against water activity have been published widely in the literature; see, for example, Iglesias and Chiriffe (1982).

Ranken *et al.* (1997b) quote the following a_w ranges for microbial stability. Normal bacteria are unlikely to grow below $a_w = 0.91$, normal yeasts below 0.88, normal moulds below 0.80, halophilic (salt-loving) bacteria below 0.75, xerophilic (dry-loving) fungi below 0.65, and asmophilic (sugar-tolerant) yeasts below 0.60. The growth of pathogenic bacteria does not occur below $a_w = 0.86$ and, for short storage periods at room temperature, $a_w = 0.70$ is generally considered safe.

11.2.4 Dryer throughput

The solids throughput is a principal factor influencing dryer selection. Broadly speaking, low throughputs favour the selection of batch dryers whereas high throughputs are best handled by continuous dryers. It is not possible to define rigorously 'low' and 'high' throughputs since these will vary widely with both the nature of the material being processed and the initial and final moisture contents. The evaporation rate, perhaps, provides the best guide.

Specifying dryer throughput is not a trivial task. The dryer must be capable of processing product at a rate equal to or greater than that of the upstream equipment or it will become a bottleneck in the system. The simplest definition of throughput (in units of e.g. $kg\,h^{-1}$) is annual production divided by total operating time. This, however, may mask a number of important considerations. For example, instantaneous throughputs may vary widely from this mean value and the dryer selected should be sufficiently versatile to handle the anticipated range. Due allowance should also be made for scheduled (e.g. production breaks, routine maintenance, cleaning) and unscheduled downtime (e.g. break-downs). The anticipated turndown (ratio of maximum to minimum throughputs) should also be considered. Some dryers are more flexible than others in handling variable feedrates.

Note that throughput may be defined either on a wet or a dry basis. In the case of the former, the total mass flowrate of solids will progressively decrease as moisture is evaporated during the course of drying. Throughput defined on a dry basis will, however, remain constant during the drying process.

11.2.5 Physical properties of the feed

The physical properties of foods that are most likely to affect dryer selection are: size and size distribution, shape, density, and bulk and surface, including flow, characteristics. Thermal properties, such as heat capacity, may have some influence on dryer size, but are unlikely to influence the selection of one class of dryer over another.

If the feed is a liquid, the physical properties that may influence dryer selection are its rheological characteristics and, where appropriate, the size and concentration of the suspended particles, their solubility, and their abrasiveness.

If the feed is a solid, this will also prompt further consideration of its physical characteristics. These include:

(a) Size. The characteristic dimension(s) of the wet feed entering the dryer not only affect the materials handling requirements of the dryer but also influence the drying rate. The design features of certain types of dryer make them more suitable for processing large, discrete particulates, including flakes and extrudates, whereas others are more appropriate for drying fine powders. The size distribution may also need to be considered.

It is also worthy of note that the size of particulate products can change during the course of drying as a result of attrition, or the breakdown or formation of agglomerates. This can influence both dryer selection and operation. Food products are frequently combustible. Those that are fine or capable of generating dust (e.g. starch, flour and sugar) are also potentially explosible. Appropriate precautions, as outlined by Abbott (1990) for example, must be employed when drying these materials.

(b) Density. Density can influence the selection of dispersed-phase dryers. The use of pneumatic conveying or fluidized bed dryers, in particular, may not be possible if the material is too dense, particularly if the particles are also large. Density, like size, may also change during the course of drying, particularly if the bulk of the moisture is removed from internal pores.

(c) Structure. The internal structure of a particulate material will have a major influence on the drying rate and, therefore, on the choice of dryer. Materials

having an open, porous structure will dry relatively fast, as will those whose moisture is restricted to the outer surface. However, food products are often characterized by cellular structures through which moisture migration is relatively slow. This limits the drying rate, particularly in cases where high temperatures cannot be tolerated because of the resulting damage to the product. Both optical and scanning election microscopy can provide considerable insights into the structural characteristics of foodstuffs.

(d) Surface characteristics. Dryer choice is also influenced by surface characteristics such as stickiness, which makes handling difficult. The problem is normally most acute with the wet feed but, in certain cases, can also occur at intermediate moisture contents.

11.2.6 Physical properties of the product

In certain instances, the physical properties of the product will differ from those of the feed in several respects, even when both are solids. It is important to note these differences, as the requirements could affect the choice of dryer.

In most cases, the size of the particulates will be specified, albeit in a rudimentary manner in some instances. The required size may be the same as that of the feed, or different. Size reduction can normally be effected by milling the dried product post-drying. Size enlargement can be achieved through the appropriate use of fluidized bed dryer/granulators.

The shape of the dried product can also influence dryer selection in certain cases. For example, spray dryers normally produce approximately spherical particles, whereas drum dryers and agitated dryers tend to produce flaky or irregular shaped product.

Bulk density is another important product characteristic, which, again, may influence the choice of dryer. It is affected not only by the particle density, which is an intrinsic property of the material being dried, but also by particle size and size distribution.

Drying may also bring about chemical changes in the product. These can manifest themselves in, for example, a change in functionality, such as with wheat gluten, or the development of colour and taste. These 'quality' requirements should be specified at the outset as many are highly dependent upon the nature of the drying process.

It is always important to remember that dehydration of food products may sometimes be best accomplished by employing a combination of processes and this option should always be considered. If an agglomerated product, such as milk powder or instant coffee powder, is required, it is common practice to employ two-stage drying in which a spray dryer is followed by a plug flow fluidized bed in which the agglomeration occurs and the residual moisture is removed in a more economical manner.

11.2.7 Quality changes during drying

The drying process can impart both desirable and undesirable changes to the foodstuff undergoing dehydration. These changes can, at least in part, be influenced by the choice of dryer and, in many cases, are dependent upon the temperature–moisture content–time history of the product. For example, freeze dried foods are, in many cases, considered to be of greatly superior quality to their air dried counterparts because they are not exposed to elevated temperatures during the dehydration process. Some of the quality parameters that need to be considered in selecting a drying process are given in the following sections.

(a) Colour and taste.   The heating of foodstuffs that occurs during drying processes may induce changes to both colour and taste. As these primary attributes are closely associated by the consumer with the perceived quality of the food, their development needs to be both understood and controlled. Moreover, these changes can continue, albeit at a slower rate, during storage of the dried product, thereby limiting shelf life.

Jayaraman and Das Gupta (1995), for example, discussed colour changes that can occur during the drying of fruit and vegetables. These authors considered both the destruction of natural pigments during dehydration and the development of colour and flavour through both enzymic and non-enzymic browning reactions. They also stressed the importance of pre-treatment processes in reducing colour changes. Thus, sulphur dioxide was reported to have a pronounced effect on protecting carotenoids in unblanched carrots during drying. Moreover, enzymic browning, which occurs naturally in both fruit and vegetables, can also be avoided by pre-sulphiting and, in the case of vegetables, an additional step, blanching. In this latter process, the vegetable is exposed to either near-boiling water or steam for typically 2–3 min in order to inactivate the enzymes. This not only minimizes colour change but also reduces the development of off-flavours and the loss of vitamin C. Non-enzymic browning occurs in a wide range of products through complex (Maillard) reactions between amino groups, as found in proteins, and carbonyl groups, present in reducing sugars, to yield insoluble brown polymeric pigments known as melanoidins. This becomes very pronounced at high temperatures and is frequently accompanied by flavour development and nutrient loss. Jayaraman and Das Gupta (1995) reported that the addition of sulphites during the pre-drying step was the only effective means at present of controlling non-enzymic browning in dried fruit and vegetable products. However, as certain individuals, albeit a very small percentage of the population, are hypersensitive to sulphite, there is a growing pressure within the industry to phase out its use.

(b) Reconstitution.   Ultimately, many dried foods are reconstituted by the addition of water. The speed of reconstitution and the appearance of the product

are both important features that need to be adequately considered in selecting a
drying process.

The reconstitutability of food products in piece form, such as sliced or diced
vegetables and meat, depends principally on the internal structure of the dried
pieces and the extent to which water-holding components (e.g. proteins and
starch) have been damaged during drying (Brennan *et al.*, 1990). These authors
indicate, for example, that the rate at which air dried vegetables reconstitute can
depend on the initial drying rate. If this is high, case hardening may occur and
the outer layers become rigid. As drying proceeds, the tissues split and rupture
internally. This gives rise to a relatively open structure, which favours rapid
reconstitution. Excessive heating, however, may reduce the water holding
capacity of the product and give rise to a poor appearance. Freeze dried foods,
on the other hand, often reconstitute rapidly because of the open, porous
structure that is characteristic of products processed in this manner.

A number of factors may influence the reconstitution of dried powders such
as soup and gravy mixes, and hot and cold beverages. These include (Brennan *et
al.*, 1990) the following physical properties.

Wettability.   Wettability describes the ability of a particle to adsorb water on
its surface, thus initiating reconstitution. This process is affected both by the size
of the particles and the nature of their surface. Small particles tend to clump
together, thereby reducing the rate at which water is able to penetrate individual
particles. This effect can be reduced by increasing the particle size or agglomer-
ating the powder. The presence of free fat on the surface can also impede
wetting; it may be possible, however, to counteract this by the use of surfactants
such as lecithin.

Sinkability.   Sinkability describes the ability of particles to sink rapidly in
water; it depends principally on their size and density. Thus, large, dense
particles sink more rapidly than small, light particles. Note that the presence of
occluded air may reduce sinkability.

Dispersibility.   Dispersibility is the term used to describe the ease with
which single particles are distributed over the surface and throughout the bulk of
the liquid. Dispersibility increases with increasing sinkability and is reduced by
clump formation.

Solubility.   Solubility describes the rate and extent to which components
dissolve in water and is a function mainly of the composition of the powder and
the temperature.

In order that a powder may exhibit good reconstitution characteristics, it is
important that a proper balance be struck between the above properties. The
drying step can have a notable impact on several of these properties, as well as

related characteristics such as particle size, density, and bulk density, which may also impact on the perceived quality of the product.

(c) Volume changes.   Drying frequently results in shrinkage of both animal and vegetable products and of colloidal material. Early in the drying process, if the rate of moisture loss is sufficiently slow, the shrinkage is related in a simple manner to the amount of water removed. However, these volume changes become progressively less marked as drying proceeds and the final size and shape of the product are fixed well before the process is complete.

At the other extreme, selection of appropriate drying conditions can give rise to a puffed product. This is normally carried out intentionally and is the principal aim of the process, with moisture loss being essentially incidental. In order to achieve puffing of, say, cooked cereal grains or fruit and vegetable pieces, it is first necessary to adjust the moisture content to a value appropriate to the particular product and then to allow this moisture to equilibrate uniformly throughout the solid. As the temperature of the material is raised in an appropriate dryer or puffing device (e.g. 'gun'), the water is vaporized inside the product and the internal pressure rises. Provided the solid is still sufficiently plastic, it will ultimately expand rapidly, thereby giving rise to the desired puffed product.

11.2.8 Drying kinetics

The rate of moisture loss that is exhibited by a given product under given conditions may have a profound influence on dryer selection. Drying kinetics have been discussed at some length in Chapter 2 and the detail will not be repeated here. Broadly speaking, the rate of moisture loss will depend on some or all of the following.

- Location of moisture (on or near the surface, or uniformly distributed within the product?).
- Nature of moisture (free or strongly bound?).
- Mechanisms of moisture migration through the solid, and rate-limiting steps.
- Dimensions of product.
- Temperature, humidity and flow rate of heated air in an atmospheric dryer.
- Nature of contact between heated air and solids in an atmospheric dryer.
- Pressure and heating-plate temperature in a contact dryer.

Note that several of these parameters may change during the course of the drying process, particularly in the case of hygroscopic materials. Elementary methods of measuring the drying kinetics are described in section 11.4. It is probably satisfactory from the viewpoint of preliminary dryer selection to determine whether the drying rate is fast, medium or slow within the range of moisture contents of interest.

11.3 Preliminary dryer selection

A rigorous analysis of the type described in section 11.2 above will provide much of the information required to make a preliminary selection of the types of dryer that may be appropriate in a given set of circumstances. A range of batch and continuous equipment appropriate for the drying of foods is illustrated in Figures 11.3 and 11.4, respectively. The dryers are categorized, firstly, into layer and dispersion dryers. These are subdivided according to the mode of heating – conduction (contact dryers), convection, and other 'special' techniques. The numbers in parentheses denote the chapter in which a detailed description of the particular dryer type can be found. The suggested preliminary selection technique involves consideration and assessment of the principal factors that influence dryer selection. These are given in the following sections.

11.3.1 Nature of the feed

If the feed is a liquid, then the choice is essentially limited to a drum or agitated dryer (low to medium throughputs) or a spray dryer (higher throughputs). If it is a solid, the choice is much wider. Difficulties can arise with a paste or gelatinous feedstock. If it is both pumpable and atomizable, a spray dryer may be the best choice. If it is not, the spin-flash dryer is a good alternative. Another possibility is to dilute it in order to make spray drying possible. Although this approach is wasteful in terms of energy utilization, product quality considerations may override this drawback. Alternatively, a pneumatic conveying dryer may be suitable at high throughputs or one of a number of contact dryers at lower

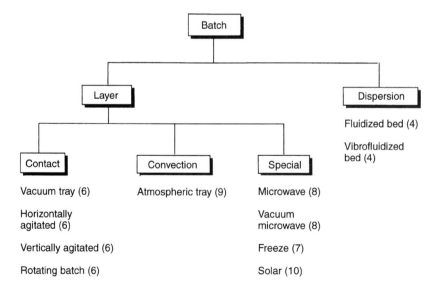

Figure 11.3 Categories of batch dryer.

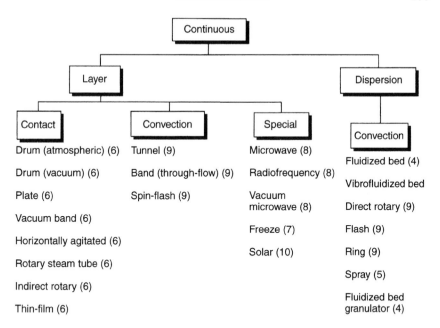

Figure 11.4 Categories of continuous dryer.

throughputs. In either of these cases, materials handling or quality considerations may dictate that the feed should be backmixed with an appropriate quantity of product prior to drying. Freeze dryers can handle both liquid and solid feeds but the product must be capable of absorbing the high processing costs associated with this method of drying.

11.3.2 Nature of the product

Several fundamental product properties can be determined by the type of dryer employed. These may include, for example, size, shape, density, colour and functionality.

If the material fed to the dryer is a solid, the feed and product will, in most cases, be of comparable size. If the size of the product is to exceed that of the feed, a granulation (agglomeration) step is necessary. A fluidized bed granulator, in which the solids are re-wetted with binder, agglomerated and dried in a single unit, is perhaps the most obvious choice of technique. The use of more basic equipment, such as an inclined disc or drum agglomerator (Capes, 1980), followed by a suitable dryer, is also a possibility. In cases where the desired size of the product is less than that of the feed, the inclusion of a milling step will normally be necessary.

Where the feed to the dryer is a liquid, a paste or a suspension, the size and shape of the product will clearly be dictated by the dryer characteristics and those of any post-drying treatment such as agglomeration or milling. Spray dryers tend to produce quasi-spherical particles, which are often hollow. Drum dryers yield flaky products and other contact dryers may produce irregularly shaped particles.

11.3.3 Throughput

Batch dryers are appropriate at low throughputs and continuous dryers at high throughputs. There is, however, considerable overlap between the two. The mode of operation (batch or continuous) of the upstream and downstream processing equipment should also be taken into account. In certain situations it may be appropriate to operate two or more dryers in parallel if the desired throughput exceeds the maximum capacity of a single dryer.

11.3.4 Mode of heating

Most dryers rely on either convective or conductive heating. The choice between the two can be determined by several factors. If relatively high drying temperatures are required, then a direct fired convective dryer may be essential. In cases where contact with combustion products is undesirable, the inlet air to a convective dryer can be heated indirectly with steam. In such cases, its temperature is limited to around 200°C or lower, depending upon the steam pressure available. This is normally quite adequate for drying many foodstuffs. Alternatively, if higher temperatures are required, a specialized heat transfer fluid may be employed as an alternative to steam. Generally speaking, direct fired dryers exhibit a somewhat greater thermal efficiency than their indirectly heated counterparts.

Since both batch and continuous contact dryers are capable of being operated under vacuum, they are particularly useful for drying heat sensitive foods at low temperature. Their size is, however, limited by the heat transfer area available. Only indirect and steam tube rotary dryers are capable of drying relatively large throughputs.

Dielectrically heated dryers are generally more expensive in terms of both capital and operating costs than thermal dryers and, consequently, are not regarded as direct substitutes. They are particularly useful as finish dryers in situations where relatively small quantities of residual moisture must be removed or where moisture levelling is required. Thermal dryers require large quantities of heat to remove the last traces of moisture. In contrast, in dielectric dryers, energy is absorbed directly by the moisture, which is located within the bulk of the product; their efficiency is therefore much higher than that of thermal dryers under these conditions, and the size of the plant is much smaller. This can be of particular value where factory floor space is restricted. Dielectric dryers may also be considered under circumstances where their use has a positive impact on the quality of the product by, for example, accelerating the drying time or varying the moisture content–temperature–time profiles.

Tables 11.1–11.4 provide information required to make the preliminary dryer selection. Table 11.1 gives details of the suitability of various layer and dispersion dryers for handling different types of feed. Further information on feed and product capabilities is given in Table 11.2 for batch dryers and in Table 11.3 for continuous dryers. Table 11.3 also includes details on typical residence times and on scales of operation. Table 11.4 summarizes typical evaporative loads that can be accommodated by different dryers. Note, however, that the higher values quoted for direct fired dryers may only apply where the inlet air temperature is much higher than that which could be tolerated by foodstuffs. Examples of the use of these tables in making a preliminary selection of suitable dryer types are given in section 11.7.

11.4 Bench scale tests

Bench scale tests form an integral part of the dryer selection process. In this section, their role is defined, their reliability as far as scale-up discussed and, finally, the methods employed are described briefly.

11.4.1 Role of bench scale tests

The selection of a dryer for new product is always risky and, once the choice is made, it is very difficult, if not impossible, to alter that decision. Food manufacturers often attempt to make at least the preliminary selection in-house. Whereas this is clearly advantageous from the viewpoint of maintaining commercial confidentiality, it does have a major drawback in that the selection is often, by necessity, made by a non-specialist, who, in many cases, will not have the necessary background knowledge or skills. This could mean that opportunities for making a more appropriate choice are lost as a result and may not be redeemed even in subsequent discussions with equipment manufacturers.

In many instances, a principal objective in selecting a dryer is not only to remove the appropriate quantity of moisture but also to achieve particular product attributes. These may have been specified as a result of market surveys or by the desire to match a competitor's product. In any event, from a commercial viewpoint, they are of great importance. The risks involved in selecting a dryer capable of delivering a product with the desired attributes can be minimized by conducting directed tests in appropriate small scale equipment. These will provide sufficiently detailed and accurate information to refine the preliminary selection, which may have been made on the basis of, say, past experience of similar materials, or literature articles not entirely relevant to the present requirements. In product development projects, it is important to identify the most likely type of dryer that will be ultimately be used, so that meaningful tests can be conducted in the laboratory. For example, there is little point in

Table 11.1 Dryer feedstock handling capabilities (adapted from Noden, 1979)

Dryer type	Operation	Liquids, liquid suspensions	Pastes, dewatered cake	Powders	Granules, pellets, extrudates, pieces
			Application rating		
Layer dryers, contact					
Tray dryer (vacuum)	Batch	(b, c)	(a)	(b)	(b)
Horizontal agitated dryer (atmospheric/vacuum)	Batch/Continuous	(b)	(b)	(b)	(c)
Vertical agitated dryer (atmospheric/vacuum)	Batch	(b)	(b)	(b)	(c)
Double cone dryer (atmospheric/vacuum)	Batch	(d)	(c)	(b)	(c)
Drum dryer (atmospheric/vacuum)	Continuous	(a)	(b)	(d)	(d)
Plate dryer (atmospheric/vacuum)	Continuous	(d)	(c)	(b)	(a)
Vacuum band dryer	Continuous	(d)	(a)(Pastes)	(d)	(b)
Rotary steam-tube dryer	Continuous	(d)	(c)	(a)	(b)
Indirect rotary dryer	Continuous	(d)	(c)	(a)	(a)
Thin-film dryer	Continuous	(a)	(d)	(d)	(d)
Layer dryers, convective					
Atmospheric tray dryer (cross-flow)	Batch	(c)	(b)	(b)	(a)
Atmospheric tray dryer (through-flow)	Batch	(d)	(d)	(d)	(a)
Band dryer (through-flow)	Continuous	(d)	(c)	(d)	(a)
Band dryer (cross-flow)	Continuous	(d)	(c)	(d)	(a)
Band dryer (air impingement)	Continuous	(d)	(d)	(d)	(a)
Tunnel dryer	Continuous	(d)	(d)	(c)	(a)
Rotary louvre dryer	Continuous	(d)	(d)	(a)	(a)
Spin-flash dryer	Continuous	(b)	(a)	(b)	(c)
Layer dryers, special					
Microwave dryer (atmospheric)	Batch/Continuous	(d)	(c)	(b)	(b)
Microwave dryer (vacuum)	Batch/Continuous	(d)	(c)	(b)	(b)
Freeze dryer	Batch/Continuous	(b)	(b)	(b)	(b)
Radio-frequency	Continuous	(d)	(c)	(b)	(b)
Air-radio-frequency (ARFA) dryer	Continuous	(d)	(d)	(c)	(b)
Induction-heated rotary dryer (Rotek)	Continuous	(d)	(c)	(b)	(b)
Dispersion dryers					
Fluidized bed dryer	Batch/Continuous	(d)	(d)	(a)	(b)
Vibro-fluidized bed dryer	Batch/Continuous	(d)	(d)	(a)	(a)
Fluidized bed granulator	Batch	(d)	(d)	(a)	(d)
Direct rotary dryer	Continuous	(d)	(b)	(b)	(a)
Pneumatic (flash) dryer	Continuous	(d)	(b)	(a)	(b)
Pneumatic (Ring) dryer	Continuous	(d)	(b)	(a)	(b)
Spray dryer	Continuous	(a)	(d)	(d)	(d)

(a) Good, (b) fair, (c) unsatisfactory, (d) not applicable.

Table 11.2 Operating characteristics of batch dryers

Selection parameter	Fluid bed	Vibro-fluidized bed	Fluid bed granulator	Atmos-pheric tray	Vacuum tray	Hori-zontal agitated	Vertical agitated	Double cone	Freeze
Feed									
Solution/slurry				X	X	X	X		X
Paste				X	X	X	X		X
Wet particles:									
free flowing	X	X	X	X	X	X	X	X	X
cohesive		X		X	X	X	X		X
Size[a]: (a) small, (b) medium, (c) large	(a)	(a–c)	(a)	(a–c)	(a–c)	(a)	(a)	(a)	(a–c)
Product									
Very low final moisture content					X	b	b	b	X
Fragile		X		X	X				X
Temperature sensitive					X	b	b	b	X

[a] Typically (a) fine powders, (b) coarse powders, grains and small pieces and (c) large pieces, flakes and extrudates.
[b] Operation under vacuum is recommended for drying food materials that are particularly temperature sensitive or for which a very low product moisture content is required.

producing test samples in a laboratory freeze dryer, when this technique, for whatever reason, stands no chance of being adopted commercially.

The small scale tests that are likely to be undertaken will, in most cases, be directed at providing information on the physical properties of the feed and product, the quality attributes of the product, the thermal stability of the material being dried, the drying kinetics and the equilibrium moisture content. It is important at the outset to conduct these measurements on representative samples. In the case of solids, coning and quartering, and sample splitting, are two widely used methods of obtaining such samples.

Many of the relevant feed and product properties are relatively easy to measure, given appropriate equipment and skills. These include, for instance, particle size distribution, density, bulk density and moisture content. Note, however, that the results of moisture content determinations can vary quite widely according to the manner in which they are conducted. Other properties, including, for example, colour, texture, surface properties and functionality, are more difficult to assess, and the results may be qualitative or, at best, semi-quantitative. If the feed is a liquid, it will probably be appropriate to characterize its rheology. However, the ultimate questions to which answers are required are often 'is the feedstock pumpable' and 'will it atomize' as these factors will determine whether the use of a spray dryer is feasible.

Table 11.3 Operating characteristics of continuous dryers

Type of dryer

Selection parameter	Well-mixed fluid bed	Plug-flow fluid bed	Vibro-fluidized bed	Flash	Ring	Direct rotary	Indirect rotary	Rotary louvre	Spray	Atmospheric band	Vacuum band	Tunnel	Drum	Horizontal agitated	Spin-flash	Freeze
Feed																
Liquid/solution/slurry									X				X	X	X	X
Paste											X		X	X	X	X
Wet particles:																
free flowing	X	X	X	X	X	X	X	X		X		X		X		X
cohesive	X	X	X	X	X	X	X	X		X				X		X
Size:[a] (a) small, (b) medium, (c) large	(a)	(a)	(a–c)	(a)	(a)	(a–c)	(a–c)	(a–c)	(a)	(b, c)	(b, c)	(b, c)	(b)	(a)	(a–c)	(a–c)
Product																
Uniform moisture content		X	X	X	X	X	X	X	X	X	X	X	X	X	X	X
Low final moisture content	X	X	X		X	X	X	X		X	X	X	X	X	X	X
Fragile			X					X		X	X					X
Temperature sensitive	X						X		X		X		X	X		X
Moisture																
Surface moisture	X	X	X	X	X	X	X	X		X		X		X	X	X
Internal moisture	X	X	X		X	X	X	X		X		X		X	X	X
Residence time (a) <10 s, (b) 10–30 s, (c) 5–10 min, (d) 10–60 min, (e) 1–6 h, (f) >6 h	(d)	(d)	(d)	(a)	(b)	(c, d)	(d)	(d)	(b)	(d)	(d)	(d)	(b)	(d)	(a–c)	(e, f)
Throughput[b] Low (L), medium (M), high (H)	(M, H)	(M)	(M)	(M, H)	(M, H)	(M, H)	(M, H)	(M, H)	(M, H)	(M)	(L, M)	(M)	(M)	(L, M)	(M, H)	(L, M)

[a] Typically (a) fine powders, (b) coarse powders, grains and small pieces and (c) large pieces, flakes and extrudates.
[b] See also Table 11.4.

Table 11.4 Typical maximum evaporation rates in some common types of dryer (adapted from Noden, 1979)

Dryer type	Operation	Typical maximum evaporative capacity (kg h^{-1})
Tray dryer (cross-flow)	Batch	55
Tray dryer (through-flow)	Batch	75
Vertical agitated dryer (vacuum)	Batch	120
Vertical agitated dryer (atmospheric)	Batch	120
Double cone dryer (vacuum)	Batch	300
Fluidized bed dryer	Continuous	910
Band dryer (through-flow)	Continuous	1820
Indirect rotary dryer	Continuous	1820
Direct rotary dryer	Continuous	5450
Drum dryer (atmospheric)	Continuous	410
Spin-flash	Continuous	7800
Pneumatic (flash) dryer	Continuous	15 900
Spray dryer	Continuous	15 900

11.4.2 Details of testing methods

(a) Physical properties. The physical properties that most commonly influence dryer selection are particle size, shape, and structure and bulk density of both the feed and product, the particle density and friability of the dried product, and the cohesive and adhesive properties of the wet solids. In certain cases, it may be appropriate to estimate other, more specialized properties such as minimum fluidizing velocity or the terminal velocity of the particles. As noted above, if the feed is a liquid, an assessment must also be made of its pumping and atomizing capabilities. Although a knowledge of the thermal properties of the solids may ultimately be required for design purposes, these are unlikely to influence dryer selection.

A variety of techniques may be used to measure particle size, and are fully described in standard texts on the subject (see, e.g. Allen (1990)). The choice of method is largely dependent upon the dimensions of the particles. Whichever method is chosen, it is important to establish standardized and reproducible techniques in order to ensure that the measurements are reliable and accurate.

For relatively small particles, techniques based on laser diffraction and on the change of electrical resistance that occurs when a particle displaces conducting liquid flowing through an orifice, are perhaps the most appropriate. Microscopy techniques can also be used but analysing sufficient data to yield a statistically reliable result can be tedious in the absence of image processing and analysis equipment. For larger particles, sieve analysis is widely used. The British Standard series of sieves, for example, covers a wide practical size range, 38 μm to 16 mm.

At the very least, a mean particle size should be extracted from these tests. An assessment of the smallest and largest particles present in any substantial amount should also be made. Alternatively, it may be possible to describe the particulate system in terms of the geometric mean size and standard deviation of the commonly observed log normal distribution (Allen, 1990).

Qualitative descriptions of particle shape and structure are normally sufficient for dryer selection purposes. This information may often be obtained by optical or scanning electron microscopy, which can also provide a wealth of additional information, particularly when trying to achieve a good match against an external standard. In such cases, the micrographs may indicate quite clearly, for instance, whether a liquid product has been dried in a spray dryer, an agitated dryer or on a drum dryer. They will also show whether the product is an agglomerate and, if so, how it has been formed.

Bulk density is defined as the mass of the solids divided by their bulk volume. Consequently, the value obtained in a given experiment depends on the packing characteristics of the particles, which may vary quite widely, depending on the technique employed. It is therefore particularly important in this case to ensure that a consistent methodology is employed.

The particle density is defined as the mass of the dry particle divided by its external (displacement) volume, which excludes the volume of any pores present within the solids. It is normally measured using a density bottle or other equivalent vessel which can be filled repeatedly to a reproducible volume.

A knowledge of the adhesive and cohesive properties of the wet solids is of importance since it is necessary both to feed them into and transport them through the dryer. Operational problems often occur as a result of the wet, sticky feed material adhering to feeders and to the internals at the front-end of a continuous dryer. As a result, some drying techniques are unsuitable for processing such solids.

Although the adhesive and cohesive properties of wet solids and pastes are very difficult to describe quantitatively, it is relatively easy to obtain sufficient qualitative information in the laboratory to satisfy the requirements for dryer selection. Thus, manipulation of the solids with one's fingers or a spatula may be all that is needed to assess their stickiness. It is often advisable to examine the material at a number of moisture contents intermediate between those of the feed and the product. This will also provide information on the friability of the solids as this could also influence dryer selection.

(b) Quality attributes. In many cases, the drying of foodstuffs can have a marked effect on attributes that are deigned to influence their quality. Common examples include taste, colour, physical appearance, structure, texture and functionality. These attributes are normally very difficult to measure quantitatively and one normally has to resort to sensory analysis techniques in order to provide a qualitative, but meaningful, description. Even in those cases where relatively sophisticated instrumentation is available as, for instance, in the measurement of colour using tristimulus methods, or crunchiness using mechanical test equipment, it is often difficult to correlate the results directly with human perception.

Functionality tests, which determine whether a particular food product is meeting predetermined standards of performance, are specific to the product in question. For example, the reconstitution of dried meat and vegetables is assessed by rehydration tests in which their moisture uptake is measured. Gluten quality, on the other hand, is determined through bread baking trials.

Again, the benefits of both optical and scanning electron microscopy in assessing many of the surface attributes of a dried product cannot be stressed too highly. In particular, photographic or electronic records of the images obtained during the course of an investigation can prove invaluable.

(c) Thermal stability. It is very important to determine the maximum temperature to which food products can be exposed without detriment. This will certainly influence the size of the dryer ultimately chosen and, because size limitations on different types of dryer vary, may also influence the selection.

In performing thermal stability tests, several points need to be borne in mind. Firstly, the chemical reactions that result in quality deterioration in foods are very complex and often exhibit temperature–moisture content–time interdependence; a single test result should not be regarded as being sufficient to give a complete picture. It is therefore advisable to carry out a number of thermal stability trials at various initial moisture contents over the range of interest. Secondly, it is not unknown for the results of such tests to be influenced by the mode of heating. Thus, layer tests on a heated plate at a specific temperature may give a different outcome from, say, tests at the same temperature carried out in a small-scale fluidized bed dryer.

A typical thermal stability test might be performed as follows. Several samples of the feed at the envisaged initial moisture content are placed on dishes in an oven set to an appropriate lowish temperature, say 50°C. The samples are removed periodically from the oven and examined for deterioration. The temperature and final moisture content of each sample should be measured and the drying time recorded. The last sample should remain in the oven until the desired product moisture content is achieved. If there is no obvious deterioration, the test should be repeated at progressively higher temperatures until obvious signs of quality impairment become apparent. The results of these tests will yield a range of possible drying temperatures.

(d) Drying characteristics. A cursory knowledge of the drying kinetics at an appropriate temperature is vital for dryer selection purposes and should be obtained as early as possible in the project. As discussed in Chapter 2, this information is best represented as a simple drying curve, in which the average moisture content of a batch of drying material is plotted against time. It can normally be obtained by periodically withdrawing and analysing samples of material contained in a drying oven or a suitable dryer simulator.

(e) Equilibrium moisture content. A dryer cannot dry a product below its equilibrium moisture content (EMC) and, consequently, the outlet moisture content must be judged in relation to this limiting value rather than zero moisture. Some dryers are more suitable than others at reducing the moisture content to a value close to the EMC.

Equilibrium moisture content is a function of both humidity and temperature. An adsorption isotherm (see Chapter 2) can be measured by allowing the solids to equilibrate at the desired temperature over a number of standard salt solutions, which generates known relative humidities in the closed air space. Such measurements are normally lengthy and it can take several days for equilibrium to become established. A much faster method, in which air of known humidity and temperature is blown through a sample holder filled with the material in question, has been described by Papadakis *et al.* (1993). These authors also examined the validity of a number of published correlations for EMC and found that none was entirely satisfactory. Several were, however, useful for interpolating EMC data.

11.5 Economic comparison of alternatives

11.5.1 Introduction

The iterative processes of preliminary selection and directed bench scale testing will undoubtedly have narrowed the number of acceptable dryer types very significantly. It is even possible that results will point towards a single preferred option. However, this is unlikely, and two, or perhaps three, possibilities will remain open at this stage of the selection process. In such cases, the next step is to make an economic comparison of the alternatives.

In practice, dryer selection is frequently made on the basis of least capital cost. As discussed by Bahu *et al.* (1983), however, this may result in false economy since, over the lifetime of a typical dryer, the total cost of the operation is dominated by expenditure on fuel. The capital cost element is normally relatively small, typically less than 10%, of the total.

11.5.2 Vendor-generated capital cost estimates

In-house generated estimates of the capital cost of a given dryer are, in many cases, accurate only to an order of magnitude and are not normally recommended as a basis on which to make a definitive selection. It is far better, in practice, to approach reliable dryer manufacturers with a request for a budget quotation against a relatively detailed specification. At this time, it will probably be necessary to provide the manufacturer with a sighting sample of the material to be dried. It is important to be quite specific as to what the quotation should include as the cost of ancillary equipment such as conveyors, instrumentation and controls, feed and product bins, and gas cleaning equipment may exceed the base cost of the dryer itself. If these items are not included in the quotation, their costs should be obtained separately as there can be significant differences between the requirements for different types of dryer. Most dryer manufacturers will, if requested, provide a cost for installing their equipment. Again, as this cost is far from insignificant, it is preferable to obtain a reliable quotation.

It is good practice to obtain a capital cost estimate for a given dryer from more than one supplier, if possible. This will satisfy the purchaser that the quoted costs represent good value for money. In cases where more than one type of dryer has been shortlisted, it may well be necessary to deal with more than one vendor, as many do not offer a comprehensive range of drying equipment.

11.5.3 In-house generated capital cost estimates

If, for commercial or other reasons, it is deemed undesirable to approach a dryer vendor, then it will be necessary to arrive at an in-house estimate of capital cost. There are two basic steps involved in this process, both of which may be subject to considerable error. The first is to determine the approximate size of the dryer, and the second, to cost the dryer on the basis of this knowledge.

Estimation of the required size of a dryer is, at best, difficult, and the results may not be particularly accurate, especially when attempting to scale-up from laboratory test results and/or published literature correlations. Layer dryers and fluidized bed dispersion dryers (Figures 11.3 and 11.4) can, however, be scaled reasonably reliably even from bench scale data, since the solids and gas flows and the interfacial area can be modelled mathematically with some accuracy. The methodology employed is beyond the scope of this chapter. However, full details are given elsewhere by, for example, Moyers (1994) for layer dryers and Bahu (1994) for fluidized bed dryers (see also Chapter 4).

Other dispersion dryers cannot be modelled with the same degree of accuracy and hence scale-up rules are considerably less reliable. The state of the art for rotary, pneumatic conveying, and spray dryers has been discussed by Papadakis et al. (1994), Kemp (1994), and Masters (1994), respectively.

The capital cost of a dryer is a function of its size, its complexity, and its material of construction. Some data on the costs have been published in the open literature (see e.g. Peters and Timmerhaus (1991) and van't Land (1991)), but most are retrained as confidential information by the industry. In using published data, it is worth noting that the costs quoted by different equipment suppliers against a given specification can vary quite widely, depending on commercial conditions pertaining at the time.

There are two possible approaches to obtaining a capital cost estimate. Where a company has relevant data in its files, such as, for example, the cost of a similar type of dryer purchased some years ago, this may be used to specify a base cost. This is the preferred option. Alternatively, if this information is not available, it will be necessary to resort to data published in the open literature as a basis for the estimate.

Having obtained a base cost, it will invariably be necessary to scale this value to take into account differences in size between the new and base-case dryers and the years in which the two dryers were purchased. Equation 11.5 accounts for such differences and may be applied in cases where the two dryers are of the same type.

$$C_{new} = C_{base} \times (c_{new}/c_{base})^n \times (I_{t,new}/I_{t,base}) \qquad (11.5)$$

Here

C_{new} = purchase cost of new dryer
C_{base} = purchase cost of base-case dryer
c_{new} = capacity of new dryer
c_{base} = capacity of base-case dryer
n = an exponent
$I_{t,new}$ = current value of appropriate cost index
$I_{t,base}$ = value of cost index at time of purchase of base-case dryer

The dryer capacity c is any suitable size related parameter. For example, dryer volume, solids throughput, air mass flowrate, and evaporation rate may all be used. The exponent n may be taken as the commonly accepted value of 0.6 in the absence of more precise information. However, a range of values have been reported in the literature and $n = 0.6$ can only be regarded as a rough approximation. A number of appropriate cost indices I_t are available. These include, for example:

1. The Marshall and Swift (M&S) index, which is based on costs in the United States for a range of engineering industries. An updated value is published monthly in the US magazine *Chemical Engineering*.
2. The Chemical Engineering index, which is again published in *Chemical Engineering*, is based on equipment and related costs in the American chemical industry.
3. The Process Engineering index, which is based on weighted engineering costs in the United Kingdom for a representative process. It is published monthly in the UK magazine *Process Engineering*.

The cost of dryers, like other process equipment, does vary from country to country. For example, corresponding costs in the USA are significantly less than those in Europe. A commonly applied rule of thumb is that a dryer will cost the same number of US dollars in America is as it does pounds sterling in the UK.

Having obtained an estimate for the cost of the basic dryer, it will be necessary to add on the costs of ancillary equipment and installation, in order to make a realistic comparison between possible alternatives. The data obtained by van't Land (1991) suggested that the cost of ancillaries is around twice the cost of the basic dryer. Installation costs are typically around 2.5 times the purchased-equipment cost but these do vary somewhat according to the materials of construction. Thus the fully installed cost of a $100 000 basic dryer is likely to be around $750 000.

11.5.4 Cost comparison

Having obtained cost data for the different alternatives, it should now be possible to select one of the candidate dryers as the clear favourite. Before making a final decision, however, it is advisable to address the following points.

1. Are the accuracy and reliability of the competing quotations comparable? If not, any possible cost advantage exhibited by one dryer may turn out to be spurious.
2. Does the most probable equipment supplier have the technical competence, experience, and back-up support necessary to design, build and install the dryer?
3. Are there any major differences in energy utilization between the different dryer types? For example, contact dryers are significantly more fuel efficient than their convective counterparts. Given that the cost of drying is dominated by fuel costs, this factor should be given due consideration.
4. Have all appropriate energy conservation measures been considered and properly costed?
5. Have any hidden costs, such as, for example, the resulting need for added boiler capacity, been taken into account?

11.6 Pilot-plant trials and final selection

Pilot plant trials of the favoured option are often desirable and should be carried out in conjunction with a reliable dryer manufacturer. Their purpose is threefold. Firstly, they will provide additional, more reliable, data with which to size and design the dryer. This is particularly important in the case of dispersion dryers, for which scale-up rules are not, in general, particularly accurate. In contrast, the scale-up of layer dryers is considerably more reliable since the solids and air flows are reasonably well defined. Secondly, they are likely to highlight any materials handling problems that may be encountered. Finally, they will give the purchaser the opportunity to assess the facilities and experience of the equipment supplier.

Pilot plant trials can have three possible outcomes. Firstly, they will yield satisfactory results. In this case, the proposed choice of dryer can be confirmed and its design finalised. Secondly, they may show that the scale-up rules employed earlier in the selection procedure, and hence the estimated size and cost of the dryer, were inaccurate. In this case, it may be appropriate to reassess the economic comparisons; a possible loop back to the 'Compare economic alternatives' step in Figure 11.1 is therefore indicated. The third possibility is that the outcome of the trials is unsatisfactory. This will require a complete reassessment of the problem and a loop back to the 'Make preliminary dryer selection' step (Figure 11.1) is appropriate under such circumstances.

11.7 Typical examples of the use of the dryer selection algorithm

The typical examples presented in this section focus principally on the first two stages in the dryer selection algorithm (Figure 11.1), namely on the formulation of the dryer specifications and the preliminary selection process. No attempt is made to refine the selection beyond this point, with the exception of the provision of general guidelines, where appropriate.

11.7.1 Example 1

800 kg h^{-1} (wet basis) of cooked, flaked cereal grains are to be dried from 30% to 18% moisture. The flakes fed to the dryer are cohesive, and exhibit a tendency to clump. Their size is 5–15 mm (largest dimension) by 1–2 mm thick. Past experience with similar products suggests that the inlet air temperature should probably not exceed 50–70°C and that the corresponding drying time will be around 25–35 min. The wet flakes are flexible and fragile but at 18% moisture, the dried flakes are rigid but not unduly susceptible to breakage. A continuous process is envisaged. What dryers are suitable?

(a) Formulation of dryer specification. At 800 kg h^{-1}, a continuous dryer will be required. The evaporation load on the dryer is 117 kg h^{-1} of moisture, which is relatively small in absolute terms. However, it is consistent with the limited drying rates associated with the low permitted drying temperature.

(b) Preliminary dryer selection. Table 11.1 lists the following continuous dryers as being 'good' for granules, pellets, extrudates and pieces, which encompass the present product: plate, indirect rotary, band (through-flow), band (cross-flow), band (air impingement), tunnel, rotary louvre, vibrofluidized bed, direct rotary.

Table 11.3 lists only two dryers that are suitable for cohesive feeds, large particles, fragile products, internal moisture, 10–60 min drying time, and medium scale operation. These are the vibrofluidized bed and the atmospheric band. The latter includes the through-flow, cross-flow and air-impingement types. Note that the plate dryer is not included in Table 11.3 as it is not widely used in the food industry.

It is at this point that technical judgement is required to refine the initial selection. Of the atmospheric band dryers, the cross-flow type can be eliminated as the drying rate is likely to be considerably slower than that in the through-circulation dryer and it offers no compensating features in this case. Several dryers are selected in Table 11.1 but not in Table 11.3. It is always worth checking that there are sound reasons for the omissions. The direct rotary dryer is not included in the Table 11.3 selections solely on the grounds of product fragility. Given the nature of this particular product, this omission probably unwarranted. It would therefore seem reasonable to include on the list of possibilities at this stage. The indirect rotary, tunnel and rotary louvre dryers are all excluded on the grounds that they are not well suited to handling cohesive feedstocks. These exclusions are justified. A plate dryer would not be used for the same reason.

The preliminary selection therefore consists of the following dryer types: vibrofluidized bed, through-circulation band, air-impingement band, direct rotary. The evaporation rate (117 kg h^{-1}) is considerably lower than the upper limits for these dryers given in Table 11.4. The relatively stringent temperature limitations should, therefore, not be a problem.

(c) Subsequent actions. A number of bench scale tests would be advisable to establish the limiting air temperature for drying and elementary drying kinetics. A knowledge of the bulk density of the feed and dried product is also required. The next step is to obtain budget quotes for each of the options and to pilot test selectively the most promising alternatives before arriving at a final decision.

11.7.2 Example 2

In the course of its manufacture, a proposed new food-thickening ingredient is to be obtained by drying a pasty material from around 65% moisture (wet basis) to 6%. At 65% moisture, the paste, which is produced on a batch basis, is neither pumpable nor atomizable. The envisaged production rate is 50 kg h^{-1} of the dried product, which is required to be a free-flowing powder having a size $<150 \text{ μm}$, and readily dispersible. After drying, the solids are subsequently packed. Suggest appropriate dryers.

(a) Formulation of dryer specification. The paste is to be dried from 65% moisture to 6%; the production rate is 50 kg h^{-1} of the dried (6%) solids. The feed rate to the dryer is therefore 134 kg h^{-1} (wet basis) and the evaporation rate 84 kg h^{-1}. At this throughput, the use of a batch dryer is probably the preferred option (see Table 11.4), particularly as the upstream processing is performed on a batch basis. However, as drying, together with any associated particle size modification of the product, is the last major processing step, the use of a continuous dryer is not precluded at this stage.

(b) Preliminary dryer selection. The only 'good' batch dryer for pastes listed in Table 11.1 is the vacuum tray dryer. The following batch dryers are listed in the 'fair' category: horizontal agitated, vertical agitated, atmospheric vertical

tray (cross flow), freeze. Table 11.2 lists the following possible batch dryers: atmospheric tray (cross flow), vacuum tray, horizontal agitated, vertical agitated, and freeze. There is therefore a reasonable consensus between the types suggested by the two sources. In all these cases, milling of the dried product is likely to be required.

The following continuous drying possibilities are listed in Table 11.1: horizontal agitated, drum, vacuum band, spin-flash, freeze, direct rotary flash, and Ring. Table 11.3 also lists vacuum band, freeze, spin-flash and horizontal agitated dryers as possibilities. In the case of the drum dryer, the feed would probably have to be diluted to make it suitable for processing. If we allow for this possibility, then we should also include that of a spray dryer. In the case of flash and Ring dryers, dried solids would have to be backmixed with the wet feed. Of these possibilities, only horizontal agitated and spin-flash dryers are likely to be realistic contenders, unless quality considerations dictate otherwise. The other continuous dryers will probably not be economical at such low throughputs (see Table 11.4).

(c) Bench scale tests.   The next step is to undertake a series of laboratory trials in order to determine an acceptable range of drying temperatures. Thus, small quantities of the feedstock should ideally be dried in a small freeze dryer, a vacuum dryer and a convection oven. The resulting samples should be tested for functionality and for dispersibility. If the results of these tests suggest that drying at relatively high temperatures is acceptable, then this will probably be the preferred option on the basis of least cost. Similarly, vacuum drying is likely to be less expensive than freeze drying. These tests should, therefore, divide the possibilities into three groups, as follows.

- Atmospheric cross-flow tray (batch), horizontal agitated (batch or continuous), vertical agitated (batch), spin-flash (continuous).
- Vacuum tray (batch), horizontal agitated (batch or continuous, operated under vacuum), vertical agitated (batch, operated under vacuum).
- Freeze (batch).

(d) Subsequent actions.   Evaluation of the options in the favoured group can now proceed. Provided that the product quality is acceptable, appropriate dryer manufacturers can be approached for budget quotations. The most promising option(s) can then be pilot tested if this is regarded as necessary by the dryer manufacturer.

11.8 Summary

In this chapter, an attempt has been made to describe the principal factors involved in the selection of dryers and to present a unified and logical approach to the task. As may be seen, arriving at a satisfactory recommendation involves the careful weighing of a large number of process and product parameters, all of which have to be properly considered and evaluated. The task is far from trivial and the reader is strongly recommended to take advice from an appropriate dryer manufacturer or an independent specialist as early as possible.

It is encouraging to note that a rule-based algorithm for dryer selection is in the process of development and refinement (Kemp and Bahu, 1995). This approach will undoubtedly provide a valuable aid to process engineers. Only time will tell whether the expert systems that ultimately emerge will make the need for a specialist input redundant. In the case of food dryers, this is, on balance, unlikely because of the complex and subtle nature of the materials being dried.

References

Abbott, J. (1990) (ed.) *Prevention of Fires and Explosions in Dryers: A User Guide*, 2nd edn, Institution of Chemical Engineers, Rugby, UK.
Allen, T. (1990) *Particle Size Measurement*, 4th edn, Chapman & Hall, London.
Bahu, R.E. (1994) Fluidised bed dryer scale-up. *Drying Tech.*, **11**(3), 329–39.
Bahu, R.E., Baker, C.G.J. and Reay (1995) Energy balances on industrial dryers – a route to fuel conservation. *J. Separ. Process Tech.*, **4**, 23–8.
Brennan, J.G., Butters, J.R., Cowell, N.D. and Lilly, A.E.V. (1990) *Dehydration in Food Engineering Operations*, 3rd edn, Elsevier Applied Science, London, pp. 371–415.
Bruce, D.M. and Nellist, M.E. (1992) *Modelling the Effect of Heated-air Drying of Grains on their Quality for Bread Baking and for Seed*, in I. Chem. E. Symposium on Food Engineering in a Computer Climate, Cambridge, 1991, pp. 47–56.
Capes, C.E. (1980) *Particle Size Enlargement, Handbook of Powder Technology*, vol. 1 (eds J.C. Williams and T. Allen), Elsevier, Amsterdam.
Iglesias, H.A. and Chiriffe, J. (1982) *Handbook of Food Isotherms*, Academic Press, New York.
Jayaraman, K.S. and Das Gupta, D.K. (1995) Drying of fruits and vegetables, in *Handbook of Industrial Drying*, Vol. 1, 2nd. edn (ed. A.S. Mujumdar), Marcel Dekker, New York, pp. 643–90.
Kemp, I.C. (1994) Scale-up of pneumatic conveying dryers. *Drying Tech.*, **11**(3), 279–97.
Kemp, I.C. and Bahu, R.E. (1995) A new algorithm for dryer selection, *Drying Tech.*, **13**(5–7), 1563–78.
Masters, K. (1994) Scale-up of spray dryers. *Drying Tech.*, **11**(3), 235–57.
Moyers, C.G. (1994) Scale-up of layer dryer: a unified approach. *Drying Tech.*, **11**(3), 393–416.
Noden, D. (1979) Selection of dryers, in *Handbook of Solids Drying*, Continuing Education Course, 17–20 September 1979, University of Birmingham, UK.
Papadakis, S.E., Bahu, R.E., McKenzie, K.A. and Kemp, I.C. (1993) Correlations for the equilibrium moisture content of solids. *Drying Tech.*, **11**(3), 543–53.
Papadakis, S.E., Langrish, T.A.G., Kemp, I.C. and Bahu, R.E. (1994) Scale-up of rotary dryers. *Drying Tech.*, **11**(3), 259–77.
Peters, M.S. and Timmerhaus, K.D. (1991) *Plant Design and Economics for Chemical Engineers*, 4th edn, McGraw-Hill, New York.
Ranken, M.D., Kill, R.C. and Baker, C.G.J. (eds) (1997a) *Food Industries Manual*, 24th edn, Blackie Academic & Professional, London, pp. 26–7.
Ranken, M.D., Kill, R.C. and Baker, C.G.J. (eds) (1997b) *Food Industries Manual*, 24th edn, Blackie Academic & Professional, London, p. 527.
van't Lund, C.M. (1991) *Industrial Drying Equipment – Selection and Application*, Marcel Dekker, New York.

12 Dryer operation and control

S.P. GARDINER

12.1 Introduction

In order to achieve optimum performance of a food dryer, it is important that the operating conditions are correctly specified and that the dryer is fitted with an effective control system. The operating conditions will naturally influence the quality of the dried product. They will also affect the running costs of the plant, the impact that the dryer may have on the environment, and, not least, its susceptibility to fire and explosion hazards. These latter aspects are considered in section 12.3.

Adequate control of the dryer is important principally from the viewpoint of maintaining the production of consistent, high quality product at least cost. The various control options available are described in section 12.4, together with appropriate instrumentation that may be used to sense the process variables.

12.2 Dryer operation

12.2.1 Introduction

Dryers have a life in service of 20 years and often more. In consequence, it is common for them to be used to dry several different products (for which they were probably not designed) during their lifetimes. For this reason alone, it is useful to have some guidelines on dryer operation. The process variables are different for different dryers, and a full consideration of each particular type is beyond the scope of this work. However, the general guidelines which follow should prove useful.

12.2.2 Operating conditions

(a) Gas velocity. This in general is run at a maximum value compatible with the product being dried. The evaporative capacity of a dispersion dryer is dependent on inlet gas flow (or velocity in a pre-determined geometry), and inlet and outlet temperatures. The water-carrying capacity of a contact dryer is determined by gas flow and outlet gas temperature. For maximum throughput, the gas velocity is set at its maximum controllable level. Factors which may cause reconsideration are those relating to attrition of product, product loss, particulate emission and, sometimes, energy saving (see section 12.2.3).

(b) Inlet gas temperature.   Energy efficiency requirements dictate that, for a given dryer throughput, the inlet gas temperature should be as high as the product, process or equipment can tolerate. Gas flow can be moderated to reduce capacity, as long as material transport is not hindered. In the food industry, product quality is an issue, and so a temperature is selected with a safety margin to avoid product degradation following a mild process upset (e.g. a reduction in feed rate or feed moisture content).

(c) Outlet gas temperature.   The most suitable dryer outlet gas temperature is determined by one of two considerations. If the feed has a high moisture content, the gas leaving the dryer is likely to be close to saturation. An outlet temperature should be selected that avoids condensation in the exhaust gas stream. Normally 5–10°C above saturation at the dryer exit is sufficient to avoid condensation in the cyclone, bag filter or exhaust ductwork, and to give a margin to cater for normal variation in plant operating conditions.

If the product is being dried to a level close to its equilibrium moisture content, then the relative humidity of the outlet gas (assuming cocurrent operation) will influence the product moisture, and the most convenient way of adjusting product moisture is to change the gas outlet (and perhaps consequentially the product outlet) temperature.

(d) Feed temperature.   Feed temperature is not normally an independent variable, but is set by the upstream process. If the plant output is limited by the evaporative capacity, a small gain can be made by increasing the wet feed temperature, and so putting more heat into the dryer in this way. This should only be considered in these circumstances, and only then if there is no adverse effect on feed handling properties.

(e) Start-up and shutdown.   Unless it is particularly well understood and controlled, start-up is the time when things are most likely to go wrong, and off-specification product is most likely to be produced. Continuous steady-state dryers normally run smoothly but, because the time response to changes can vary widely between gas and solid, most operators will avoid unnecessary stoppages by operating at sub-optimal conditions.

Each dryer type has its own preferred start-up and shutdown method, but a few general rules are common. Normally, the gas flow is started and established first. Heating and temperature control of the gas is then established, and discharge devices are switched on and confirmed before the introduction of any feed. The wet solids (or liquid feed) are introduced at low rate initially, and the rate increased to normal as quickly as possible consistent with retaining control of the operating conditions. Since the principal product property of interest is the moisture content, and there may well be no on-line measurement, the time to develop the full operating rate can be extended. However, there is usually a correlation between one of the measured conditions (e.g. gas outlet temperature) and product moisture, so that an immediate approximate view is available.

On shutdown, the first step is to cut the wet feed and progressively reduce the gas inlet temperature. (This may be an immediate cut in heating medium in the case of short residence time dryers.) Gas flow is normally only cut after all product has been removed.

Dryers with liquid feeds may be started up and shut down by switching feed from or to water. This is wasteful of energy, but ensures that product does not see the extreme conditions which may arise as a result of this unsteady-state situation.

(f) The role of the operator. A continuous dryer operator is essentially a machine minder. There is usually a modicum of automatic control to avoid the necessity of continually watching and adjusting the conditions. From time to time a sample is taken for moisture measurement, and this may result in a minor adjustment to a control set-point. The operator needs to tour the dryer(s) regularly to check that conditions on the plant are consistent with what the instruments are indicating, and to look for potential problems. One of the most common duties of a dryer operator is to ensure that material transport is effective. Feed or discharge blockages or restrictions are not uncommon, and an experienced operator can often anticipate them and avoid more serious stoppages.

Many processes are run on a shift basis, and this means that more than one operator is responsible for a particular dryer. Given that the degree of automation and automatic control is often minimal, the operator has plenty of scope to affect the running of the dryer. As mentioned elsewhere, with every pair of hands you get a free brain, but these brains rarely agree on the optimum operating and control strategy for the dryer. Operators are the first line trouble-shooters, and so the supervisor must educate and train them on cause and effect, and try to ensure a uniform approach to operation and control.

12.2.3 Energy savings

(a) Introduction. Thermal drying is used in many processes where the final product is a solid. From an energy standpoint, drying is best eliminated, but this is not often possible. Significant amounts of energy are used in removing water from the intermediate or final products in a wide range of industries, and thermal drying is often one of the final stages in a process. Approximately 8% of UK industrial energy usage is consumed by drying, and approximately 12% of energy usage in the process industries (Table 12.1). These figures apply to the position in 1978. The present position may have altered somewhat, but not significantly. It is clear that the energy used in drying food materials is significant and therefore represents a significant and often reducible element of process cost.

Table 12.1 Energy usage in drying in UK industry sectors

Industry/subsector	Total energy usage (PJ y^{-1})	Drying energy usage (PJ y^{-1})
Food and agriculture	286	35
Chemicals	390	23
Textiles	128	7
Paper	134	45
Ceramics and building materials	127	14
Timber	34	4
Total	1103	128

(b) Practical energy conservation measures. In attempting to save energy, a strategy to follow is, firstly, determine what you are using; and, secondly, follow the conservation measures that cost the least. The following is a short summary of the techniques to employ.

1. Monitor consumption. 'If you don't measure, how do you know when you've improved?' This is one of the foundations of quality management, and particularly true of energy management. It involves reliable and accurate instrumentation for flow (gas or steam, two of the commonest energy sources for drying). This instrumentation is available, but often not fitted.
2. Maintain a good housekeeping standard. Ensure obvious leaks are repaired promptly, and keep insulation in good condition.
3. Carry out a mass and energy balance. This is not as straightforward as it sounds, and requires a good standard of instrumentation. Portable instruments for investigative purposes are adequate. With improved data acquisition techniques now available at much lower cost, it is feasible to consider on-line mass and energy balances as a management tool to keep costs under control. On-line techniques for humidity and solids moisture measurement are available, but reliability and sustained accuracy under the more arduous conditions of plant operation are generally inadequate. With a complete picture of energy consumption, it is possible to identify and attack the most significant areas in which potential savings can be made.
4. Use the heat balance to identify and repair the less obvious leaks.
5. In convection dryers, the major heat loss is usually in the exhaust gas. Minimize the gas flow, and reduce the exhaust temperature as far as possible. The extent to which this is possible will be determined by the product flow rate and the moisture specification.

   Recovery of heat from exhaust gas, and the use of recycle techniques to ensure minimum flow, need to be planned into a dryer from the design stage in order to achieve maximum effect, and may be difficult to justify on retrofit.
6. Minimize start-ups and shutdowns. These periods usually represent the least energy efficient period of dryer operation. Planning of production schedules and maintenance can minimize the planned interruptions. The duration of a

start-up or shutdown depends to some extent on the dryer type, but can be minimized by use of responsive automatic control techniques, often based on predictive modelling. These techniques often require on-line measurement of humidity and/or solids moisture content. The availability of suitable techniques depends on plant and product, but reliability at a reasonable price is often not obtainable.

7. Minimize the drying load. This splits into two parts, feed moisture and product moisture. If it is possible to reduce the feed moisture mechanically without penalty on product quality, then this is at least an order of magnitude cheaper in terms of energy costs than thermal evaporation. Sometimes dramatic changes can be achieved by, for example, increasing feed particle size. Measurement of feed moisture is desirable to monitor the condition closely, but on-line feed moisture measurement is rarely installed.

   Product moisture level is normally part of the selling specification, and the product quality control specification. On-line product moisture measurement is rare, so the plant operating target for product moisture is usually set some way below the quality control specification, so that normal plant variability does not give out-of-specification product. Cheap, robust moisture measurement would allow operating targets to be set much closer to specification limits, saving energy and possibly increasing plant output.

8. Use direct firing where product quality constraints allow, and there is no conflict with the overall site steam balance on an integrated site. This option requires significant investment on an existing dryer, and may not be justified. Sources of waste heat (e.g. gas turbine exhaust) should also be considered.

9. Consider heat recovery from the exhaust gas (interchange with cold, fresh inlet gas before the heater is the usual way). For an existing dryer this option can rarely be justified on likely payback criteria, but the calculations are worth doing, and it should always be considered for a new design.

12.3 Safety, health and the environment

12.3.1 Introduction

In all processes, some activity in the safety, health and environmental areas are essential for effective operation. In the drying of foodstuffs, the principal safety issue is one of fire and explosion, and avoidance of any condition that can give rise to a food safety risk (Chapter 2). In general, the materials used are not toxic, so operator health from exposure to such materials is not an issue. Environmental problems in food drying are two-fold. Particulate emissions are becoming increasingly intolerable to the public, and odours can create a nuisance, even though they present no danger.

Foodstuffs dryers may be fitted with temperature alarms and interlocks to prevent unsafe conditions, particulate alarms and interlocks to avoid dust emissions to the atmosphere, or flame failure devices to prevent flammable gas escapes. Such controls are increasingly required in order to provide doc-umentary evidence to the regulatory authorities, and on-line measurement, diagnostics and reporting has become the norm in the chemical industry, because of regulatory control. The food industry is also becoming further regulated in the area of environmental matters, and such measures will probably become the norm in this sector too.

12.3.2 Safety

(a) Dust explosions. Some of the most devastating dust explosions have taken place in food processing plant, particularly those processing starch powder and flour. Such explosions not only cause loss of life (possibly) and property, but by destroying the plant cause loss of business until the processing capability is restored. This can take months, and so consequential losses are severe.

The safety record of particulate drying operations in the UK has been an average of one explosion and 30 fires per year. The Institution of Chemical Engineers has produced a safety guide *Prevention of Fires and Explosions in Dryers* (Abbott, 1990), and the reader is referred to this document for a more detailed treatment.

If a combustible material is dispersed in air, and a source of ignition is present, an explosion will occur if the following conditions apply.

- The dust concentration is within the explosive limits (0.01–0.06 kg m^{-3} lower level, upper level around 4 kg m^{-3} but seldom measured and of no practical significance). Dust is normally reckoned to be 75 μm or less.
- There is sufficient oxygen present to support combustion (normally 3–15% oxygen by volume).
- The ignition source has sufficient energy (around 25 mJ or more).
- There is sufficient contact time.

In the most severe cases of devastation, the majority of the damage is caused by a secondary explosion. The primary explosion occurs within the equipment, which then ruptures due to the sudden pressure rise. As a result, dust deposited on surfaces in the building generally is disturbed, which then explodes to cause more widespread damage.

In the UK dusts are classified as Group A, explosible; or Group B, non-explosible, as a result of a series of tests performed under controlled conditions (Table 12.2). The tests are performed at ambient temperature, but the results are temperature dependent. If the material is likely to be exposed to a temperature

Table 12.2 Dust explosion classification tests

Equipment	Dust dispersion	Ignition source
Vertical tube	Vertically up	Electric spark/hot wire
Horizontal tube	Horizontally	Hot wire (1300°C)
Inflammator	Vertically down	Electric spark/hot wire

above 110°C, a high temperature vertical test is performed if no ignition is seen at ambient temperature. If a material is classified as Group A, precautions must be taken to prevent, vent, suppress or contain the explosion. Absence of an ignition source is not regarded as a basis for safe operation. A flow chart indicating dust safety is given in Figure 12.1.

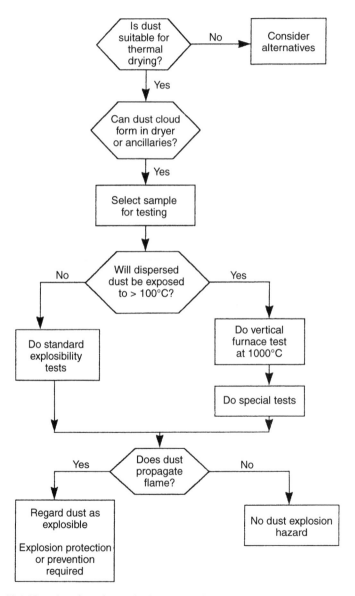

Figure 12.1 Flow chart for safety evaluation proposed by the Institution of Chemical Engineers.

(b) Ignition of layers or deposits. When a layer of combustible material reaches a temperature at which the heat generated by oxidation exceeds the heat losses, it will ignite. This may just be a smouldering fire, or flames may be produced. This can then act as a source of ignition for a dust cloud, either in the dryer or downstream after being transported from the dryer.

A simple test of a layer of material on a metal plate, heated in an oven with air circulation, is sufficient to predict the layer ignition temperature for band dryers and oven dryers. The actual ignition temperature is dependent on layer thickness, the ignition temperature being lower for thicker layers. In a spray dryer, where deposits progressively build up, the tests can be used as a basis for deciding the maximum safe run length.

The pure dry material in a test may give different results from a layer built up from wet deposits, and the presence of combustible additives can affect the result. In particular, contamination of a layer of material with oil will result in a significant lowering of the layer ignition temperature.

(c) Bulk ignition. Powder in bulk will ignite at a lower temperature than that required for layer ignition, as the bulk material insulates against heat loss. The larger the bulk, the lower the ignition temperature. Because the high temperature is at the centre of the bulk, it may be some time before a problem is detected. In a drying plant, a problem can result from powder being discharged from a dryer directly into a storage hopper at too high a temperature. A product cooler may be necessary if the dryer outlet temperature is above the maximum safe storage temperature for the material. The maximum safe storage temperature can be determined by holding small quantities at various temperatures in vacuum flasks. The temperature at which a temperature increase is detected is about 50°C above the maximum safe storage temperature.

(d) Vapour explosion. In some drying operations, evaporation of a flammable solvent in addition to water takes place. This is generally not the case in the food industry, although alcohol evaporation is a possibility. Flammable vapours have much lower (<1 mJ) ignition energies than dust clouds, and rates of pressure rise and maximum pressures are generally higher. Inert gas operation is usually the answer in such cases. Little is known about the influence of flammable dust on the explosion characteristics in these cases.

(e) Fuel cloud. Direct gas firing of food materials dryers is uncommon, but indirect gas-fired heaters are used. An escape of gas from a burner is usually prevented by flame failure devices in the burner, and this is the basis of safety, rather than explosion protection.

(f) Ignition sources. Ignition sources include:

• sparks from welding and cutting;
• sparks from friction or impact (e.g. tramp metal);

- frictional overheating (e.g. bearings, distorted machinery);
- exothermic decomposition of product;
- incandescent particles from direct heating systems;
- sparks from electrical equipment;
- electrostatic discharges.

Although absence of a source of ignition is not regarded as an adequate basis of safety, it is prudent to eliminate as many sources as possible. Thus, sensible operational rules should prevent spark generating operations when the dryer is functioning. Proper maintenance regimes should avoid creation of electrical or friction sparks, and good operating and design practice should minimize the risk due to electrostatic causes, exotherm and incandescent particles.

(g) Explosion protection. Safety can be achieved through prevention by inerting, through protection by venting or suppression, or through mechanical design for containment of the maximum pressure. Inerting involves ensuring that there is insufficient oxygen present to support combustion, and is achieved in one of two ways. For complete inerting, drying takes place in the presence of nitrogen (usually) rather than air. Carbon dioxide or water vapour in the form of superheated steam are alternatives, but uncommon. In the case of direct fired dryers, a proportion of the products of combustion are recycled to reduce the oxygen concentration below the minimum level to support combustion (11% v/v for CO_2 inerting, and 8% v/v for N_2 inerting). In either case, continuous on-line oxygen concentration monitoring is recommended, and alarms and interlocks set to act well below the levels quoted above.

Venting relieves the pressure build-up by releasing the gases in the equipment directly to atmosphere, thereby reducing the excess pressure to a safe level, and is one of the cheapest forms of protection. Vents usually consist of blow-out panels or hinged flaps. As the design pressure of most drying equipment is near to atmospheric pressure, the required vent area can be large, and fitting it into the dryer design can be problematic. Venting must be to a safe area (not inside the plant), and vent ducts should be straight and as short as possible and, in any case, no longer than a few metres. The subject is complex and the reader is recommended to consult a specialist in the area before finalizing a design. More detail is available in an IChemE publication *Guide to Dust Explosion Prevention and Protection: Part 1 – Venting* (Schofield, 1987).

Explosion suppression works by detecting a pressure rise in the early stages of an explosion and then very rapidly injecting a suppressant material into the space to stop the propagation of the flame. The suppressant material may act by forming a heat sink to reduce temperature rise in the flame front, or by free-radical scavenging, or both. This form of protection is rarely used in food processes because of contamination problems from the suppressant.

Explosion protection by containment means designing the dryer and connected equipment to withstand the maximum explosion pressure (10 bar g) without damage. This is acceptable without significant cost penalty for small equipment, but many food processes operate on a large scale, and design for 10 bar g would increase capital cost greatly.

Start-up (particularly) and shutdown are periods when conditions in a dryer change rapidly and there is the greatest potential for unsafe situations to develop. For this reason, the start-up and shutdown procedures should be examined closely to ensure that no hazardous conditions are created.

12.3.3 Environmental issues

Environmental performance is increasingly a focus for regulatory and public scrutiny, and the requirements for monitoring in this area are placing demands on sensor suppliers that are resulting in rapid development of off- and on-line techniques. This area is likely to be one of significant advance in the next few years.

(a) Particulate emissions. All dryers emit a wet air stream to atmosphere, although, in the case of vacuum dryers, its volume is very small. In all other cases, it constitutes the principal environmental impact by virtue of dust content and condensation of moisture vapour to form a plume. Particulate concentration in atmospheric discharge is limited to typically 20 mg m^{-3} by the UK Environmental Agency for prescribed processes under the environmental protection legislation. Other countries have different levels, but the pressure to reduce emissions is steadily increasing.

Dispersion dryers typically have cyclones or bag filters as their primary dust collectors, and material removed from the exhaust gas stream is recycled directly into the dry product. For most dusts, cyclones alone will not give the separation efficiency required to meet particle emission standards, and they are either being replaced by bag filters (particularly in new plant designs) or supplemented by scrubbers. Impingement or venturi scrubbers can be very efficient in removing fine particulates, which can then be recovered by recycling to the dryer feed after a primary solid/liquid separation stage. Scrubbers also are helpful in odour removal where this is a problem.

Bag filters are usually of the reverse-jet cleaning type. The dust laden air passes through a fabric filter in the shape of a narrow cylinder, and the particles collect on the outside of the bag. Without isolation, a section of bags (say 1/6 of the total) is subjected to a rapid pulse of high pressure air injected at the top of the cylinder on the clean side through a solenoid valve. The reverse air flow, and the fabric flexing it induces, cause the solids to be dislodged from the outer surface. The bag then resumes its filtration duty. Low face filtration velocities are normal (60–100 m min^{-1}) to prevent the particles clogging the fabric. Occasionally, bag failure can cause an emission of particulates, which must be rapidly detected.

Routine monitoring of stack emissions for particulate concentration is encouraged by the authorities, and some of the methods available are described below. In both types (impingement probes and optical devices), data logging and manipulation to give periodic averages and total emissions are available.

Impingement probes can be mounted directly in a duct and give either continuous quantitative monitoring, or simply a high level alarm to warn of, for example, a filter failure. These are claimed to be suitable for $0.1–1000$ mg m^{-3}, between -10 and $500°C$. Output can be expressed directly in mg m^{-3} from a knowledge of the stack flow rate. Quantitative monitoring of very fine (sub-micron) particles might be expected to give rise to difficulty, as they will be carried by the air stream round the probe, rather than impinging upon it. The price for a simple alarming system is quoted at $\sim£2000$, and up to £5000 for continuous quantitative measurement.

Optical devices are based on either light-extinction or light-scattering methods. Both can give an accurate measurement, but depend on the optical system being kept clean. Most systems offer an air purge to prevent dust ingress, and diagnostics built into the system can warn of the need for cleaning. Calibration is needed by iso-kinetic sampling to translate obscuration or scattering into a dust concentration. Typical prices range from £1500–4000, depending on accuracy and robustness. Data manipulation costs are extra.

(c) Odour control. Although odours have been emitted from food processing plants for a long time, it is only comparatively recently that pressure to abate them has arisen because, although somewhat unpleasant in high concentrations, they are not noxious.

Methods used for odour abatement are:

- reducing air flows to atmosphere wherever possible;
- scrubbing to absorb (physically or chemically) the offending material;
- biological scrubbers, where the odour-bearing air is contacted with a wet packing carrying bacteria which break down the odour-causing material, have been successful in a number of cases;
- activated carbon adsorption;
- incineration (if all else fails, because of the high expense).

12.4 Dryer control and instrumentation

12.4.1 Operating and control strategy

In deciding on how to control a dryer, it is firstly important to take an overview of the operating strategy of the plant as a whole. There may be an upstream process which feeds, possibly via intermediate storage, to a dewatering, drying and packing section. This section of the plant may also include size classification, milling and some form of recycle. Each of these sections may be operated and controlled separately, but it is common for a process operator's responsibility to range over a range of operations. The overall concept of the control system must be simple enough for an individual operator to understand. In the end, success depends on the ability of the operator to understand and 'feel' in control of the plant; the more complex the strategy and its implementation, the more care must be taken in designing the operator interface.

The operating strategy of a plant will result from the consideration of the following factors, which then determine the objectives of process control:

- inventory management and production control; multi-product, multigrade, clean-in-place;
- extent of manual intervention; less means higher instrument maintenance cost; every pair of hands come with a free brain;
- plant utilization targets and maintenance strategy; solids plants and dryers have lower utilization factors than fluids plants;
- product quality control and assurance;
- material, manpower and energy costs;
- management information;
- process diagnostics; troubleshooting;
- process safety, health and environmental performance.

12.4.2 Alarms and interlocks

Process alarms are intended to give a warning that some part of the process or equipment is behaving abnormally. An on-line sensor is required, which, when a pre-set level is reached, triggers the alarm action. When this occurs, it is a warning either that there has been a failure of the control system, or that some part of the process normally set constant has varied unexpectedly. Sometimes it is the result of a faulty sensor; it may also indicate that the pre-set value should be changed. Assuming that the alarm is not spurious, the process operator is required to take some action. Some common dryer alarms are given in Table 12.3.

Table 12.3 Possible causes of alarm

Symptom	Possible cause
Gas flow low	Filter blockage, dryer blockage, fan stopped
Gas inlet temperature high	Control failure, low gas flow
Gas inlet temperature low	Loss of heating medium, fouling
Gas outlet temperature high	Gas by-passing, loss of feed
Material temperature high	Loss of control
Filter pressure drop high	Blocked filter
Filter pressure low	Element missing
Cyclone pressure drop low	Cyclone solids exit blocked
Pressure high (vacuum dryer)	Vacuum pump fault leaks
Pressure low (atmospheric dryer)	Blocked filter; risk of suck
Motor stopped	Inadvertent switch-off, motor failure, fuse blown, switch gear problem, overload and high temperature trip

The alarm condition may have a predictable consequence and, if not attended to promptly, may cause serious damage to product or plant. In these cases, particularly if the speed of the operator's response is not fast enough, an action is triggered automatically. This is one form of interlock.

Interlocks are intended to prevent undesired combinations of conditions arising through maloperation, or loss of control. Typical examples in a dryer are:

• no feed unless the gas flow or fan is on;
• no feed unless the outlet rotary valve is running;
• motor shutdown on high motor temperature (bi-metal strip, no indication);
• no heat source unless the gas flow or fan is on;
• moving machinery stops if access doors are opened.

The modern way of handling large numbers of alarms is through a computer system, even if the interlocks associated with the alarm are hard-wired, as is often the case with safety interlocks. The computer will present information to the plant operator in a way that allows him to tackle the most urgent problems first and, in the case of an emergency shutdown, which causes multiple alarm conditions, will suppress all the consequent alarms, and display only the one(s) that initially caused the problem.

12.4.3 Dryer control strategies

The drying plant can be controlled in two basic ways, depending on the type of overall process, the type of drying equipment, and the constraints imposed by the product.

(a) 'Wild' feed rate and moisture content. In the first mode, the dryer has to accept the variations in feed rate and moisture content which are fed, and the operator has to manipulate the drying conditions to deliver the product within the desired quality limits. This is the most common mode of operation historically, and gives the dryer operator most scope, since this mode probably determines the overall feed rate. A variation of this strategy is to feed the dryer at a constant wet feed rate, expecting it to cope with variations in feed moisture content. Dryers designed for these modes of operation are usually oversized, and under-run for an easy life. This is not a condemnation of the strategy, but an observation of the result of its deployment.

A typical example of this strategy is shown in Figure 12.2. Here, the moisture content of the dry product is measured by sampling on a periodic basis, with off-line measurement in a laboratory. The measurement is compared with the target value and, if it is too high, the set point of the inlet gas temperature controller is increased. The amount of adjustment is determined by the skill and experience of the operator, and is not precisely defined. Thus different operators will respond in different ways, and moisture control may be less than optimal.

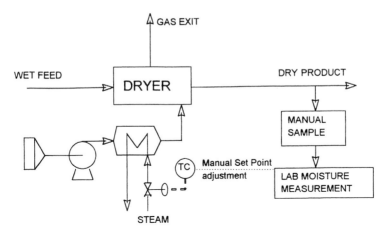

Figure 12.2 Wild feed rate and moisture content, with gas temperature control.

Because moisture measurement takes place intermittently and infrequently (hourly at best, because of the labour availability in the laboratory, and possibly only once or twice per shift), the operator looks for better control methods. Typically, a correlation between a temperature measurement in the dryer and the product moisture content is recognized. In this case the dryer temperature will be used as a control substitute for moisture content, since the measurement is a continuous one. Fluctuations or drift in this temperature will be used to adjust the inlet temperature set point between moisture measurement results, and actual moisture measurement values then used to update the empirical qualitative relationship between dryer temperature and moisture content.

This method, undefined and in the operator's brain, may be successful to the extent of reducing the frequency of moisture measurement. The system is vulnerable to loss of skills if the operator leaves, and to different operators behaving differently. These skills are jealously guarded, and sharing such information with other shifts may be resisted. It is sometimes possible to develop an empirical model of the relationship between the moisture content and the dryer operating parameters. In these cases, more advanced forms of automatic control can be employed, removing some of the dependence on operator skill.

(b) Feed rate control. In the second mode, the dryer operating parameters, such as gas flow and inlet temperature are set at the maximum sustainable, within the constraints of either the equipment or (in the case of temperature) the product. The feed to the dryer is then modulated to keep the product within acceptable limits. This mode of operation is most commonly used when the dryer is the bottleneck in the process, and when the time lag of the process is short. Some method of accumulating wet feed, or a backward integration of the process to vary the upstream rate, is implicit in this mode of control, which incidentally maximizes thermal efficiency.

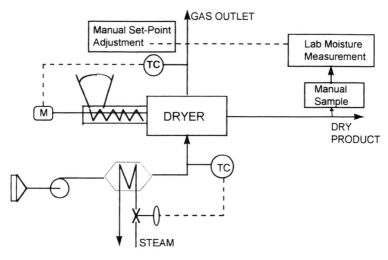

Figure 12.3 Feed rate control.

An example is shown in Figure 12.3. The inlet gas temperature is controlled at a set value by adjustment of the heating medium supply, but plays no part in correcting deviations from the target moisture content. As before, moisture content is measured off-line, intermittently, and compared with a target. This information is used to alter the set point of the outlet gas temperature controller, which changes the speed of the wet feed screw.

One of the main disturbances to dryer operation is uncontrolled variation in feed moisture content, which, because it is difficult to measure, causes problems in dryer control. In dryers with long transport lags, it may be possible (and, if possible, is almost certainly desirable) to reduce feed moisture variation to a minimum.

(c) Control of batch dryers. In batch drying, temperatures are changing as the batch progresses through its drying cycle. The control strategies outlined above for continuous drying are inappropriate, and a different approach is required.

Historically, samples would be taken at intervals during the drying process, and moisture measurement compared with target until the latter was reached. In a repeatable process, the time to reach target will be the same for each batch and, under these circumstances, a simple timer can be used to determine the end point. Since disturbances in the process do occur, time alone is not normally sufficient. A correlation is needed between moisture content and some continuously measured operating parameter. This may be the temperature of the dryer contents, or the outlet gas temperature and humidity, depending on the type of dryer and material being dried. In any case, considerable savings in processing time can be made by having a good correlation, since stopping to take a sample, and restarting, can be time-consuming and costly, particularly for vacuum dryers. In most cases, the drying gas flow is set manually by a damper, and not varied; inlet gas temperature, or jacket heating fluid temperature, is controlled to the maximum that the product can withstand, and material temperature is monitored as an indication of moisture content. Where the material temperature is difficult to measure, a generous fixed time is set.

(d) Quality control. In some plants, product quality control measurements can be made automatically, on- or off-line. This is rarely true of drying plant, since sample presentation represents a problem. The type of moisture analyser used for on-line measurement and process control, where it is practicable (e.g. infrared reflectance), may not be satisfactory for quality control purposes.

Since the majority of drying plant does not have continuous on-line moisture measurement, the customer seeks reassurance that the material dried between samples also conforms to his purchasing specification. This can be achieved by further off-line sampling and testing before despatch (expensive), or techniques such as statistical process control (SPC) can be employed.

The principle of SPC is that if a property varies cyclically, it is possible to contain the cycling within the specification limits by taking action when variation from target exceeds a narrower limit. If the rules of action and the limits are clearly defined, and evidence of their being followed can be provided, then statistical confidence in achieving specification for all material produced is high. The benefit of this approach is, firstly, that material is likely to be within specification and, secondly, that it is cheaper than testing at a high frequency.

In providing evidence of SPC, critical process variables must be recorded in some form. A plant control system can provide the means of both recording the data and presenting it in a documentary form to satisfy the customer. As the concept of supplier–customer partnership extends, so electronic data interchange (EDI) can reduce the need for documentation.

Most drying operations are not just about reducing moisture; other quality parameters are both measured and controlled. Typical of such properties are colour, bulk density, dustiness (particle size distribution), flowability, dispersibility and even trace contaminants. This is by no means an exhaustive list, but each product has its combination and hierarchy of properties. Each property needs to be measured, and assurance must be given to the customer that it falls within his acceptable limits. Very few of these non-moisture properties of a solid can be measured on line, and the correlation between the properties and the dryer operating variables is even less well defined than that for moisture. Thus apart from the documentation aspects, plant control linkages with laboratory based measurements are few.

12.4.4 Types of process control system

The objectives in most drying operations are to produce dried product, conforming to some given moisture content at minimal cost and maximum throughput.

As long as the processes upstream of the dryer continue to produce wet material at a constant rate and of constant moisture content, these objectives may be met without resort to automatic control. When, however, variations in these quantities or changes in product specification due to the requirements of customers or of subsequent processes are important factors, then the need for compensating control systems becomes more pressing.

Control systems may take one of two basic forms. These are usually known as open loop and closed loop systems, respectively, and are described in the paragraphs that follow. Details of more advanced methods of process control are described thereafter.

(a) Open loop control.   Open loop control systems use an estimate of the required control action needed to achieve a desired result. No automatic check is made to determine the effects of the control action and, in particular, whether it actually does achieve the desired result. For example, a batch drying system has open loop control when the duration of each batch is predetermined (Figure 12.4) with no attempt to measure the finished product dryness. The features of a control system of this sort are:

- performance is determined by the accuracy of the estimated control action (drying time in the above example);
- dynamic instability because of control action cannot occur.

(b) Feedback control.   Feedback control is a closed loop system in which a dependent variable (e.g. product wetness) is continuously measured, and the difference between the actual value and the desired value (set point) of the variable is derived. This error signal is used to initiate control action in a direction which tends to decrease the error signal, that is to equalize the actual and desired values of the controlled variable. This is illustrated in Figure 12.5

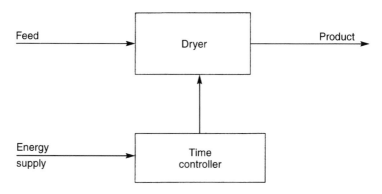

Figure 12.4 Open loop control.

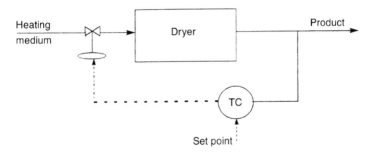

Figure 12.5 Feedback control.

for a continuous dryer, where product temperature is shown controlled by heat supply. The features of a feedback control system are as follows.

- The system will work in principle even if the actual relationships between the feed (input) and product (output) variables are unknown, though time lags and delays always existing in plant may cause instability, i.e. oscillations of the controlled variables. This means that a dynamic model of the plant is almost certain to be needed for effective control system design.
- In plant with significant transport time delays between the input and output variables, and this includes most dryers, feedback control does not begin to correct a feed disturbance until it has passed through the system to the output, where it is detected and initiates control action. This delay in compensating for disturbances can lead to very unsteady control; a possible solution is to combine feedback and feedforward control (see below) in an attempt to make the best use of the characteristics of each.

Feedback control can, of course, be applied to situations other than the overall feedback control of the plant. For instance, when any plant variable such as a temperature or flow rate is to be held constant, a local feedback controller would normally be employed. This uses the desired value of the variable as its set point, the sensor signal as a measured value, and the difference between the two as an error signal to actuate a control valve.

(c) Feedforward control. Feedforward control uses measurements of the feed to control the dryer to a required product property. It can be either open or closed loop in nature.

The features of feedforward control are therefore: (1) the process conditions are altered by control action to compensate for feed disturbances before those disturbances can upset the process; and (2) the effectiveness of control is entirely dependent on a good estimate of the dynamic relationship between feed and output product variables, i.e. on a partial model of the process.

(d) Advanced control. This term can be applied to all but the simplest form of control systems, and can include for instance cascade control, where a deviation

from a set point in one variable can be used to change the set point of a second controller. This is useful in damping oscillations in the second variable that might occur if the control action was direct. However, the term advanced control is normally used to describe model based predictive control, and the more complex areas such as fuzzy logic and neural networks.

(e) Model based predictive control.   As the term implies, a dynamic under-standing of the relationship between the product properties and the process variables is needed for this approach to succeed. If such a model exists, a disturbance in the feed conditions will cause a disturbance in the output which can be predicted in both magnitude and time. To prevent the output disturbance, changes can be made to the process conditions that will reduce the effect of the initial disturbance such that the specified limits of the output properties are not exceeded. This type of control is most useful where the transport delays in the dryer are large (e.g. in cascading rotary dryers). Much work has been carried out in this particular area, for cement kilns and sugar beet.

The model used for control can be either based on sound theoretical prin-ciples, or be entirely empirical. In the latter case, it should not be extrapolated beyond the range of conditions over which it is known to apply. If theoretically based, then it can be used for extrapolation with more confidence.

(f) Fuzzy logic.   The term Fuzzy Logic, based on the fuzzy set theory in mathematics, was first proposed and described by Zadeh in 1965. Driving this theory is the idea that traditional true or false logic cannot deal with cases that contain a number of exceptions. Although it originated in USA, the major applications have been development in Japan, mainly in consumer products such as cameras, camcorders and washing machines. More recently, the benefits achievable in the process industries are being recognized, and several control companies are offering fuzzy control as part of their range. The concepts were first applied to the control of steam engines in 1974, and later to cement kilns.

In classical logic, a rule can be introduced only on the basis of quantified data (e.g. if $T < 10$, then ...). In fuzzy logic, the boundaries of action are blurred by the concept that, for instance, if $T = $ LOW, then ..., without initially defining what LOW means. However, for the system to work, some function must be implied, and many rules are needed for each output.

The advantages in following this approach is that it is not necessary to define the model by which the process operates in precise mathematical terms. Indeed, many drying operations have defied all attempts to so define them. They are still more easily controlled using operator expertise than rigid rules.

The advantages of fuzzy logic control are that it can control processes that previously could only be controlled manually, the principles are easy to under-stand, and unpredictable disturbances can be handled. Its disadvantages are that it is hard to back up theoretically, and no methods are available to determine the rules.

There are no published cases in which fuzzy logic has been used to control a dryer, but the process is one which meets the criteria for application:

- a mathematical description of the process is not available to formulate (non-linear, time-variant);
- the control strategy cannot be expressed in mathematical formulae;
- process can be controlled by the intuition and experience of skilled operators;
- many inputs are needed to control one or more outputs.

The fact that cement kilns are very similar in nature to cascading rotary dryers suggests that the latter would be susceptible to the approach.

(h) Neural networks.   Neural networks are made up of a number of inter-connecting software or hardware building blocks called processing elements, and vaguely resemble the architecture of the vastly more complicated network of neurons in the human brain. A fuller description is given by Samdami (1990). A feedback network can learn from its own experience, without being explicitly programmed for each new input. Thus, in a system where relationships between input and output variables are unknown, a neural network can determine the relationships on the basis of experience, and constantly update them as new data arrives. Interpolation within the extremes of experience is thus reliable, but extrapolation is dangerous because the correlations identified have no theoretical basis.

The application of neural networks is at an early stage of development in the process industry, and no cases have been published of applying neural networks to solids drying. However, they are suitable for modelling processes with long time constants, and those with complex interactions, so drying of products in long residence time dryers where product quality parameters are linked to a number of drying variables is a subject which is suitable for analysis. This is particularly true if the process is subject to a high degree of 'noise'.

A major disadvantage to neural networks is their complexity. Improving their performance is thus very difficult, no matter how well the user understands the process. Further reading on the subject is available from Wasserman (1989) and Chitra (1993).

12.4.5 Process measurements

The main process parameters in drying are:

- temperature of the inlet and outlet gas, liquid heating medium, wet and dry solid;
- flow of the drying gas, solids;
- pressure, i.e. the low differential pressures from atmospheric, and vacuum;
- moisture in the inlet and outlet gas and solid.

In addition, product properties important in handling, packaging or end-use application are often measured. These vary widely between products, and could commonly include:

- bulk density of particulates;
- particle size distribution, or perhaps just a measure of fines/dustiness;
- dispersibility (for products whose end use involves re-wetting);
- flowability.

For process control alone, the most important attributes of any measurement are reproducibility, accuracy and robustness at an affordable price. Once a measurement is being communicated outside the plant or company, then the method must be stated, or agreed. To avoid having to agree a range of methods for a single property, each to meet the needs of different customers, the use of an appropriate ISO, American or British Standard may be preferred.

Whether on- or off-line, an instrument will only measure the property of the material presented to it. Sampling of a stream for off-line measurement, and positioning of a sensor for on-line measurement are critically important for representative results.

(a) Temperature. Temperature is the most frequently used measurement in the control of dryers. Since temperature sensing equipment is relatively inexpensive compared with process analysers, it is worth selecting on performance and reliability criteria alone.

Resistance thermometers, thermocouples, thermistors and automatic pyrometers are all used to measure temperatures industrially. While resistance thermometers are more accurate and give a signal less prone to interference, because of its high level, than thermocouples, the latter are less expensive, smaller, more rugged and show a quicker response. Thermistors have much greater sensitivities than the other detector types and are best employed in situations where temperatures must be regulated precisely over a very limited range.

Pyrometers are used where remote measurement is necessary or where contact between the material being measured and the sensor is difficult or erratic. Thermocouples remain the most widely used industrial temperature sensor, with the other types meeting specialist requirements.

(b) Pressure. Pressure measurements can be important in drying in two areas. Firstly, on atmospheric dryers, pressure balancing between inlet and outlet fans at an opening in the dryer can ensure that leaks of hot air from the system, or cold air into it, are minimized. A small suction is often maintained at such a point to prevent escape of product. Pressure difference across a filter can give early warning of the need to replace the element, and gas flow is often inferred from pressure tappings across an orifice plate. All these applications require pressure difference measurement from atmospheric which is small, say

0–100 mbar. Secondly, on vacuum dryers, a low pressure allows drying to take place at low temperature; often this is the reason for selecting vacuum drying. Here the most common range of operation is 0.01–0.1 bar. Only in freeze drying are absolute pressures of <0.01 bar needed.

Manometers are used for local indication of low pressure differences, and Bourdon gauges for local pressure differences >0.75 bar. Diaphragm and bellows sensors can be designed for a wide range of pressure differences.

Wherever possible, pressure tappings should be located in clean, non-condensing areas of the process. In drying this is not always possible, and it is often necessary to measure pressure in the presence of particulates and condensable vapours.

Condensation can be tackled by one of a number of techniques:

- filling the impulse lines with an incompressible seal fluid (and compensating for any difference in vertical liquid height);
- arranging the transmitter above the tappings, with self-draining impulse lines (of large enough diameter) and no pockets;
- heat tracing the impulse lines to prevent condensation;
- placing the sensing diaphragm flush with the process surface.

If solids ingress is a risk, then the tappings should be mounted vertically upwards from the process surface. If this is not enough, then a clean gas purge of the tapping point can be arranged. Again, direct flush mounting of a diaphragm element can also be considered.

(c) Gas flow. Absolute measurement in a dryer is a complex matter, since the conditions for best sensor performance are seldom met. Long straight sections are rarely found in duct work and changes in section are more rapid than recommended for accurate measurement because of the desire to economize on space. For on-line process control purposes, precision is unimportant, so long as any change in the measurement is related in a consistent manner to the absolute value. There are three principal methods of continuous flow monitoring in dryer ductwork:

- pitot tube;
- differential pressure meter;
- vortex meter.

Alternative methods of flow measurement tend to be excluded by considerations of fluid density, scale and cost. In general, because precision is not essential, but reproducibility is, it is better to locate the flow sensor on the clean dry gas side, even if the duct layout is non-ideal, rather than in the exhaust stack where risk of condensation is greater.

(d) Solids flow. Solids flow is the fundamental measurement of any solids processing plant, since it is required in order to determine the plant output.

However, it is rarely measured continuously, on-line. Indeed, it is not uncommon for the only measurement of throughput on a solids handling plant to be the number of bags or bins packed. If there is intermediate storage before packing, then some other means of throughput measurement is needed. This may be instantaneous (rare), or time-averaged to some degree. For dryer control, a direct measurement is not normally required, since the control strategy usually dictates that the throughput is allowed to vary, and is controlled elsewhere in the process.

(e) Humidity. The measurement of moisture in gases is covered in detail by Carr-Brion (1986). Methods described are suitable for either laboratory or on-line measurement. This section will concentrate on commercially available on-line methods. Most humidity sensors rely on a change in the electrical properties of the detector element with change in moisture content. The mechanical hygrometer works on the principle that the dimensions of an organic material (e.g. hair) change with moisture content, which in turn responds to changes in atmospheric moisture. The response to changing humidity is slow, and the environment for measurement needs to be clean. However, this device is cheap. Calibration is not easily traceable for quality assurance purposes. Accuracy is claimed as ±5% (Foxboro, 1984), and is unaffected by gas velocity.

Psychrometers measure the wet and dry bulb temperatures of the operating gas stream, allowing calculation of humidity, dew point and other psychrometric parameters. The measurements are dependent on gas velocity and, if this varies significantly, a correction factor must be applied. Relative humidity levels of 20–80% over a dry bulb temperature range of 0–80°C are regarded as a suitable operating envelope for a wet box, in-line sensor system (Foxboro, 1983). Clean water is advisable, as is a flow-through system to prevent errors due to concentration build-up of soluble species. Regular maintenance to check for water level and contaminants is essential. This type of sensor is losing ground in on-line installations to electrical measurement techniques.

In recent years there has been a large expansion in suppliers of instruments based on the electrical measurement of humidity. Currently, there are over 40. For both portable and on-line measurement, the majority of suppliers offer capacitive sensors. These usually consist of a hygroscopic material (e.g. aluminium oxide or a polymeric substance) whose electrical impedance or capacitance, or a combination of the two, varies with relative humidity. Many suppliers claim an unrealistic RH range (e.g. 0–100%). In a review article, Pragnell (1993) of SIRA points out severe non-linear effects below 15% and above 85% RH, and concludes that an accuracy of ±4% of reading is probably the limit for instruments of this type. In a study of five different manufacturers' instruments over a 2 y period, Pragnell reports that all drifted significantly, up to 5% RH in a year. For this reason, all electrical humidity instruments should be frequently and regularly calibrated. Many manufacturers recognize the problems of drift, and offer calibration tools, usually as standards which can be reused several times before replacement is necessary.

Despite the limitations described above, electronic humidity instruments have a fast response, can be easily read, often with a digital output, and are not sensitive to vibration in handling. Portable instruments are compact and cheap, with price falling in real terms as electronic hardware prices fall. Many instruments combine a measurement of RH with a temperature, so that dew point can also be calculated and displayed.

Capacitance sensors are used for monitoring humidity in sugar storage silos to prevent caking, with reliable performance being reported over several years. However, this is at ambient temperature, and medium RH, so not a demanding duty, except for the dust, which is eliminated by the use of a sintered metal filter (available as standard on most instrument probes).

Dew point detection relies on a sudden change in a property, such as capacitance or optical reflection, when condensation takes place. A temperature sensor on the element gives a direct measurement of dew point temperature, as the element is cooled, often by thermoelectric Peltier effect. Optical sensors are accurate and reliable, so long as contaminants do not build up on the wet surface, leaving a deposit when the surface is reheated. Filtration of a dusty air stream is essential. The drift which affects capacitive humidity instruments is absent from optical types, and developments to minimize mirror contamination include automatic chill–reheat cycling. A typical chilled mirror sensor, with process gas temperature measurement to enable RH to be calculated costs around £3500.

Infrared analysers will typically measure a number of components in a gas stream, of which water can be one. A cheaper variant is available which measures water only. The specific wavelength used (e.g. between 2 and 12 mm) depends on the components to be analysed. A recent development has been an in-line analyser mounted directly in the gas duct, removing the need for a separate sample line. A mirror system is used to double the path length while keeping the probe length down to 450 mm (Procal Analytics Ltd). Dust is excluded by a sintered metal shroud, and an automatic purge of dry air can be supplied to ensure that no condensation occurs on the optical system. An operating temperature range of 0–300°C is quoted.

(f) Moisture in solids. Moisture measurement of solids is usually performed off-line, and related by experience to a process variable, or a combination of process variables, with on-line monitoring, and possibly control.

On-line measurement of moisture is commercially available by a number of methods: capacitance, microwave and near-infrared. Neutron moderation is also available, but is unsuitable for foodstuffs, and anyway is extremely expensive. On-line nuclear magnetic resonance (NMR) techniques are currently being developed, which have the potential of measuring other physical and chemical parameters in addition to moisture content.

Capacitance instruments. Both on- and off-line instruments are available, and are based on the change in the capacitance of the sensor with moisture content. All are subject to all the comments in section 12.4.5(e). On-line experience is limited. Calibration against a standard laboratory method is necessary for each material to be measured, but once calibrated the method is easy to use and gives quick results. For process control purposes, this method is likely to be acceptable with regular calibration, except at low moisture contents. Most industrial experience has been obtained on foodstuffs where moisture contents of several per cent are quite acceptable.

Infrared instruments. When a beam of infrared light is incident on a moist material, some of the light is selectively absorbed by the moisture, so that the reflected light from a moist material differs from that of a dry one. The extent of the difference is a measure of moisture content at the reflecting surface. In order to cope with different materials, with a range of particle sizes and moisture contents, a number of different narrow bands of wavelength are used simultaneously. The technique can be used both on- and off-line. In laboratory use, a number of different materials are likely to be analysed on the same machine, each of which will have a different infrared/moisture calibration. Instruments can store a number of different calibrations (up to 50), and the operator must ensure that the right reference calibration is being used. Even for on-line use, this feature is essential unless the application is for a single grade of a single product.

For on-line use, the sensor head must be positioned 75–300 mm from the sample, on, for example, a conveyor belt. A variation of 0.3% of the reading can result from ±50 mm variation in pass height. Dust in the atmosphere, or on the optical system will affect the measurement. Alternatively, an automatic sampling system can present a sample to a local analyser.

Moisture contents from 0–90% can be measured, but for greater accuracy at low moisture a range from 0–0.2% is the minimum. Accuracy is claimed as ±0.5% FSD, and repeatability at 0.2% FSD.

Manufacturers of on-line instruments include Infra-Red Engineering and Quadra-Beam. Components other than moisture can also be measured: for example, the fat content of a food material.

(g) Particle size/size distribution. Particle size and distribution are a complex subject and cannot be treated in any detail here. The methods routinely used on plant are sieve analysis, which is used off-line on a sample from the process, and light scattering. In a wet system the Coulter method may be used to measure light scattering; while in a dry system the Malvern method may be used.

Sieve analysis is adequate for sizes above 50 μm and, if dustiness is a concern, then typically material below 75 μm may be called 'fines'. The amount of dust (say below 20 μm) may be small as a proportion by weight, but as a proportion by number of particles is much higher.

References

Abbott, J. (ed.) (1990) *Prevention of Fires and Explosions in Dryers*, Institution of Chemical Engineers, Rugby, UK.

Carr-Brion, K. (1986) *Moisture Sensors in Process Control*, Elsevier, London.

Chitra, S.P. (1993) Use neural networks for problem solving. *Chemical Engineering Progress*, April, 44–52.

Foxboro (1984) *45P Indicating Relative Humidity Transmitter*. Specification sheet PSS 1–5C1 B, Foxboro Co, Foxboro, MA, USA.

Pragnell, R. (1993) Relying on humidity. *Process Engineering*, Feb., 41–2.

Samdani, G. (1993) Fuzzy logic more than a play on words. *Chemical Engineering*, Feb., 30–3.

Schofield, C. (1987) *Guide to Dust Explosion Prevention and Protection: Part 1 – Venting*, Institution of Chemical Engineers, Rugby, UK.

Wasserman, P.D. (1989) *Neural Computing Theory and Practice*, Reinhold, New York.

Index

Page numbers appearing in **bold** refer to figures and page numbers appearing in *italics* refer to tables.

Absolute humidity *13*, *15*
Adiabatic saturation lines 12–16
Adiabatic saturation temperature *13*
Adsorption isotherms, *see* Equilibrium moisture curves
Advanced control systems, *see* Process control of dryers, types of process control system, advanced
Agglomerate production in spray drying, *see* Spray drying, production of agglomerates
Agitated pan dryer, *see* Contact dryers, types of, vertically agitated
Agricultural crop dryers, *see* Through-flow dryers for agricultural crops
Air filtration in spray dryers, *see* Spray dryers, design features for hygiene and safety; drying air filtration
Alarms and interlocks 282–3
Alfalfa, chopped, drying of 5
Apple, drying of *197*, *208*
Ascorbic acid, drying of, *see* Vitamin C, drying of
Aspergillus niger, minimum water activity for growth *20*
Atomization in spray drying processes, *see* Spray drying, atomization methods
a$_w$, *see* Water activity

B. cereus spores, minimum water activity for growth *20*
B. subtilis spores, minimum water activity for growth 20
Baby foods, drying of *82*, *91*, *116*, 121
Bacteria, minimum water activity for growth *20*
Bag filters, *see* Particulate emissions, dust collectors
Baker's yeast, drying of *82*
Baking quality of wheat, *see* Crop quality and drying
Bananas, drying of 217, *229*
Band dryers
applications 207, *208*
construction of 206–7
description of 205–7
multi-pass 207
selection criteria *257*, *258*, *259*

single-pass 206–7
variability in moisture content over band 4
see also Fluidized bed dryers, types of, Jetzone
Barley, drying of *37*, 229
Basics of fluidization, *see* Fluidization, basic principles of
Batch drying curves, use of in fluidized bed dryer design 86–8
Batch fluidized bed dryers, *see* Fluidized bed dryers, types of, batch
Beans, French, drying of *229*
Beef, cooked, drying of 151
Bench-scale tests, use of in dryer selection
role 256–60
testing methods
drying characteristics 262
equilibrium moisture content 262–3
physical properties 260–1
quality attributes 261–2
thermal stability 262
Binding energy, *see* Enthalpy of wetting
Biscuits, post-baking of 4, 174–5
Biscuits, baking of 175, *208*
Bound moisture, *see* Moisture, bound
Bound moisture, removal of in freeze drying, *see* Freeze drying, process stages, desorption
Bran, drying of *208*
Bread volume, *see* Crop quality and drying, bread volume, relative
Breadcrumb, drying of *208*
Breakage of grains, *see* Crop quality and drying, stress cracks and broken kernels
Breakfast cereals, drying of *208*
Brewers' grains, drying of *116*
Bulk density of spray-dried powders, *see* Spray drying, meeting powder specifications in, bulk density
Butter, drying of *91*
Buttermilk, drying of *91*

C. botulinum, minimum water activity for growth *20*
C. botulinum spores, minimum water activity for growth *20*
Cabinet dryers, *see* Tray dryers

Canadian Prairies, harvest conditions   *41*
Canola, drying of   *37*
Capillary-flow model   26
Capital cost estimates of drying equipment
    cost comparisons   266
    in-house generated   264
    vendor-generated   263
Carbohydrates, drying of   91
Cardamon, drying of   *229*
Cascading rotary dryers, *see* Rotary dryers,
        cascading
Casein, drying of   *82*, 185, 190
Caseinates, drying of   *91*
Cassava, drying of   *229*
Cereal products, drying of   174, 175, *197*,
        267–8
    see also, Breakfast cereals, drying of
Characterization of dryer feeds, *see*
        Specification of the drying process
Characterization of dryer products, *see*
        Specification of the drying process
Characteristic drying curve   **25**
Cheese, drying of   11, *82*, 91
Chemical changes in foods during drying   27,
        28
Chicken, cooked, drying of   151–2
Chillies, drying of   *82, 229*
Chips, drying of   175
Chitosan gel, drying of   192
Chocolate crumb, drying of   *116*, 118, 124,
        208
Citric acid, drying of   *82*
Citrus fruit concentrates, drying of   173
Citrus pulp, drying of   185
Cladding of spray dryers, *see* Spray dryers,
        design features for hygiene and
        safety, cladding
Classification of dryers   9–10
Cocoa products, drying of   192, *197*, 216
Coconut, drying of   5, *82, 208*, 216
Coconut milk, drying of   *91*
Coffee, drying of   *82, 91, 116*, 149
Coffee beans, drying of   213, 216, *229*
Coffee whitener, drying of   *91*
Combined hot air/dielectric drying, *see*
        Dielectric drying, combination dryers
Commercial grain dryers   43–4
    see also Crop drying costs
Conductively heated dryers, *see* Contact dryers
Confectionery, drying of   *116*, 124
Constant-rate drying period   *13*, 20–3, **22**
Contact dryers
    applications of   3, 115, *116*
    classification of   *116*
    comparison with convective dryers   114–5
    definition of   114
    design of   131–2
    factors influencing selection of   115–7
    selection criteria   *257, 258, 259*

Contact dryers, types of
    drum   121–3
    horizontally agitated   124–6
    indirectly heated rotary   126–7
    plate   119–20
    rotating batch, vacuum   123–4
    thin-film   120
    vacuum band   117–18
    vacuum tray   117
    vertically agitated   127–8
Contact drying, theoretical overview
        of   128–30
Containment of explosions, *see* Safety
        considerations in drying, explosion
        protection
Continuous fluidized bed dryers, *see* Fluidized
        bed dryers, types of, continuous
Control of dryers, *see* Process control of dryers
Convective dryers, comparison with contact
        dryers   114–15
Conveyor dryers, *see* Band dryers
Cooked cereals, drying of, *see* Cereal products,
        drying of
Cookies, post-baking of   175
Copra, drying of   *229*
Corn products, drying of
    grits   *37, 208, 229*
    meal   124, 116
    starch   186, *187*
    syrup   91, *121*
Cost estimates of dryers, *see* Capital cost
        estimates of drying equipment
Cost of dielectric dryers   176
Cost of drying agricultural crops, *see* Crop
        drying costs
Crab meat paste, drying of   192
Crackers, post-baking of   175
Cream, drying of   *91*
Crispbreads, post-baking of   175
Critical moisture content   *13*, **22**
Crop dryers, *see* Through-flow dryers for
        agricultural crops
Crop drying costs
    direct drying costs   59–60
    risk costs   60–1
    total cost   61–2
Crop quality and drying
    bread volume, relative   58
    factors affecting   56
    germination ratio   56, 58
    gluten, relative   56, 58
    maximum allowable temperature   56, 62, 57
    quality models
        exponential model   57–8
        linear model   58
        probit model   58–9
    specific quality attributes affected by
        drying   56
    stress cracks and broken kernels   59

Crumb, drying of 208
Cyclones, *see* Particulate emissions, dust
    collectors

Dairy products, drying of 1, *82*
Definition of drying 7
Dehydrated foods, consumption of 1
Design methods 8–9
  see also individual dryer types
Desorption 2
Desorption, in freeze drying, *see* Freeze
    drying, process stages,
    desorption
Desorption isotherms, *see* Equilibrium
    moisture curves
Deterioration of foods
  control of *27*
  rate of 19–20, **21**
Dewpoint 13, 15–6
Dextrose, drying of 91
Dielectric dryers
  cost of 176
  selection criteria *257*
  use of in postbaking, *see* Biscuits,
    postbaking of
Dielectric drying
  applications of
    combination dryers 175
    microwave atmospheric dryers 174
    microwave vacuum dryers 173
    radiofrequency dryers 174–5
  advantages and limitations of 155, 172,
    176
  choice between microwave and
    radiofrequency dryers 172
  combination dryers 170, 175
  dielectric heating in 170
  dielectric loss factor 165–7, **168**
  drying rates in 169
  effect of a dielectric on electric field
    strength 164–5
  heat and mass transfer in 168–9
  moisture levelling in 167–8
  moisture movement in 169–70
  power dissipated within dielectric materials
    microwave drying 163–4
    radiofrequency heating 161–3
  safety of 161
Dielectric heating, permitted frequency bands
  EMC (electromagnetic compatibility)
    regulations 155–6
  ISM (industrial, scientific and medical)
    frequency bands 155–6
Dietary fibres, drying of 185
Diffusion coefficients 2
Diffusivity of moisture in foods 24
Discoloration of grains, *see* Crop quality and
    drying, maximum allowable
    temperature

Distiller's dark grains, drying of 185
Distributors in fluidized bed dryers, *see*
    Fluidized bed dryers, features of
    construction, distributor
Dog biscuits, drying of *208*
Double-cone dryers, *see* Contact dryers, types
    of, rotating batch, vacuum
Drum dryers, *see* Contact dryers, types of,
    drum
Dry-bulb temperature *13*
Dryer operation
  energy savings
    energy use in drying in UK industrial
      sectors 273, *274*
    practical energy conservation
      measures 274–5
  operating conditions
    feed temperature 272
    gas velocity 271
    inlet gas temperature 272
    operators, role of 273
    outlet gas temperature 272
    start-up and shut-down 272–3
Dryer selection
  approach to 241–2
  checklist 243, **244**
  criteria 11
  use of expert systems in 4, 270
  worked examples of use of selection
    algorithm 367–9
  see also Bench-scale tests, use of in dryer
    selection; Capital cost estimates of
    drying processes; Pilot-plant trials,
    use of in dryer selection; Preliminary
    selection of dryers; Specification of
    the drying process
Dryer types 7–8
Dryeration, *see* On-farm crop dryers,
    combination heated- and unheated-air
    dryers
Dryers, control of, *see* Process control of
    dryers
Drying and cooling times in agricultural dryers
  in on-farm combination heated- and
    unheated-air dryers 37–8, *42*
  in on-farm heated-air dryers 36–8
Drying costs, agricultural crops, *see* Crop
    drying costs
Drying, definition of, *see* Definition of drying
Drying kinetics
  in contact dryers, *see* Contact drying,
    theoretical overview of description
    of 20–6
  in fluidized bed dryers, *see* Fluidized bed
    dryers, design methods, detailed
    design
Drying mechanisms 2
  see also Drying times predicted by different
    drying-rate models

Drying process specification, *see* Specification of the drying process

Drying rate, definition of   21

Drying rates in dielectric dryers, *see* Dielectric drying, drying rates in

Drying times predicted by different drying-rate models   23

dryPAK software, *see* Psychrometric charts

Dust collectors, *see* Particulate emissions, dust collectors

Dust explosions in dryers
  causes of   276
  during start-up and shut-down   280
  dust safety classification   276–7
  powder explosion classification tests   276, 277
  secondary explosions   276
  UK statistics   276
  see also Explosion protection in dryers; Safety considerations in drying

Economic evaluation of drying processes, *see* Capital cost estimates of drying equipment

Economic evaluation of solar dryers, *see* Solar dryers, economic evaluation of

Egg, drying of   *91*, 152

EMC, *see* Equilibrium moisture content

Energy conservation in fluidized bed dryers, *see* Fluidized bed dryers, operation, energy conservation

Energy consumption in agricultural drying
  by on-farm combination heated- and unheated-air dryers   42–3
  by on-farm heated-air dryers   37–8

Energy saving techniques, *see* Dryer operation, energy savings

Engineering properties of air-water mixtures, *see* Psychrometry

Enthalpy of air-water mixtures   *15*

Enthalpy of wetting   18–19

Enzymatic deterioration of foods during drying   *28*

Equilibrium moisture content   13, 16–19, **22**

Equilibrium moisture curves   16–18

Examples of use of dryer selection algorithm, *see* Dryer selection, worked examples of use of selection algorithm

Exotoxins, formation of, *see* Toxicological safety

Explosion hazards
  in fluidized bed dryers, *see* Fluidized bed dryers, operation; fire and explosion hazards
  in spray dryers, *see* Spray dryers, design features for hygiene and safety, fire and explosion protection

Explosion protection in dryers, 279–80

External heat/mass transfer control during drying   20–3

Factors affecting dryer selection, *see* Preliminary selection of dryers, factors affecting

Falling-rate drying period   20–3

Feed systems for pneumatic conveying dryers, *see* Pneumatic conveying dryers, feed systems for

Feedback control, *see* Process control of dryers, types of process control system, feedback

Feedforward control, *see* Process control of dryers, types of process control system, feedforward

Fibre, gluten and corn steep liquor, drying of   187

FILTERMAT® dryer, *see* Spray dryers, types of, cocurrent drying chamber with integrated belt

Fire hazards in dryers, *see* Safety considerations in drying
  see also Fluidized bed dryers, operation, fire and explosion hazards; Spray dryers, design features for hygiene and safety, fire and explosion protection

Fish products, drying of   82, *152, 197*

Flaked cereal grains, drying of   267–8

Flash dryers, *see* Pneumatic conveying dryers, flash dryers

Flour, drying of   *82, 197*

Fluid bed dryers, *see* Fluidized bed dryers

Fluidization, basic principals of   65–8

Fluidized bed dryers
  design methods
    detailed design   86–8
    scoping design   84–6
  features of construction
    distributor   68–70, 99, **100**
    freeboard   67–8, 71
    gas cleaning equipment   71
    plenum chamber   70–1
  feedstocks   65
  operation
    control   81
    drying times   65, 82
    energy conservation   81
    fire and explosion hazards   81
    gas velocity range   67–8, 82
    particle size range   65, *258*
    throughput range   65
  selection criteria   *257, 258, 259*
  test procedures   82–4

Fluidized bed dryers, types of
  batch   72–3
  continuous
    centrifugal   80
    granulators/coaters   79, **80**

internally heated 76
Jetzone 80
mechanically agitated, *see* Spin-flash
    dryer
multi-stage 3, 74–6
plug flow 74, **75**
vibro-fluidized 76, **77**
well-mixed 73–4
Fluidized bed drying applications 81–2, **257,
    258**
Fodder, drying of 217
Food colours, drying of *116*, 192
Food concentrates, drying of *208*
Food flavours, drying of *82*
Food properties 1–2
Forage, drying of *205, 229*
Free moisture content, *see* Moisture, free
Freeboard in fluidized bed dryers, *see*
    Fluidized bed dryers, features of
    construction, freeboard
Freeze dried products, packaging of 153–4
Freeze dryers, selection criteria *257, 258, 259*
Freeze drying
    advantages and disadvantages of *134, 137*
    applications of 133, 137
    aroma retention in 4
    benefits of 134
    comparison with other methods 134–6
    cost of 148–9
    definition of 133
    history of 133
    phase equilibria in 134, *136, 138, 139*
    process overview 133–7
    product criteria for 134, 139
    product shape retention in 134
    stages of 133–4, **135**
Freeze drying equipment
    batch 145, **146, 149**
    continuous 147–8, **150**
    ice condensers
        defrosting of, in batch plants 147
        double, patented 147–8
        multiple, use of
        need for 140
    racking in 145
    trays 145
    vacuum locks in continuous dryers 147
Freeze drying procedures for individual
        products
    beverages
        coffee 149
        tea 149
    difficult-to-dry products
        high-fat products 4, 153
        high-sugar products 153
    egg 152
    fruit and vegetables
        fruit juices 153
        mushrooms 152

strawberries 153
vegetables, milled 153
meat products
    beef, cooked 151
    chicken, cooked 151
    ingredients for use in soups and prepared
        dishes 151
    lamb, cooked 151
    meat, cooked 151
    meat, raw 150
    offal 152
    pork, cooked 151
seafood
    fish 152
    shrimp 152
Freeze drying, process stages
desorption 144
freezing
    collapse temperature 138
    effect on the drying process 139–40
    eutectic diagrams 138, **139**
    glass transition temperature 138, **139**
    methods 138–40
rehydration and use 145
storage after drying 144
sublimation
    drying front 140–1
    heat transfer in 140–3
    mathematical models of 141
    methods 142–3, **144**
    microwave heating in 140
    operating conditions 142
    pretreatment, effect of 141
    temperatures 141
    vacuum pumps 140
    vacuum requirements 140
    vacuum spoiling 141–2
French fries, drying of 175, *208*
Frequency bands permitted for dielectric
        drying, *see* Dielectric heating,
        permitted frequency bands
Fructose, drying of *91*
Fruit juices, drying of *91, 116*, 118, 153, 173
Fruits, candied, drying of 5
Fruits, drying of *116*, 174, 211, 216
Fungi (Xerophilic), minimum water activity for
        growth *20*
Fuzzy logic, *see* Process control of dryers
        types of process control system,
        fuzzy logic

Garlic, drying of *208, 229*
Gas flow, measurement of, *see* Process
        measurements, gas flow
Gelatine, drying of *116, 208*
Geldart classification of powders 67
Germination of grains, *see* Crop quality and
        drying, germination ratio
Glass transitions 27–8

Glucose, drying of   *116*
Gluten, drying of   185, 190
   see also Crop quality and drying, gluten
      relative
Grain, drying of   1, 32–3, 82, 217
Grain dryers, commercial, *see* Commercial
      grain dryers
Grain properties, *see* Properties of grain
Grains, brewers', drying of   *116*
Grapes, drying of   *208, 229*
Gravy mix, drying of   *116*, 178
Gum, drying of   192

Harvest conditions on Canadian Prairies   *37*
Hay, drying of   *229*
Heat transfer in contact dryers, *see* Contact
      drying, theoretical overview of
Herbs, drying of   *82*, 211, 216
High-fat products, drying of   3, *91*, 153
   see also Spray-dryers, types of, suitable for
      drying high-fat products;
      Spray-dryers, types of, unsuitable for
      drying high-fat products
High-sugar products, drying of   3, 153
Horizontally agitated dryers, *see* Contact
      dryers, types of, horizontally agitated
Humid heat   *13, 15*
Humid volume   *15*
Humidity, absolute, *see* Absolute humidity
Humidity, relative, *see* Relative humidity
Humidity, measurement of, *see* Process
      measurements, humidity
Hydrolysed products, drying of   *91*
Hygiene in pneumatic conveying dryers, *see*
      Pneumatic conveying dryers, hygiene
      in
Hygroscopicity of spray-dried powders, *see*
      Spray drying, meeting powder
      specifications in, hygroscopicity

Ice-cream mix, drying of   *91*
Indirectly heated dryers, *see* Contact dryers
Indirect rotary dryer, *see* Contact dryers, types
      of, indirectly heated rotary
Inerting as a means of explosion protection,
      see Explosion protection in dryers
Ingredients for use in soups and dehydrated
      prepared dishes, drying of   *91*, 151,
      185
Insulation of spray dryers, *see* Spray dryers,
      design features for hygiene and
      safety, cladding
Interlocks, *see* Alarms and interlocks
Internal heat/mass transfer control during
      drying   22
Isotherms, sorption, *see* Equilibrium moisture
      curves

Jellies, drying of   *208*

Jetzone dryer, *see* Fluidized bed dryers, types
      of, Jetzone

Kilning, *see* Malt, drying of
Kinetic data   2
Knocking hammers, use on spray dryers   101
Knudsen diffusion effects, *see* Receding-front
      model

Lactose, drying of   *91*
Lamb, cooked, drying of   151
Latent heat of vaporization of water   *15*
Liquid-diffusion model   24
Lucerne hay, specification of solar dryer
      for   222–3

Maize germ, drying of   185
Maize proteins, drying of   185
Malt, drying of   44–6
Malt extract, drying of   *91*
Maximum allowable temperatures in grain
      drying, *see* Crop quality and drying,
      maximum allowable temperature
Meat products, drying of
   cooked meat   151
   cured meat   11
   meat/cereal strip   175
   meat extracts   *116*, 118
   raw meat   150
   sundry meat products   175, *208*
Medicinal plants, drying of   *229*
Microbiological deterioration of foods, control
      of   *27*
Microwave dryers, cost of   176
Microwave drying, *see* Dielectric drying
Microwave equipment
   applicators
      leaky waveguide   160–1
      multimode cavity   160
      slotted waveguide   161
   heating systems
      circulator   **159**, 160
      magnetron   159–60
      waveguide   159–60
   see also Radiofrequency equipment; Safety
      of dielectric heating equipment
Military-use products, drying of   4, 133
Milk products, drying of
   liquid milk   1, *82, 91*
   milk, flavoured, mixed, skim   *91*
   milk sugar   *82*
   milk, sweetened condensed, whole   *91*
Milk replacer, drying of   *91*
Millability of grains, *see* Crop quality and
      drying, maximum allowable
      temperature
MIVAC dryer   174
Model-based predictive control, *see* Process
      control of dryers, types of process

control system, model-based predictive
Modelling of solar dryers, *see* Solar dryers, design of, modelling and simulation
Moisture, bound *13*, **18**
Moisture, free *13*, **18**
Moisture, unbound, *see* Moisture, free
Moisture content
  in drying process specification 245–7
  measurement of, *see* Process measurements, moisture in solids
  of spray-dried powders, *see* Spray drying, meeting powder specifications in, moisture content
Moisture levelling in dielectric dryers 167–8
Molds, minimum water activity for growth *20*
Mother liquor, drying of *91*
Mountaineering, dried products for use in 4, 133
Moving-bed dryers, *see* On-farm crop dryers
  see also Commercial grain dryers
Multi-stage dryers, *see* Spray dryers, types of, multi-stage
Mushrooms, drying of 152

Neural networks, *see* Process control of dryers, types of process control system, neural networks
Nitrogen oxides, contact with, *see* Toxicological safety
Nitrosamines, formation of, *see* Toxicological safety
Non-edible proteins, drying of 185
Non-uniformity of final moisture content in hopper-bottom bin dryers 41
Nougat, drying of *208*
Nozzle atomizers in spray dryers, *see* Spray dryers, atomization methods, pressure nozzle atomizers
Nuts, drying of *82, 197, 208*

Oats, drying of *229*
Objectives of drying 7
Odour control 281
Offal, drying of 152
Oil recovery from grains, *see* Crop quality and drying, maximum allowable temperature
Olive-oil residues, drying of 185
On-farm crop dryers
  combination heated- and unheated-air dryers 42–3
  heated-air dryers
    batch internal recirculating types 34
    cross-flow types 35–6
    drying and cooling times *37*, 38
    holding capacity *37*, 38
    intermittent types 32–33

    maximum temperature in 37
    mixed-flow types 34–5
    performance of 36–8
    specific energy consumption in *37*
    throughputs *37*, 38
  types of unheated-air dryers
    flat-bottomed bins 38–9
    hopper-bottomed bins 39–41
    performance of 41
  see also Crop drying costs; Process control of dryers, agricultural dryers; Through-flow crop dryers, design of
Onions, drying of 174, *208, 229*
Open loop control, *see* Process control of dryers, types of process control system, open loop
Organisms producing slime on meat, minimum water activity for growth *20*
Osmotic dehydration 5

Packaging of freeze dried products 153–4
Paddle dryers, *see* Contact dryers, types of, horizontally agitated
PAHs, contact with, *see* Toxicological safety
Particle size, measurement of, *see* Process measurements, particle size and size distribution
Particle size and structure of spray dried powders, *see* Spray drying, meeting powder specifications in, particulate size and structure
Particulate emissions
  dust collectors 183–4, 280
  monitoring of 281
Pasta, drying of 2, 174
Pastilles, drying of *208*
Peanuts, drying of *208*, 217, 226, *229*
Pectin, drying of *116, 208*
Pepper, drying of *229*
Percent humidity *15*
Permeate, drying of *91*
Pharmaceuticals, drying of 173
Phase transitions 27–8
Physical changes in foods during drying *27, 28*
Physical properties of dryer feed and product, *see* Specification of the drying process
Pilot-plant trials, use of in dryer selection 266
Plant extracts, drying of *116*
Plate dryers, *see* Contact dryers, types of, plate
Plenum chambers in fluidized bed dryers, *see* Fluidized bed dryers, features of construction, plenum chamber
Pneumatic conveying dryers
  applications 179, 185
  comparison of different systems *187, 188, 189,* 190–1

design of
    computerized methods   190
    operating conditions   186, 188–90
    thermal efficiency   186–8
    wet-bulb depression   186
    worked example   186, *187*, **188**, **190**
    environmental control   185
    feed systems for   183
    flash dryers   179, **180**, **181**
    hygiene in   184–5
    product collection in   183–4
    Ring dryers   178, 179–83
    selection criteria   *257, 258, 259*
Poisseuille flow, *see* Capillary-flow model
Polycyclic aromatic hydrocarbons, contact
        with, *see* Toxicological safety
Pork, cooked, drying of   151
Potato products, drying of
    instant potato   *82, 116*, 121, 185
    potato proteins   185
    potato starch   187
    slices for snacks   175, 197
    sundry products 174, *208, 229*
Powder handling in spray dryers, *see* Spray
        dryers, powder handling in
Powder specifications in spray drying, *see*
        Spray drying, meeting powder
        specifications in
Power dissipation in dielectric dryers, *see*
        Dielectric drying power dissipated
        within dielectric materials
Preliminary selection of dryers
    factors affecting   243, 253–6
    selection charts and tables   **253, 254**, *257,
        258, 259*
Preserves, drying of   82
Pressure, measurement of, *see* Process
        measurements, pressure
Primary drying, in freeze drying, *see* Freeze
        drying, process stages, sublimation
Process control of dryers
    agricultural dryers   46–7
    control strategies
        batch dryers   285–6
        description of   281–2
        feed rate control   284–5
        'wild' feed rate and moisture
            content   283–4
    difficulties in   5
    fluidized bed dryers   81
    quality control   286
    types of process control system
        advanced   288–9
        feedback   287–8
        feedforward   288
        fuzzy logic   289–90
        model-based predictive   289
        neural networks   290
        open loop   287

Process measurements
    gas flow   292
    humidity   293–4
    moisture in solids   294–5
    particle size and size distribution   295
    pressure   291–2
    principal process parameters   290–1
    solids flow   292–3
    temperature   291
Product collection in pneumatic conveying
        dryers, *see* Pneumatic conveying
        dryers, product collection in
Properties of grain
    drying rates   53
    equilibrium moisture content   53
Protein, vegetable, drying of   91
Protein quality in wheat, *see* Crop quality and
        drying, maximum allowable
        temperature
Proteins, drying of   178, 185
Pseudomonas spores, minimum water activity
        for growth   *20*
Psychrometic charts   12, **14**, 186, **188**, **189**
Psychrometry   11–16
Psychrometric properties of air-water
        mixtures   12
Puffed food products   3, 82
Pulses, drying of   *56*

Quality changes in foods during drying   2, **21**,
        26, *27*
    see also Specification of the drying process,
        quality changes during drying
Quality of dried crops, *see* Crop quality and
        drying
Quality of grains, *see* Crop quality and drying

Radiofrequency dryers, cost of   176
Radiofrequency drying, *see* Dielectric drying
Radiofrequency equipment
    applicators
        fringe or stray-field   158, **159**
        staggered through-field   158–9
        though-field   158
    dryers
        conventional equipment   156
        50-ohm equipment   156–8
    see also Microwave equipment; Safety of
        dielectric heating equipment
Receding-front model   26
Rehydration of freeze-dried products, *see*
        Freeze drying, process stages,
        rehydration and use
Relative humidity   *13*
Rice
    drying of   *82, 208*, 214, *229*
    specification of solar dryer for   214
Ring dryers, *see* Pneumatic conveying dryers,
        Ring dryers

Risk costs, agricultural crops, *see* Crop drying costs, risk costs
Rotary atomizers in spray dryers, *see* Spray drying, atomization methods, rotary (wheel) atomizers
Rotary dryers
  applications 197
  cascading
    air flow in 193
    description of 193–6
    design of 197–9
    flights in 193, **194**, **195**
    temperatures in 195–6
  rotary louvre 196–7
  selection criteria *257, 258, 259*
Rotary louvre dryers, *see* Rotary dryers, rotary louvre
Rotating batch, vacuum dryers, *see* Contact dryers, types of, rotating batch, vacuum
Rusk, drying of 208

Saccharin, drying of *116*
Safety considerations in drying
  bulk ignition 278
  dust explosions, *see* Dust explosions in dryers
  explosion protection, *see* Explosion protection in dryers
  fuel cloud ignition 278
  ignition of powder layers and deposits 278
  ignition sources 278–9
  vapour explosions 278
Safety of dielectric heating equipment 161
Salmonella, minimum water activity for growth *20*
Saturation humidity *15*
Sauce mixtures, drying of *82*
Sausage rusk, drying of *208*
Scrubbers, *see* Particulate emissions, dust collectors
Seaweed, drying of 174
Secondary drying, in freeze drying, *see* Freeze drying, process stages, desorption
Seeds, drying of *54, 82*, 219–20
Selection of dryers, *see* Dryer selection
Shrimp, drying of 152
Simulation of solar dryers, *see* Solar dryers, design of, modelling and simulation
Slime-producing organisms, minimum water activity for growth *20*
Solar dryers
  design of
    definition of the drying process 228, *229*
    functional parts, description of 210
    materials of construction 230
    modelling and simulation 232–3
    selection of dryer type 229–30
    structural design of solar collector 230–2

economic evaluation of 226–7
  operation and process control strategies
    dryers with recirculation 236
    less-sophisticated dryers 234
    operational aspects 233–4
    solar-assisted dryers
      energy effectiveness of 234–5
      intermittent operation 235–6
      with water storage tank and auxiliary heater 236–7
  types of
    classification 210–11
    solar natural
      cabinet dryers 211–13
      cabinet dryers fitted with a chimney 213–14
      cabinet dryers fitted with a chimney and heat storage 214–15
    semi-artificial
      greenhouse types 217
      solar rooms 217
      tunnel dryers 215–16
    solar-assisted dryers
      combined with adsorbent units 224–5
      combined with heat pumps 225–6
      with heat storage, justification for 220
      integrated high-performance 200–4, **222**, **223**
      with rock-bed heat storage 224
      for seeds and herbs 218–20
      with water heat storage 220, **221**
Solar drying
  applications of 214, 219–20, 222, *229*
  history of 209
  importance of 4, 209
Solids flow, measurement of, *see* Process measurements, solids flow
SOLVER function in EXCEL, use of 48
Sorbitol, drying of *91*
Sorghum, drying of *229*
Soup mixes, drying of 4, *91*
Soy isolate, drying of *91*
Soy sauce, drying of *91*
Soya, drying of 174
SPC, *see* Process control of dryers, quality control
Specific absorption rate (of electric power by dielectrics) 161
Specification of the drying process
  drying kinetics 252
  feed, form of 243–4
  moisture contents of feed and product 245–7
  physical properties of feed 248–9
  physical properties of product 249
  quality changes during drying
    colour and taste 250
    reconstitution 250–2
    volume 252

throughput 247–8
upstream and downstream processing
     operations 245
Spent grains, drying of *197*
Spices, drying of 211
Spin-flash dryer
     description of 78–80, 191–2
     selection criteria *257, 258, 259*
Spouted bed dryer 66, **67**
Spray dryers
     description of 3, 90–2, **93**
     design features for hygiene and safety
          drying air filtration 110
          fire and explosion protection
               inertizing 111
               in multi-stage dryers 112–3
               with products containing fat 111–2
               suppression systems 112–3
               venting 112–3
               warning signs during operation 112
          cladding 110–1
     hazards 3
     history of 90
     operation of 6
     particle formation and drying in 6, 96–8
     powder handling in 99
     selection criteria *257, 258, 259*
     spray-air contact in 94–6, **97**
     thermal efficiency of 97–8, *100*
     throughputs 90
     types of
          cocurrent drying chamber with integrated
               belt 102–3, **107**
          cocurrent drying chamber with integrated
               fluid bed 102–3, **107**
          cocurrent drying chamber with nozzle
               atomizer 101, **102**, **106**
          cocurrent drying chamber with rotary
               atomizer 100, **101**
          mixed-flow drying chamber with
               integrated fluid bed 104–6
          multi-stage 98, 99–100, 102–6
          suitable for drying high-fat
               products 101–2, 103–4, 104–6
          unsuitable for drying high-fat
               products 100, 103
     wall deposit formation in 101, 103, 104
Spray drying
     applications of
          high fat products, *see* Spray dryers, types
               of, suitable for high fat products
          products, table of *91*
          sugar-containing foods 103–4, 104–5
     atomization methods
          comparison of methods 94, *96*, 105–6
          pressure nozzle atomizers 94, **95**
          rotary (wheel) atomizers 92–4, **95**
          two-fluid nozzles 92
     meeting powder specifications in

bulk density 109
hygroscopicity 109
moisture content 108
particulate size and structure 108
production of agglomerates
     in multi-stage spray drying systems 100,
          105
     in spray dryer with integrated belt 104,
          108
     in spray dryer with integrated fluidized
          bed 98, 103, 105, 108
     in spray drying chamber 97, 100
Starch, drying of *91, 116, 197*, 276
     corn 186, *187*
     potato *187*
Starch gelatinization 28
Statistical process control, *see* Process control
     of dryers, quality control
Steam-tube rotary dryers, *see* Contact dryers,
     types of, indirectly heated rotary
Stein breakage tester 59
Storage of freeze-dried products, *see* Freeze
     drying, process stages, storage after
     drying
Stoves, *see* Tray dryers
Strawberries, drying of 153
Stress cracking of grains, *see* Crop quality and
     drying, stress cracks and broken
     kernels
Structural design of solar dryers, *see* Solar
     dryers, design of, structural design of
     solar collector
Sublimation, in freeze drying, *see* Freeze
     drying, process stages, sublimation
Sugar, drying of *197*
Sugar beet, drying of 5
Sugar-containing foods, spray drying of, *see*
     Spray drying, applications of,
     sugar-containing foods
Superheated-steam dryer 5, 12
Suppression, of explosions, *see* Safety
     considerations in drying, explosion
     protection
Sweet potato, drying of *229*
Sweeteners, drying of *91*

Tea, drying of 149, *82, 116, 229*
Temperature, glass transition, *see* Glass
     transitions
Temperature, effect of in multi-stage
     spray-drying systems 98
Temperature, measurement of, *see* Process
     measurements, temperature
Temperatures, maximum allowable in grain
     drying, *see* Crop quality and drying,
     maximum allowable temperature
Thermal efficiency of spray dryers, *see* Spray
     dryers, thermal efficiency of

Thermodynamic properties of air-water
    mixtures, *see* Psychrometry
Thin-film dryers, *see* Contact dryers, types of,
    thin-film
Thickening ingredient, drying of   268–9
Through-circulation band dryers, *see* Band
    dryers
Through-flow dryers for agricultural crops
  flow configurations   31
  types of   32
  see also Commercial grain dryers; On-farm
    crop dryers; Malt, drying of
Through-flow crop dryers, design of
  sizing the equipment
    heat transfer equations   48–9
    Thompson's procedure   50
    time- and space-dependent drying   50–2
    well-mixed transient dryer   49–50
    steady-state continuous drying   47–8
  verification of design calculations
    heated-air dryer   54–5
    unheated-air dryer   55–6
Tobacco, Virginia, drying of   *229*
Tomato paste, drying of   *91*, *116*
Total sugar, drying of   *91*
Toxicological safety   11
Transport properties of air-water mixtures, *see*
    Psychrometry
Tray dryers   4, 200–1, 257, 258
  see also Tunnel dryers
Tunnel dryers
  description of   201–4
  design of   204–5
  performance of   204–5
  selection criteria   *257, 259*
  varieties of
    cocurrent-flow types   201
    combination-flow types   203–4
    countercurrent-flow types   203
    transverse-flow types   204
  see also Tray dryers
Two-fluid nozzles in spray dryers, *see* Spray
    drying, atomization methods;
    two-fluid nozzles

Unbound moisture, *see* Moisture, free
Unloading flat-bottom bins, problems
    associated with   39

Vacuum band dryers, *see* Contact dryers, types
    of, vacuum band
Vacuum dryers, *see* Contact dryers

Vacuum tray dryers, *see* Contact dryers, types
    of, vacuum tray
Vegetable products, drying of
  vegetable extracts   *116*, 118
  vegetable protein   *91*
  vegetable purees   *91*
Vegetables, drying of
  in band dryers   *208*
  in dielectric dryers   174, 175
  in freeze dryers   153
  in solar dryers   211, 216, *229*
  osmotic dehydration of   5
Venting, of explosions, *see* Safety
    considerations in drying, explosion
    protection
Venturi scrubbers, *see* Particulate emissions,
    dust collectors
Vertically agitated dryers, *see* Contact dryers,
    types of, vertically agitated
Vibro-fluidized bed dryers, use of in
    multi-stage spray drying
    systems   99, 105
  see also Fluidized bed dryers, types of,
    vibro-fluidized
Vitamin C, drying of,   5
Vitamins, drying of   82

Wall deposit formation in spray dryers, *see*
    Spray dryers, wall deposit formation
    in
Water activity
  definition of   *13*, 19, 247
  minimum for microbial growth and spore
    germination   *20*
  plot versus moisture content for various
    foods   **20**
  plot of relative reaction rates versus   **21**
  prediction of   28
Wet-bulb temperature   12–13, 23
  lines of constant   12–15
Wet scrubbers, *see* Particulate emissions, dust
    collectors
Wheat, drying of   37, *54, 55, 56, 57, 229*
Wheat gluten, drying of   185, 190
Wheat residues, drying of   *197*
Whey, drying of   *91*

Yams, drying of   *229*
Yeast, baker's, drying of   *91*
Yeast residues, drying of   185
Yeasts, minimum water activity for
    growth   *20*